PHOTOSYNTHESIS IN RELATION TO MODEL SYSTEMS

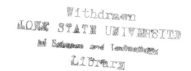
Photosynthesis in Relation to Model Systems

edited by

J. Barber

Imperial College of Science and Technology,
Department of Botany, Prince Consort Road,
London SW7 2BB, United Kingdom

ELSEVIER SCIENTIFIC PUBLISHING COMPANY
AMSTERDAM — NEW YORK — OXFORD — 1979

ISBN 0-444-41596-3 (Series)
ISBN 0-444-80066-2 (Vol. 3)

Published by:

Elsevier/North-Holland Biomedical Press
335 Jan van Galenstraat, P.O. box 211
Amsterdam, The Netherlands

Sole distributors for the U.S.A. and Canada:

Elsevier North-Holland, Inc.
52 Vanderbilt Avenue
New York, N.Y. 10017

Library of Congress Cataloging in Publication Data
Main entry under title:

Photosynthesis in relation to model systems.

(Topics in photosynthesis ; v. 3)
Includes bibliographical references and index.
1. Photosynthesis. 2. Photosynthesis--Simulation
methods. I. Barber, James, 1940-
QK882.P557   581.1'3342   78-23915
ISBN 0-444-41596-3

PRINTED IN THE NETHERLANDS

Foreword to Volume 3

This third volume in the series of Topics in Photosynthesis appears at a particularly appropriate time and with a particularly appropriate combination of subjects.

For some years now, there has been an increasing realization that alternative energy sources to the fossilized photosynthetic products of the recent past will have to be found; and the natural turn is to current photosynthetic systems for that purpose. While the actual green photosynthetic systems can and will be used as direct collectors and converters, they perhaps have a greater use as models for synthetic systems which might do more specific tasks. It is this particular relationship between our understanding of the photosynthetic process as such and the construction of model systems to accomplish one or another part of the total process to which this book is devoted. It reviews practically every approach that has grown up in the last decade or so toward this end. The construction of models to simulate photosynthesis, or some parts of it, has been going on for a long time. The purpose of this construction has been two-fold: one, to help understand the physical basis of the photosynthetic quantum conversion process and two, more recently, to possibly construct totally synthetic systems to achieve one or another of the objects of the photosynthetic process itself. It is in this latter role that the book will play its greater part. Anyone seeking to enter the field now will find this book a complete introduction to the various aspects of the subject that will allow him to make a choice most suited to his own interests and talents. There is hardly any doubt but that the use of our growing knowledge of the fundamental processes of photosynthesis will one day be the principal way in which our energy and materials sources will be achieved.

There still remains at least one approach to the natural process that has not yet been reported in the literature with any success. This is to simulate the two photosystems in two separate phases. One may expect this to happen in the very near future. Finally, the use of the separated charges, or stored energy, for the direct reduction of carbon dioxide to one-carbon reduced materials, has also yet to be reported.

Melvin Calvin

Preface

As with earlier volumes of this series the aim has been to produce a book centred around a specific theme. In this volume I have collected together a number of authoritative articles which high-light those aspects of the photosynthetic process which may have implications in the development of practical devices for solar energy capture and storage. Realizing that the book is likely to appeal to nonspecialists as well as to specialists, a deliberate effort has been made, where appropriate, to briefly explain the basic features of photosynthesis as well as cover the more specialized topics which are relevant to the theme of the volume. For this reason the book enables readers with various backgrounds and levels of training to appreciate the special features of the photosynthetic apparatus which would be worth consideration when trying to develop new technologies designed to help eleviate the energy supply problem which will inevitably arise as the reserves of fossil fuels diminish.

The first three chapters act as service chapters to the remaining portion of the book since they pin-point the basic energetic and structural features of the primary photosynthetic processes. In Chapter 1, Alexander Borisov covers some general aspects of the properties of solar radiation and goes on to emphasise the features of photosynthetic organisms which make them efficient solar energy converters. He has presented his arguments from the view of a physicist and emphasises the unique properties of the photosynthetic light harvesting antenna system and associated reaction centres which could form the basis for the construction of a photoelectric device. Chapter 2 is by Philip Thornber and myself and its purpose is to give a detailed description of how higher plants and other photosynthetic systems organise their pigments so that they can efficiently capture and transfer light energy to specialized reaction centres. It is explained how proteins help to keep the chromophores of the pigments at distances and orientations advantageous for energy transfer but avoid wasteful quenching processes. Chapter 3 is an excellent contribution by Robert Blankenship and William Parson. They discuss in detail various properties of photosynthetic reaction centres using the bacterial system as a specific example. This chapter covers the thermodynamics of charge separation and emphasises the ability of the system to resist back reactions. In Chapter 4, Professor Tien reviews mainly his own work using

vii

planar black lipid membranes and liposomes which have been made light sensitive by the incorporation of chlorophyll complexes extracted from chloroplasts. The next chapter, Chapter 5, is by V.P. Skulachev who describes experiments conducted by his group in Moscow in which photosynthetic bacterial chromatophores and reaction centres have been incorporated into artificial phospholipid membranes. Chapter 6 also deals with the making of artificial membrane systems which are light sensitive and able to produce photoelectric potentials. However in this case the pigment considered is bacteriorhodopsin and Thomas Schreckenbach has given a thorough review of both the properties and potential of this unique pigment-protein complex.

Although there is a great deal of interest in the possibility of using artificial membranes incorporating various light sensitive complexes, there is also the other alternative of stabilizing natural pigmented membranes against ageing. In Chapter 7, George Papageorgiou has reviewed the chemical background and progress being made in attempts to preserve the energy conserving properties of isolated thylakoid membranes. The following chapter (Chapter 8) is by Tony Harriman and myself and deals with the photochemical splitting of water into its elemental constituents. The initial part of the chapter outlines what is known about photosynthetic water splitting while the remaining part thoroughly reviews efforts being made to mimic this process using artificial photochemical systems. Chapters 9—11 deal with light induced hydrogen production. In Chapter 9 Professor Krasnovsky has given a general survey mainly based on work from his own laboratory touching on both artificial and in vivo systems. Chapter 10 by Rao and Hall and Chapter 11 by Hallenbeck and Benemann give comprehensive reviews on H_2 production from isolated chloroplasts and algae respectively. All three chapters emphasise and speculate about possible developments which could lead to a practical arrangement capable of generating significant amounts of hydrogen gas from illuminated photosynthetic systems. The final chapter (Chapter 12) of the book by Losada and Guerrero gives an interesting account of energy flow in biological cells and also emphasises that light mediated nitrate reduction to ammonia by photosynthetic organisms is a useful energy conserving process which could give rise to a useable fuel.

My task of editing this book has been made easier not only by the high quality of the contributions but also by the encouragement and help of several colleagues. In particular I would like to thank Cara Cherrett and my wife, Lyn, for assisting me with many of the tedious jobs involved in producing a volume of this type. Finally I wish to thank Professor Melvin Calvin for writing the Foreward. Professor Calvin was awarded the Nobel Prize in 1961 for elucidating the biochemical pathway giving rise to photosynthetic carbon fixation. For many years he has argued that the future development of man may require the manipulation and mimicking of the photosynthetic processes in order to take more advantage of the solar radiation which falls on our planet; a concept which is the basis of this volume.

J. Barber

List of Contributors

J. BARBER Department of Botany, Imperial College, London S.W. 7, England

J.R. BENEMANN Sanitary Engineering Research Laboratory, University of California, Berkeley, CA 94720, U.S.A.

R.E. BLANKENSHIP Department of Biochemistry SJ-70, University of Washington, Seattle, WA 98195, U.S.A.

A.Yu. BORISOV Moscow State University, Moscow, U.S.S.R.

M.G. GUERRERO Deparatamento de Bioquímica, Facultad de Ciencias y C.S.I.C., Universidad de Sevilla, Spain

D.O. HALL Plant Sciences Department, King's College, 68 Half Moon Lane, London SE24 9JF, England

P.C. HALLENBECK Sanitary Engineering Research Laboratory, University of California, Berkely, CA 94720, U.S.A.

A. HARRIMAN The Royal Institution, 21 Albemarle Street, London W.1., England

A.A. KRASNOVSKY A.N. Bakh Institute of Biochemistry of the USSR, Academy of Sciences, Moscow, U.S.S.R.

M. LOSADA Departamento de Bioquímica, Facultad de Ciencias y C.S.I.C., Universidad de Sevilla, Spain

G.C. PAPAGEORGIOU Nuclear Research Center Demokritos, Department of Biology, Aghia Paraskevi, Athens, Greece

W.W. PARSON Department of Biochemistry SJ-70, University of Washington, Seattle, WA 98195, U.S.A.

K.K. RAO Plant Sciences Department, King's College, 68 Half Moon Lane, London SE24 9JF, England

TH. SCHRECKENBACH Institut für Biochemie, Röntgenring 11, 8700 Würzburg, F.R.G.

V.P. SKULACHEV Department of Bioenergetics, A.N. Belozersky Laboratory of Molecular Biology and Bioorganic Chemistry, Moscow State University, Moscow, U.S.S.R.

J.P. THORNBER Department of Biology and Molecular Biology Institute, University of California, Los Angeles, CA 90024, U.S.A.

H. TI TIEN Biophysics Department, Michigan State University, East Lansing, MI 48824, U.S.A.

Contents

(for more detailed lists of contents the reader is referred to the first page of each chapter)

Photosynthesis in relation to model systems, edited by J. Barber
© Elsevier/North-Holland Biomedical Press 1979

Chapter 1

Photosynthesizing Organisms: converters of solar energy

A.Yu. BORISOV

Moscow State University, Moscow, U.S.S.R.

CONTENTS

1.1. INTRODUCTION

The energy crisis on our planet is becoming more and more grave. In the near future only solar energy and energy derived from nuclear fusion will be able to meet the global energy requirements of mankind. Until now the main effort and expense has been directed towards the development of thermonuclear devices. Although great progress has been achieved in this field, the problem is still far from being solved. But is this choice justified? Or maybe we owe this success to the fact that for the last few decades enormous sums of money have been spent on studies of the atomic nucleus, particularly for military purposes? If we forget what has been done in this respect and compare the two directions, solar energetics will undoubtedly seem more promising. In fact, visible solar radiation, when absorbed, hardly ever causes destruction of substances and "softly" generates EMF of the order of 1 volt, whereas nuclear reactions yield substances in a plasmic state at immense temperatures, with energy quanta being equal to megavolts, which means explosions that have to be harnessed. Moreover the use of these thermonuclear devices is always fraught with the danger of radioactive contamination of environment.

But there is one more thing that should be considered. The Earth's surface receives $3 \cdot 10^{24}$ J of solar energy per year. The bulk of it is absorbed in the Earth's biosphere and heats it, The thermal balance of our planet is to a great extent determined by absorption of solar radiation. As a consequence the temperature of the Earth, averaged over all geographical regions within a year, is about $280°$ K. However for solar radiation, this temperature would only be determined by geothermal heat streams and by irradiation into cosmic space and would be lowered to a value of at least $250°$ K. The estimated consumption of energy by mankind must reach, by the years 2010—2020, $2 \cdot 10^{21}$ J per year, i.e. about a thousandth of the solar energy heating our planet. If all this energy was to be generated by thermonuclear devices, the average temperature of the Earth's surface may increase by $10^{-3} \times 250° = 0.25°$. Despite the insignificance of this increase compared to yearly temperature fluctuations in certain regions of the Earth, it exceeds fluctuations of the average yearly temperature all over the planet and such heating — if it lasted for a long time — might bring about significant changes in climate, geographical environment, living conditions of plants and animals etc. In fact, a great number of anomalies observed in the recent years in our surroundings (e.g. increase of the concentration of CO_2 in atmosphere and decrease in the level of ozone, pollution of land, air and water, dying water basins, extinction of many species of animals and plants, rapid growth of harmful bacteria adaptable to newly formed ecological niches and many other things) can be attributed to the fact that mankind has at his disposal powerful sources of energy. Had the 4 billion people inhabiting our planet led a "patriarchal" mode of life, they would have attained nothing of the sort by themselves. Moreover the population explosion in this century has been paralleled by a decrease in the rest of the animal biomass, so that the balance of the plant and animal kingdoms has on a whole been retained. It seems to be high time to establish an international organization (probably, under the auspicies of the United Nations Organisation) that would work out the tactics and strategy of the development and stabilization of the World's energy systems. Such an organisation should foster the people's efforts to search for the most rational ways of energy planning with the view to restraining the chaotic growth of the power capacities that are at present improvidently exploited by nations in their own interests to the overall detriment of the whole of mankind, not forgetting the needs of our grand-children and even our children.

The situation seems to be less tragic when we speak of solar energetics. The sun heats the Earth and utilization of part of this energy would hardly change the thermal balance of the planet. No CO_2 or other harmful carbon products would be evolved into atmosphere. Therefore, in my opinion, our future energy needs must to a great extent be satisfied by utilisation of solar radiation. It is possible that mankind's survival depends on whether this choice is made.

1.2. SOLAR RADIATION

Beyond the gas envelope of the Earth the intensity of solar radiation is about $1.3 \text{ kW} \cdot \text{m}^{-2}$. Its spectrum (curve $S(\lambda)$ Fig. 1.1) approximates to the irradiation of an absolutely black body at $5000-6000°\text{K}$. Here the basic question arises about the "quality" of this kind of energy. The thermodynamic approach to this problem was undertaken by several authors for particular photochemical and photoelectric systems. In the work of

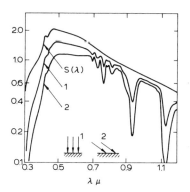

Fig. 1.1. The spectra of solar radiation. $S(\lambda)$, in space; (1), attenuated after penetrating one air mass; (2), after penetrating two air masses.

Leontowitch (1974) the general thermodynamic approach to this problem was elaborated.

Taking into account the spectrum and the intensity of solar radiation, the high degree of parallelism of solar rays, the possibility of realization of non-reflecting absorbers etc. Leontowitch came to the conclusion that the thermodynamic yield of solar energy conversion could be as high as 93%; a value which is several times more than the level currently achieved. The situation is slightly worse at the Earth's surface. Even for normally falling rays (the inclination at $0°$, curve 1) the radiation energy is weakened 1.5-fold due to light absorption and scattering by the atmosphere. Curve 2 in Fig. 1.1 represents attenuation of solar radiation inclined at $60°$ which means that the light path through the Earth's gas envelope is doubled. The cloudiness and pollution of air would also cause additional reduction of solar light reaching the Earth's surface. Furthermore, light scattering and refraction decrease the degree of parallelism of solar rays causing inadequate energy condensation. All this makes it difficult to obtain thermodynamic evaluation of solar energy conversion. But there is an easier way to solve the problem. It is known that physical solids can be heated up to $\sim 3000°$K if incident solar light is condensed by a parabolic mirror. Consequently, according to the basic law of thermodynamics:

$$\varphi_s = \frac{T_2 - T_1}{T_2} = \frac{3000° - 300°}{3000°} = 0.90$$

Even if averaged over the daytime in accordance with the radiation intensity, the mean value of φ_s will be not less than 0.80—0.85 for the middle latitudes.

Consequently, we must conclude that solar radiation is energy of rather high quality. Batteries incorporating fine inorganic crysals are already used for conversion of solar into electrical energy, mainly in space vehicles, and the maximum realized energy conversion is as high as 20%. Several other projects for solar energy utilization are now being developed very intensively especially those involving geliothermal systems providing electricity, semiconductor photoelectric systems based on silicon and some other materials, and also some minor ones using wind, ocean waves, temperature gradients in oceans etc.

But the ultimate solution of the energy problem should also include its transport over large distances (thousands of kilometres) and storage. Therefore a search for new ways of utilizing solar radiation, especially involving, electrical energy generation (to which our present day technology is mostly adapted) or photocleavage of water and subsequent utilization of hydrogen, are of great importance.

In my opinion, the solution of these problems may be facilitated by unraveling the molecular mechanisms of the primary processes of photosynthesis in which the above two goals are to a certain degree coped with.

1.2.1. Accumulation of solar radiation by world photosynthesis

The total amount of solar energy falling yearly on the perimeter of our planet is immense—$3 \cdot 10^{24}$ J. About 60% of it is absorbed by the biosphere of the Earth, the rest diffuses in cosmic space. World photosynthesis conserves in biomass about $3 \cdot 10^{21}$ J per year, i.e. the efficiency is $\sim 0.1\%$. Many cultivated plants conserve from 0.2 to 0.3% of the light falling on their leaves. Under optimal conditions (carefully arranged inter-row distances, watering and fertilization, temperature control etc.) agricultural crops may reach an efficiency of 1%, while for optimal organisms, like the alga *Chlorella*, or in some tropical plants, like sugar cane, the efficiency can reach 2—3% (see the works of Hall (1976), Calvin (1976) for more details). These figures show that improvements of all agricultural systems and even cardinal reorganization of the present-day rural economy, should be considered carefully. It also seems that biomass production from large scale algal culturing could become important as an economic chemical processing fodder etc. in the near future. Thus all agricultural, physiological, biochemical and genetic efforts resulting in an increase of photosynthesis production (food, fiber, dyes, medicines etc.) are of the greatest importance. But in photosynthesis Nature has also created an original photoelectric system which is coupled to a very complicated set of subsequent biochemical processes. Below arguments are given to show that the efficiency of solar energy conversion by this system is considerably greater than would be imagined at first sight.

1.2.2. Photoelectric systems

Let us consider the branch of solar energetics associated with the conversion of light into various forms of electrical energy. The following formula for efficiency of solar energy utilization will fit all the systems of this type:

$$\varphi_s = \frac{\int_{\lambda_m} S(\lambda) \cdot \overline{T(\lambda)} \cdot A(\lambda) \cdot \varphi_e(\lambda) \cdot \lambda/\lambda_m [1 - \Delta W_s(hc)^{-1}\lambda_m] \, d\lambda}{\int^{\infty} S(\lambda) \cdot \overline{T(\lambda)} \, d\lambda} \tag{1}$$

$S(\lambda)$ is the spectrum of solar radiation in space (curve 1, Fig. 1.1), $\overline{T(\lambda)}$ is the atmospheric transmittance in a given region averaged over a year; $A(\lambda)$ is the absorption of the system; λ_m is maximal wavelength of the quanta used by the system; λ/λ_m is the "devaluation" coefficient for shorter wavelength quanta; $\varphi_e(\lambda)$ is the spectral dependence of the quantum yield of photocurrent generation; ΔW_s is the portion of energy losses in the course of stabilization of new energy carriers (the pairs of charges of opposite signs); $hc \cdot (\lambda_m)^{-1}$ is the energy of long wavelength quantum.

Clearly $\int S(\lambda) \cdot \overline{T(\lambda)} d\lambda$ represents incident solar energy for systems located at the Earth's surface. Southern deserts with maximal numbers of sunny days and low air pollution provide mean yearly insolation exceeding $150 \text{ w} \cdot \text{m}^{-2}$ which are obviously most attractive for such systems.

Based on economic calculations, the minimal value of φ_s for conventional photoelectrical systems should exceed 6—8%. Out of many photoelectric systems known (for example, photovoltaic cells, photochemical redox reactions, artificial membranes adsorbing some subcellular particles or incorporating dye molecules, photoelectrochemical cells etc.), only present day semiconductor photocells based on superpure inorganic crystals meet this requirement. It should be noted that all the above mentioned systems are not very much different in λ/λ_m; $A(\lambda)$ and $[1-\Delta W_s(hc)^{-1}\lambda_m]$ factors. Consequently, factor $\varphi_e(\lambda)$ plays a decisive role in formation of their φ_s values. All other systems have $\varphi_e < 10\%$ and, respectively φ_s not exceeding several per cents.

Below it will be proved that one more efficient photoelectric system exists which has its φ_e close to unity and $\varphi_s \cong 10\%$ i.e. having values quite acceptable for conventional photoenergetic systems. This is the photo-electric apparatus of photosynthesis.

The above statement does not contradict the information given in Section 1.2.1 about photosynthetic organisms. It simply means that the portion of converted solar energy is considerably higher at the primary stages of photosynthesis than at the later stages.

Below the φ_s value is estimated for photosynthesis.

1.2.3. Energetics of the primary stages of photosynthesis

(a) $A(\lambda)$ Chlorophylls, like other dyes, exhibit rather narrow resonance-type absorption bands with halfwidth equal to several hundreds of cm^{-1}. However, due to several optical tricks invented by Nature in the course of evolution, the portion of solar radiation absorbed by plants reached 35—40%. These optical achievements are: (i) the diversity of spectral forms of chlorophyll (for example two basic pigments of plants Chl a and Chl b exist in vivo in more than ten forms with their absorption peaks dispersed in the region of 665—730 nm); (ii) specifically arranged surfaces of leaves decreasing the portion of reflected light; (iii) the specific characteristics of leaf cover operating as a trap of solar light even in the yellow-green optical region.

(b) λ/λ_m This factor diminishes φ_s especially in blue-green optical region because for plants $\lambda_m = 700$ nm (and correspondingly 840 and 870—890 nm for green and the most of purple bacteria). Fortunately, the portion of incident quanta is not very abundant there.

(c) ΔW_s The redox potential for oxidation of reaction centres is known to be +(0.45—0.50) volts (see, for example Kok et al., 1964). It was demonstrated in the work by Kok et al. (1965) that photosystem I is capable of photoreducing several viologen dyes with middle potentials reaching −0.60 V. Consequently, we may conclude that approximately one electron-volt of energy can be stored at the primary photoelectric stages from each quantum of electronic excitation entering the reaction center (i.e. λ_m = 700 nm and hc. $(\lambda_m)^{-1}$ = 1.75 eV).

Taking into account all above figures and using conditionally φ_e = 1 for the longest wavelength (λ_m) we arrive at the important conclusion: at the photoelectric stages of photosynthesis (i.e. separated charges of opposite signs stabilized in the millisecond time region and available for subsequent diffusion limited reactions) the portion of converted solar energy (in the form of electrical dipole) may be as high as 13—18%.

The next section will be devoted to estimation of φ_e values in the actual photosynthetic apparatus.

1.3. DELIVERY OF PHOTOINDUCED EXCITATIONS FROM ANTENNA MOLECULES TO REACTION CENTRES IN PHOTOSYNTHESIS

1.3.1. Quantum yield

There have been many studies to determine the quantum yields of photosynthetic primary charge stabilization. Usually they have been measured by following the photooxidation of reaction centres or cytochromes which are tightly coupled with them.

The most precise measurements were performed with preparations of the reaction centres (RC) isolated from carotenoidless mutant *Rh. sphaeroides* R-26. According to the data (Wraight, Clayton, 1974), in these particles, which are completely devoid of antennae pigments, φ_e = 1.02 ± 0.04, i.e. is almost equal to unity.

In other particles which have been investigated (chromatophores, chloroplasts and their subunits), and also with intact cells, light is primarily absorbed by the antenna pigments and in these cases φ_e is found normally to be less than unity. This is because the determination of the φ_e value is subject to much greater errors, due to uncertainties in optical measurements associated with light scattering. Besides, in any experimental suspension there are always dying cells with decaying pigment and young ones with an underdeveloped antenna and RC. In normal cells, especially in subcellular particles isolated from them, there can exist defective chlorophyll domains, or some of the RCs may be in an inactive state. Hence considerable divergence in the results can be obtained in different laboratories even when using identical preparations from the same organisms. However numerous data give φ_e in the range from 20 to 80% (see, e.g. Clayton, 1962; Vredenberg and Duysens, 1965; Vredenberg, and Slooten, 1967; Sybesma and Beugeling, 1967). The main difference between all these various experimental systems and RC particles is that the former have a well developed molecular antenna. It is known that in the majority of purple bacteria there are on the average 30—200 antenna chlorophyll molecules per RC (plus almost the same amount of carotenoid molecules); in algae and higher plants this ratio is 150—400 : 1, and in green bacteria it is even up to 1,000—1,500 : 1 according to Olson et al., (1976).

Thus our major conclusion is the following: *the value of the quantum yield for the primary separation of opposite charges in photosynthesis (generation of microcurrent in the photosynthetic electron transport chain) is determined by the processes of energy transfer from light-harvesting antenna to reaction centres.*

But for our purpose, which is to reveal the maximum possibilities of photosynthesis as expressed by a high value of φ_s, the optimal $\varphi_e(\varphi_e^{max})$ values obtained in well organized, fully active chlorophyll domains, rather than its mean values, are of importance. To find them, Barsky and Borisov (1971), Borisov and Il'ina (1973), Barsky et al. (1974), have developed an original method using only relative measurements. They succeeded in determining φ_e^{max} with an accuracy of not less than ±(5—10%) *).

The idea of the method may be presented as follows. Let us designate the concentrations of active and inactive RC as $[P_a]$ and $[P_i]$, respectively. Then the major assumption is that the rate (V_{ph}) of useful trapping of singlet electronic excitations induced by light in the antenna is proportional to the portion of active RC:

$$V_{ph} = K_{ph} \cdot \frac{[P_a]}{[Pa] + [Pi]} = K_{ph}(1 - x) \qquad (2)$$

where $x = P_i/[P_a] + [P_i]$ is the proportion of inactive RC.
Numerical example. Supposing at $x = 0$ (all RC are active) the rate of photosynthesis is 9 times as high as the total rate of useless losses of electronic excitations (V_Σ);

$$V_{ph}^{max} = V_{ph}(x = 0) = K_{ph} = 9V_\Sigma$$

Then the φ_e^{max} will apparently be

$$\varphi_e^{max} = \varphi_e(x = 0) = \frac{V_{ph}^{max}}{V_{ph}^{max} + V_\Sigma} = 0.9$$

---

* The specificity of the method is well illustrated by the following example. Supposing we have a suspension of cells or particles in which half of the antenna chlorophyll is not active ($\varphi_e' = 0$), and the second half has $\varphi_e'' = 0.9$. Then the classical method of determining φ_e by the ratio of moles $P + h\nu \rightarrow P^+$ to moles of chlorophyll-absorbed quanta of light will evidently give φ_e (average) = 0.45. Our relative method in this case will again give $\varphi_e^{max} = 0.9$, so we shall still be uncertain whether a fraction of inactive chlorophyll exists.

For the cases $x > 0$ the formula for $\varphi_e(x)$ is essentially the same as (2), e.g. at $x = 0.5$

$$V_{ph}(x = 0.5) = K_{ph}(1 - 0.5) = 4.5V_\Sigma$$

and hence,

$$\varphi_e(0.5) = \frac{4.5V_\Sigma}{4.5V_\Sigma + V_\Sigma} = 0.818$$

Theoretical dependences of normalized $\varphi_e(x)$ are shown in Fig. 1.2 for a number of initial values of φ_e^{max}, in particular, the curve with the largest trough corresponds to $\varphi_e^{max} = 0.9$.

The family of the curves in Fig. 1.2 allows the following conclusions to be made.

(1) In terms of the above model of photosynthetic unit (PSU) organization the value of the trough of the $\varphi_e(x)$ function is a simple measure of the maximum quantum yield of photosynthesis φ_e^{max}.

(2) If within $\varphi_e^{max} \ll 1$ the curves are almost identical, at $\varphi_e^{max} > 0.7–0.8$, relative measurements of $\varphi_e(x)$ performed with an accuracy of $\pm 3–5\%$ provide the analogous accuracy in determination of φ_e^{max}.

It should be noted that Vredenberg and Duysens (1963) were the first to make major assumption $V_{ph} = K_{ph}(1–x)$ proceeding from the experimental data on the dependence of the growth of fluorescence yield on the proportion of inactive RC in purple bacterium *Rh. rubrum*. Then similar data were obtained for one green (Sybesma and Vredenberg, 1963) and three purple bacteria (Clayton, 1967). Provided that losses for fluorescence are propor-

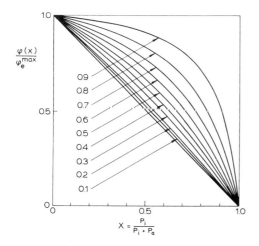

Fig. 1.2. The dependence of the relative yield of primary charge separation in photosynthesis (φ_e) on the portion of reaction centres in active state ($1–x$) according to equation 2 (see text). Numbers from 0.1 to 0.9 represent the initial values of φ_e^{max}.

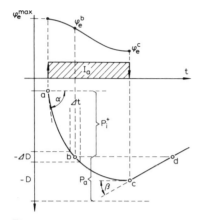

Fig. 1.3. The photoinduced kinetics of reaction centre photooxidation followed by an optical density decrease (ΔD). I_a, constant illuminating light; φ_e^b, φ_e^c, relative values of φ_e respectively in b. and c. points of ΔD curve; P_a, P_i^+, the relative concentrations of active and inactive (photooxidized) reaction centres.

tional to V_Σ, equation (2) is easy to derive. True, Clayton (1967) stated that these dependences can be obtained only in aerobic conditions and was the first to note limitations in using fluorescence as a measure for V_Σ.

The method described in the works of Barsky and Borisov (1971), Barsky et al. (1974) and Borisov and Il'ina (1973) implies measurements of the kinetics of RC photooxidation at purposely chosen redox potentials of the medium with nearly all RC being prereduced and their primary donors pre-oxidized. Experiments are usually carried out with chromatophores from bacteria, and with chloroplasts or subchloroplast particles from plants. Permanent excitation with light (I_a in Fig. 1.3) results in photooxidation of RC according to the following equation:

$$-\frac{\Delta P}{\Delta t} = \frac{\Delta P^+}{\Delta t} = I_a \cdot \varphi_e(t) - K_{DP}[P^+] \cdot [D] \tag{3}$$

where P, P^+ are concentrations of RC in normal and photooxidized states, respectively; I_a is absorbed light in einsteins; $\varphi_e(t)$ is the time dependence of the function $\varphi_e(x)$ shown in the upper part of Fig. 1.3; x in this particular case is equal to the proportion of oxidized RC, i.e. $x = P^+(P^+ + P)^{-1}$; K_{DP} is the second order rate constant for the process of dark regeneration of RC ($P^+ + D \xrightarrow{K_{DP}} P + D^+$); [D] is the effective concentration of the limiting donor system. The light state of the kinetics is represented by the a-b-c curve, in point a the light is turned on, at point c turned off.

(1) Apparently, at $t = 0$ at point a $P^+ = 0$; $x = 0$. Hence we derive from equation (3).

$$\frac{\Delta P^+}{\Delta t} = I_a \cdot \varphi_e^{max} \quad \text{or} \quad \varphi_e^{max} = I_a^{-1} \cdot \frac{\Delta P^+}{\Delta t} \sim tg\alpha$$

i.e. the initial rate of photooxidation of RC is directly proportional to the inclination of the curve at point a ($tg\alpha$).

(2) In the state of equilibrium (point c) we evidently have $\Delta P^+/\Delta t = 0$. Hence

$$\varphi_e(t = c) = I_a^{-1} \cdot K_{DP} \cdot [P_c^+] \cdot [D]$$

For determination of the $\varphi_e(t = c)$ value it is sufficient to turn off exciting light and measure the initial inclination of the rate of dark regeneration of RC ($tg\beta$ in Fig. 1.3). Apparently,

$$\frac{\varphi_e(t = c)}{\varphi_e^{max}} = \frac{tg\beta}{tg\alpha} = \frac{\dfrac{K_{ph}(1-x)}{K_{ph}(1-x) + K_\Sigma}}{\dfrac{K_{ph}}{K_{ph} + K_\Sigma}} = \frac{(1-x)(K_{ph} + K_\Sigma)}{(1-x)K_{ph} + K_\Sigma}$$

(3) In the intermediate points (e.g. point b) all the three terms in equation (3) differ from zero. Hence

$$\varphi_e(t = b) = I_a^{-1} \cdot \frac{\Delta P^+}{\Delta t} + K_{DP} \cdot [P_b^+] \cdot [D] \quad .$$

This means that the real derivative of the curve at point b should be supplemented with the derivative of dark regeneration at point d lying in the same ordinate as b.

$$\varphi_e(t = b) \sim tg(x = b) + tg(x = d)$$

By obtaining a normalized dependence of $\varphi_e(x)$ and comparing it with the theoretical curves of Fig. 1.2, one may estimate the φ_e^{max} value. It should be noted that the experimental points lying closer to the diagonal are the most informative. Hence, there is no point in using the $\varphi_e(x)^{-1}$ function which ordinarily gives a straight line, since in this case it is not clear which points provide the major information.

The reader should note however that all the above conclusions are derived exclusively for the multicentral type of PSU organization (Vredenberg-Duysens model). It is probably unreasonable to ascribe this model to the enormous variety of photosynthesizing organisms and certainly other PSU models have been put forward from a number of different studies using a range of photosynthetic material. Thus, the main assumption used in equation (2) is not valid in terms of these models.

However, in the work of Barsky et al. (1974) an important conclusion was substantiated, that for any given value of φ_e^{max} all other PSU models have a less pronounced trough in the coordinates of Fig. 1.2 than that for the limiting multicentral one. Consequently, the estimation of φ_e^{max} made according to the multicentral type of PSU (theoretical curves in Fig. 1.2) must give underestimated values for any other PSU organisation which may actually exist in vivo.

If this estimation give $\varphi_e^{max} = 0.90 \pm 0.05$, this means that the true value of this important parameter characterizing the maximal ability of energy trapping by active RC from light harvesting Chl's is $0.85 < \varphi_e^{max} < 1$ (see Barsky et al., 1974 for details).

The application of the above method to a number of photosynthesizing organism showed (Barsky and Borisov, 1971; Borisov and Il'ina, 1973; Barsky et al., 1975) that their φ_e^{max} reaches 0.9. Hence it proves that in photosynthesis Nature invented and realized an efficient photoelectric system which converts light into electricity with an energy yield φ_s not lower than 10%.

Of course, in photosynthesis this transient photoelectric stage is rapidly turned into longer lived and more stabalised states so that energy is partially lost and partially converted into the form of the potential of chemical bonds. The losses must not be too great if the ultimate goal is the synthesis of some valuable energy rich material. Here we can afford to have energy conversion with a yield of ~10%, so that ~1% of incident solar energy would be stored in the material (and even less in materials of high value). This is true for natural products of photosynthesis and also would be true for any artificial products that may be manufactured photochemically in the future.

However φ_s should not be considerably less than 10% if our main concern is energy itself. It is most unlikely that we should succeed in obtaining electical energy directly from chloroplasts or photosynthetic cells. We either have to shunt the biological electron trasport chain via an efficeint energy-consuming reaction (best of all, by H_2O-cleavage for accumulation of hydrogen) or ... we have to try and construct a new type of technological photoelectric cell based on some special organic dyes. Hence, the principles governing the building of the photosynthetic pigment complex structures are worthy of a most thorough study, particularly from the point of view of the possibility of their being extrapolated to artificial photoelectric systems.

1.3.2. Conclusions

The author hopes that after reading this section one arrives at the conclusion that in the primary cycle of photosynthesis Nature has "invented" an efficient photoelectric system the exact details of which are as yet unknown to us. Its greatest advantage is that it can provide a high (~0.9) quantum yield of solar energy conversion without the apparent need for superpure crystallinity of a light harvesting substance as required by man-made inorganic semi-conductor photocells. The efficiency of a photosynthesis-type photoelectric system may reach 10%. Thus, a new line of solar converters may be advanced which are highly promising and will undoubtedly justify all the effort made in this direction. There is no doubt that invaluable information can be obtained from investigations of ultrafast primary stages of photosynthesis. On the other hand, time is ripe to start model work the goals of which should be to synthesize artificial "reaction centres" (obviously not with chlorophylls) and the manufacture of a "molecular dye antenna" on the basis of thin polymeric films.

In the next section we will discuss the improvements in the molecular antenna that can help raise the efficiency of energy conversion up to its upper limit which is estimated for such systems as ~15% (Bolton, 1978).

1.4. SIMULATION OF LIGHT COLLECTING SYSTEM OF PHOTOSYNTHESIS

From the numerous qualities of the natural photosynthetic apparatus we must choose a few major characteristics which should if possible be reproduced and developed in model systems. These are, undoubtedly, the qualities which allow a photoelectric system to be built with φ_e close to unity without the need to use light-absorbing materials having highly crystalline structure. In my opinion, this should be a physical mechanism of resonance current-free migration of energy, previously unknown in photoelectric systems, that ensures effective delivery of energy from a light-absorbing material to energy-converting reaction centres.

Let us compare photosynthetic antenna with that of semiconductor photocells. Electrons and holes produced by light in the "antenna" of inorganic semiconductor photobatteries are liable to recombination while diffusing to the borders of "p—n" transition which acts as the energy trap. However, impurities or dislocations present in a crystalline antenna can become the centres of charge trapping, which results in a significantly greater possibility of wasteful charge recombination and thus in the increase of the quantum yield of energy losses (decrease in φ_e). Similarly, in photosynthesis quantum losses occur in the "antenna" material during the course of energy delivery via an inductive excitation transfer mechanism to the reaction centres.

Consequence I. Resonance migration of energy is effective in aromatic molecules which have an extensive network of double π-electron bonds. Hence, it is dyes that should be used for building of light-absorbing and energy-transporting antenna of model systems.

So, Naure has equipped photosynthesizing organisms with an unique photoelectric system assimilating up to 80—90% of antenna-absorbed light quanta. The question arises, whether these principles are valid for constructing model photoelectric systems. In my opinion, the organization of primary photosynthetic processes has the following advantages as far as attempts of simulation are concerned.

(a) If, when being employed, part of antenna molecules is subjected to photodegradation, with their optical spectra being shifted to the short-wavelength region (the process which is quite possible for some classes of dyes), they will still be able to perform their antenna function but already in a different region of the spectrum and, which is most important, will produce no parasitic centres where electronic excitation energy is lost, as is the case with photosensitive semi-conductor crystals.

(b) If some of the centres bringing about energy conversion are subjected to photodegradation so that the efficiency of excitation trapping becomes several times lower than that of active reaction centres (which is quite possible see Heathcote and Clayton, 1977; Godik and Borisov, 1977), the efficiency of current generation by the whole system will hardly decrease even when 30 to 50% of all the centres undergo destruction.

(c) For absorption of 70—80% of light, a quantum should encounter $(1—2) \cdot 10^3$ dye molecules. Since light should be absorbed over a wider spectral range, about 600—900 nm (in this respect photosynthesis is far from being optimal), this figure reaches $(1—3) \cdot 10^3$ molecules. Fig. 1.4 shows a simple model of such a system demonstrating successive absorption of light by layers of a dye-stained substance, with the shift of long-wavelength absorption progressively increasing. Accordingly, light-induced singlet excitations migrate towards a long-wavelength form of pigment (λ_m) and are trapped there by some dorm of transformation centres. Every layer with an absorp-

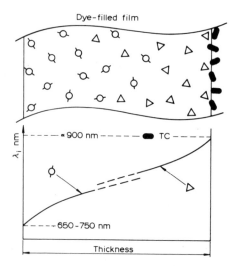

Fig. 1.4. The schematic representation of the multilayer antenna impreganted with two dyes from both sides. The absorption maxima of molecules are progressively shifted to longer wavelengths due to changes in the matrix characteristics in such a way as to favour excitation "downhill" transfer to transformation centres on the right hand side.

tion peak, λ_i contains about 200 arbitrary monolayers of a dye or about $(2—3) \cdot 10^{14}$ molecule/cm^2. It is preferable that not more than 1 or 2 dyes should be used and that their spectral properties should gradually change across the film due to changes in the physicochemical characteristics of dye-impregnated matrix. In the scheme presented in Fig. 1.4, as well as in some other models, the above-mentioned figure of about 10^3 molecules is close to the ratio of the concentrations of the antenna molecules and transformation centres since the maximum quantity of the latter corresponds to one monolayer on the right of the film. In direct sun light a dye molecule can absorb up to 5—10 quanta per second. Hence, the transformation centre that serves on an average $\sim 2 \cdot 10^3$ antenna molecules traps up to $2 \cdot 10^4$ times per second. This situation is quite realistic. For example, in photosynthesis, regeneration of active reaction centres at the level of charge exchange with the primary donor-acceptor system lasts $10^{-4}—10^{-5}$ sec.

Consequence 2. Good model systems require molecules with a lifetime of at least several years in direct sun light. Extensive experience proves that this requirement is hard to realize, especially for dyes absorbing in the red and near infrared optical region. We shall discuss this important question in Section 1.4.1.

1.4.1. Mechanism of energy migration

In accordance with the discussions above, light-absorbing antenna molecules must ensure that energy is delivered to transformation centres with a high quantum yield by one of the physical mechanisms of current free energy migration that is insensitive to the purity and order of the surrounding molecules. Potentially exchange-resonance energy migration at triplet levels would be very effective. The lifetime of these states may reach 10^{-3} sec at room temperature, whereas the time between two successive jumps, e.g. with benzene derivatives, is 10^{-12} sec and between nucleic acid bases $10^{-10}—10^{-11}$ sec. Hence, within a lifetime of a triplet, $10^7—10^9$ migration acts can occur. However, compared to singlets, the probability of molecules in this state undergoing irreversible photodestruction becomes high. For this reason it is short-lived singlet excitations that are more promising.

(1) The minimum period of time between two successive jumps involving singlet excitations is about 10^2 times shorter than in triplets. Hence, the singlet mechanism requires much less time for the energy of electronic excitation to be delivered from antenna to transformation centre and consequently it is better protected from irreversible photodegradation.

(2) Singlet migration is more long-range. The critical distance (R_0) of triplet energy migration is typically 10—12 Å, since donor and acceptor electron orbitals should overlap. For singlet energy migration, R_0 may reach in certain dye molecules 100 Å, and, with successful mutual orientation, even 130 Å (Knox, 1975).

(3) The short critical distance for triplet—triplet energy migration signifies that light-absorbing antenna must consist almost completely of dye molecules. Thereby, a number of difficulties arise (e.g. formation of nonfluorescent associates that rapidly quench electron excitations into heat). On the other hand high R_0 values for singlet energy migration, make it possible to use a dye-impregnated matrix (e.g. thin polymeric film), which may prove more feasible technolgically.

(4) There is another, not less important, point to consider. The quantum yield of photoformation of singlet states is equal to unity while for triplets is less than unity. Only a few dyes have quantum yields of triplet formation exceeding 0.9, which significantly limits our choice.

(5) The conversion from singlet to triplet state entails loss of a certain portion of the initially absorbed energy. This can be partially justified, since it is a penalty for stabilization of the new state in time.

However if this conversion takes place then many of the molecules lose more energy than is necessary, and this process causes additional heating of the system, which will decrease the efficiency of the useful processes and increases the rate of photodestruction.

(6) Finally the most important question is how to make dye molecules stable in sun light. The number of quanta absorbed by a dye molecule over a period of several years amounts to several tens of millions. If the singlet lifetime is of the order of 10^{-9} sec the overall time which a molecule spends in this excited state would not exceed several tens of millisecond. In contrast with triplets the situation is much worse because their lifetime is usually longer. Therefore to avoid photodestruction it would be necessary to synthesize dye molecules with the energy of their first triplet state less than 0.9 eV, a value which would be insufficient to activate oxygen to its highly reactive singlet state and thus avoid photooxidation. The alternative to this is to provide singlet energy migration within the shortest possible time so as to reduce drastically the quantum yield of triplet formation.

Overall the above discussion leads to the conclusion that artificial model systems should use singlet mechanisms for energy migration in the antenna.

1.4.2. Light-absorbing substance

To be ideal for this role, the molecular species must meet the following demands.

(1) The initial materials must be available and its production cheap.

(2) Their integral absorption must be high, with wide absorption bands. The dipole conversion moment corresponding to (0.0) transition from the main to the first singlet excited state should be of the order of 5D or more.

(3) The rate constant for intersystem crossing must be as low as possible, since as mentioned above it is via the triplet state that photodestruction usually proceeeds. It is noteworthy that in photosynthesis Nature has

"invented" special pigments, i.e. carotenoids, which are contained in all photosynthesizing organisms and function to discharge chlorophyll triplets and thus protecting chlorophyll from irreversible photoreactions.

1.4.3. Spectral characteristics

Naturally, the antenna molecules must absorb light in the spectral area where the maximum of solar radiation penetrates the gas envelope of the Earth. The inclination of the Sun's rays changes during the day and from season to season. For middle lattitudes, its average annual value is about $20°$, but considering that the bulk of the Sun's rays fall onto the Earth in summer and at midday, then the inclination angle should be regarded as being equal to $30-40°$. Curve 2 on Fig. 1.1 show the spectral distribution of solar energy for such a case with its main portion (~72%) being in the region 400—1200 nm. However the considerations listed below significantly limit its usage by photoelectric systems involving dye molecules.

(1) Light creates in the antenna pigments short-lived singlet excitations. The lifetime of the independent molecules is measured in nsec, but in a highly condensed matrix there arise centres with a higher probability of quenching onto which excitations from the rest of the molecules migrate. For example, in antenna pigment—protein complexes from purple bacteria the lifetime of bacteriochlorophyll fluorescence appeared, according to the data Godik and Borisov (1977) to be equal to about 1—1.5 nsec. To obtain a high quantum yield of primary energy transformation (trapping by transformation centres), it is necessary that the rate of this process (K_{tr}) should be about an order higher than that for wasteful processes, i.e. $>10^{10}$ sec^{-1}. Thus the lifetime of excitations in the antenna, should be correspondingly, of the order of 10^{-10} sec or somewhat less. Indeed $\tau_{fl} \sim 10^{-10}$ sec have been proved to exist in vivo by psec laser experiments (Searle et al., 1977; Campillo et al., 1976; Paschenko et al., 1977) as well as by my own data obtained with a time-lever method (see Borisov and Godik, 1972; for the details Borisov et al., 1977).

(2) In contrast to the above, carriers involved in the energy transformation reactions should be stabilized in time to the level of at least 10^{-4} sec, i.e. 10^6-fold. For the reverse processes to be reasonably weak, the corresponding rate constant (K_r) is to be 10^6-10^7 times as low as that of the forward one. If we assume that these constants are bound with each other via Boltzman's function

$$K_r = (10^3 - 10^4)\text{sec}^{-1} = K_{tr} \cdot e^{-\Delta W_s/kT} \simeq 10^{10} \cdot e^{-\Delta W_s/kT}$$

and consider that the system will function at a temperature of $300-330°$K, we shall have minimum energy losses (ΔW_s) in each quantum that has reached the transformation centre, equal to 0.5 eV. Hence, for molecules absorbing radiation at 1250 nm, where the quantum energy is 1 eV, only half of the energy can be fed to an electric or electro-chemical chain (these will entail new, specific, losses). The situation with shorter wavelength quanta is much worse. In the process of migration of excitations from the corresponding short wavelength forms to pigment that absorbs at 1250 nm, all their excessive energy will be quenched into heat; e.g. for quanta at 500 nm ($h\nu$ = 2.5 eV) these losses will be as high as 80%. To determine the optimal storage efficiency of this system the maximum should be derived from equation (1). If we vary the position of λ_m in equation (1), we shall obtain maximum of φ_s that functions in the area of 800—900 nm.

(3) It is evident that condensation of solar radiation should produce noticeable gain in φ_s value (see Ross and Hsiao, 1977 for particular case of photochemical systems). However, ΔW_s and $\varphi_e(\lambda)$ in equation (1) are temperature dependent. Additional heating of the photoelectric device stimulates wasteful processes. Besides, higher temperatures accelerate the wear of the cell so that some reasonable compromise should be realized in real systems.

1.4.4. Conclusions

(i) To decrease energy losses the most short-wavelength value should be taken for λ_m i.e. $800 < \lambda_m < 900$ nm.

(ii) The range of spectral sensitivity of these systems should not be too wide. According to our calculations, the ratio of the extreme wavelengths should not be higher than 1.3—1.4.

(iii) The above considerations allow the conclusion to be made that it is reasonable to have at least 2 photosystems successively absorbing light, e.g. over the ranges of 400—600 and 600—900 nm, and to use more long-wavelength rays in thermal devices. The question concerning the increase in φs in cells having two photosystems has been explicitly settled in Bolton's summarizing work (1978). However below we shall prove that for solar cells based on excitation migration in organic dyes only long-wavelength photosystems ($\lambda \geqslant 600$ nm) have the correct properties for construction of a realistic model system.

1.5. ENERGY MIGRATION

Let us consider a simplified model. Supposing that we have a system of N identical, closely positioned dye molecules and the energy of electronic excitation migrates from one of them to another via the inductive resonance mechanism. In addition, there is a molecular energy trap. So, with a light quantum being absorbed by one of the N antenna molecules, a quantum of electronic excitation may be either trapped as a result of random migrations, or deactivated en route in one of the antenna molecules. Supposing that the antenna molecules are characterized by a mean lifetime of singlet excitation τ_{fl}. The question arises, under what conditions will the quantum yield of the trapping of electronic excitations be $>90\%$? It is known from the theory for homogeneous two-dimensional pigment ensembles * (Knox, 1968), that the number of migration acts (n_m) for a ($I-e^{-1}$) portion of excitations to be trapped, is

$$n_m = 0.72 \text{ N log N} + 0.26 \text{ N}$$

Apparently, for at least 90% of the photoinduced excitations in the antenna to be trapped, the time for n_m migration acts to occur should be not more than 1/9 of τ_{fl}:

$$n_m \cdot \Delta\tau_j = (0.72 \text{ N log N} + 0.26)\Delta\tau_j \leqslant \frac{\tau_{fl}}{9} \qquad (4) \text{ **}$$

where $\Delta\tau_j$ is the mean time between two successive migration acts which is a function of an average intermolecular distance. It was mentioned above that, to provide light absorption over the 600 to 900 nm range, about $2 \cdot 10^3$ antenna monolayers are needed, i.e. for the model shown in Fig. 1.4 N $\sim 2 \cdot 10^3$ and correspondingly, $n_m \sim 5 \cdot 10^3$. If we take for pure antenna $\tau_{fl} \sim 10^{-9}$ sec, we shall obtain the lifetime of excitation in active complex $<10^{-10}$ sec and from the ratio, $\Delta\tau_j < 4 \cdot 10^{-14}$ sec. This value of $\Delta\tau_j$ characterizes typically excitonic energy migration at the energies of inter-molecular interaction of the order of 0.1 eV (Förster, 1960), which is only realized if complex organic molecules are in close contact. Such close contact is inherent in crystalline structures, but should be avoided if possible since it leads to the formation of wasteful quenching centres.

In random arrangements of organic molecules, non-fluorescent quenching associates are frequent. Thus in solutions, when the dye concentration is increased, the mean number of migration acts of singlet excitations at first increases but then drops due to effective quenching by non-fluorescent associates and excimers (Beddard and Porter, 1976). Interesting data has been obtained in my laboratory involving the adsorption of chlorophyll a onto

---

* The formula for three-dimensional ensemble is not greatly different.
** The approximate character of formula (4) should be emphasised. (a) It is valid for an absolute trap. If, say, the trapping constant of a transformation centre is equal to that of reversed energy migration the number n_m needed would be increased (less than twice) and $\Delta\tau_j$ inversely reduced. (b) On the other hand, if the trap consists of several molecules (e.g. in bacterial photosynthesis we have two P_{870} molecules and apparently the same is true for P_{700} of plants and P_{840} of green bacteria) the trapping rate should be correspondingly greater. (c) If absorbed light creates excitons delocalized over a number of antenna molecules they may be trapped distantly. It is equivalent to the situation with the trapping efficiency increased and, the number n_m decreased considerably.

lipid surfaces and onto the surface of crystalline powder MgO (unpublished data). When the surface monolayer was incompletely filled, the chlorophyll failed to form associates since the adsorption energy was markedly greater than that of association. Under these conditions the pigment was characterized by a significant fluorescence yield and by $\tau_{fl} \sim 4 \cdot 10^{-9}$ sec. But when the surface concentration of chlorophyll in the monolayer was increased, then there was a sharp decrease in the values of the yield and the lifetime of fluorescence due to migration of energy to quenching associates.

So, this way of antenna assembly suggestes that a careful selection of dye molecules which do not have the tendency to associate must be made, or if interactions did occur then they should be specific like the antenna models discussed in the works by Norris et al. (1975) that seem to exclude strong quenching of excitations.

However, there is an alternative way. It is known from studies in the field of photosynthesis that the quantum yield of energy migration from antenna to RC may reach a value of 0.9, with the long-wavelength form containing up to 50—60 bacteriochlorophyll molecules per RC (Barsky and Borisov, 1971; Barsky et al., 1975). Hence the mechanism is efficient at $N \sim 50-60$, corresponding to $n_m \sim 60-80$. Assuming that the intrinsic radiation lifetime of the dye to be 10^{-8} sec and the quantum yield of triplet formation not to be more than 0.5, we obtain for free molecules a lifetime of $\sim 5 \cdot 10^{-9}$ sec. With energy migration being efficient, quenching in parasitic centres will make it several times lower ($\tau_{fl} \sim (1-2) \cdot 10^{-9}$ sec;). Substituting the values of τ_{fl} and $n_m \sim 60-80$ in equation (4), we obtain:

$$\Delta\tau_j \gtrsim (1-2) \cdot 10^{-12} \text{ sec}$$

What is the relation between this $\Delta\tau_j$ and the average intermolecular distance? It follows from the theory of Förster (1960), which now has considerable experimental support, that for $\Delta\tau_j \ll \tau_{fl}$

$$\Delta\tau_j = \left(\frac{R}{R_0}\right)^6 \cdot \tau_r \tag{5}$$

where R_0 is the critical distance of energy migration with respect to the intrinsic radiation lifetime of the given pigment τ_r.

In the work of Kenkre and Knox (1974), formula (5) was substantiated up to time intervals when after the act of absorption of a light quantum the coherence in the induced excitation disappears due to its interaction with molecular rotations and oscillations. From this it follows that for the above dye molecules, R may be calculated using equation (5) up to the values of $\Delta\tau_j$ of the order of $10^{-12}-10^{-13}$ sec. In our case it gives

$$R/R_0 = 0.20 - 0.22$$

But besides molecular characteristics and environment parameters, the parameter R_0 is highly dependent on the wavelength (λ_{max}) of the red peak of absorption (Förster, 1960).

$$R_0 \sim \frac{\lambda_{max}}{2\pi n \sqrt{\epsilon(\nu) \cdot F(\nu) \, d\bar{\nu}}}$$

If the minimum values of R_0 for short-wavelength dyes are of the order of 15—20 Å, then for pigments with a absorption bands in the region of 800—900 nm and good overlapping of the absorption $|\epsilon(\lambda)|$ and fluorescence emission $|F(\lambda)|$, R_0 can be as large as 100 Å. Hence for these dyes efficient energy migration (hundreds of acts within about 100 picoseconds) may proceed without direct contact at distances of about 20 Å.

So, with a "loose" matrix of, for example, a synthetic polymer or micro-cellular powder which could be permeated with dye(s) in a high concentration(s) with few quenching associates being formed, the required artificial photosystem antenna absorbing in the spectral region of 600—650 to 850—900 nm might be realised, according to the heterogeneous sandwich model shown in Fig. 1.4.

The main demand for the construction of this model matrix is that there must be a certain intermolecular distance for the antenna molecules and if possible this should be localized around the most favorable values of R calculated with the help of equations (4) and (5). Moreover the use of several dyes is probably not reasonable from the technological point of view. Therefore it seems much more promising if the spectral heterogeneity of the antenna in the model system can be induced by changing the physico-chemical characteristics of the thin matrix impregnated with only two dyes (or even just one) as is shown diagrammatically in Fig. 1.4.

Pigments (or their spectral forms) should be successively coupled at the level of energy migration with one another down to the terminal long-wavelength form which acts as a trap with respect to the others. This offers great possibilities for the basic requirement that the antenna absorbs light over a large spectral interval. In fact it is known that one photosynthetic RC may efficiently trap electronic excitations from 50—60 molecules of a long-wavelength form of chlorophyll. If each of these 50—60 molecules may, in their turn, trap excitations on an average from ~ 50 more shorter-wavelength antenna molecules with the same efficiency, the time of migration and, correspondingly, the total energy losses in the system will increase by not more than a factor of two, and the size of antenna will increase ~ 50-fold and reach about $2 \cdot 10^3$ molecules per transformation centre (Borisov, 1976). Studies of molecular antenna in green bacteria (Fuller et al., 1976) indicate that this type of organisation and energy transfer takes place in these organism where, according to Olson et al. (1976), there are 80 bacteriochlorophyll molecules and $\sim 10^3$ bacterioviridin molecules per RC.

1.6. TRANSFORMATION CENTRES

Let us formulate the conditions for energy coupling of photosystem pigment antenna with energy transformation centres (TC). They are not necessarily similar to photosynthetic RC which are discussed in depth in Chapter 3 of this book. First of all the properties of a suitable TC should be characterized.

The major function of a TC is to isolate an electron from the photo-excited molecule and to stabilize this state by spacially separating the charges in the time of 10^{-4} to 10^{-2} sec. To a certain extent these functions are fulfilled in bimolecular charge transfer complexes. A simplified picture of this phenomenon is presented in Fig. 1.5. On excitation either by direct light absorption or by migration of electronic excitation, the electron of molecule C is transferred from initial orbital c_0 to excited cb* spread over the entire complex. This is essentially equivalent to the transfer of half an electron charge from molecule C to molecule B. However, the space over-lapping of orbitals cb* and c_0 causes a quick back transfer within the

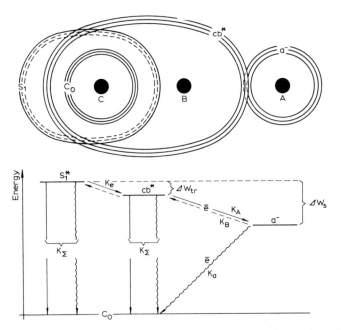

Fig. 1.5. The schematic diagram of two step charge transfer complex functioning as molecular transformation centre. C- and B-molecules (or complexes) form the initial charge transfer complex; c_0 and cb* are the ground and excited orbitals of π-electron from C; A is the acceptor molecule(s) with its vacant orbital a^- overlapped with cb*. In the lower part of the figure the corresponding energy diagram is shown.

characteristic fluorescent time $\sim 10^{-8}$ sec with energy being lost. Therefore one of the requirements of creating a TC is a double complex so that charge transfer can occur (Borisov, 1973) along the lines shown schematically in Fig. 1.5. Acceptor molecule A adheres to the complex CB on the side most remote from the initially excited molecule C. Owing to good spacial overlap between orbital cb* and a vacant orbital a^- electron reduces molecule A. Now electron transfer $a^- \to c_0$ is strongly prohibited because the corresponding orbitals are spatially distant, and stabilization in time of the state with separated charges $C^+ A^-$ may take seconds.

What are the energy characteristics of the corresponding orbitals?

(1) Apparently, at the level of energy migration, TC should be well coupled with the long-wavelength form λ_m of the antenna pigments. For slow energy migration, the TC absorption peak should be shifted to the region of the fluorescence maximum of the antenna pigment form λ_m, i.e. the TC absorption band may be even more shifted to longer wavelengths within $200-300$ cm^{-1}. With more intensive excitonic molecular interactions between the pigment form λ_m, and TC, it would be preferable that the absorption peaks of both should be almost the same.

(2) On migration of electronic excitation to TC there arises a short-lived state cb*. It may be transferred into the major state c_0 with the help of trivial intermolecular mechanisms (e.g. fluorescence, conversion to triplet state, conversion into heat) within the characteristic fluorescence time $\sim 10^{-8}$ sec. Alternatively, after overcoming the energy barrier ΔW_{tr}, cb* may give birth again to a singlet excitation in the antenna (note that in this case the resulting luminescence of the antenna dye should be regarded as delayed fluorescence even if its lifetime is in the picosecond range). The transition cb* $\to a^-$ needed for the stabilization of the separated charges is at the expense of the loss of a portion of energy $\Delta W_s \sim 0.5$ eV. Since the value ΔW_s is significantly lower than that for transition cb* $\to c_0$ (~ 1.3 eV) the quantum yield of the cb* $\to a^-$ transition is close to unity.

(3) The intermediate state cb* plays a very important role. It is this state that allows the high quantum yield of the total transition, to be realized, i.e. antenna singles excitation $\to$ state $B^+ A^-$

In the work of Borisov (1976), it was pointed out that to obtain high quantum yields, the energy gap ΔW_{tr} should be from about 0.05 to 0.1 eV. The reason for this is that energy migration in the λ_m form is very quick, with the rate constant being $> (\Delta \tau_j)^{-1} \sim 10^{12}$ sec^{-1}. Therefore for effective trapping of singlet excitations from λ_m, the rate constant of transition into cb* should be $> 10^{12}$ sec^{-1}. Two conditions should be observed therefore.

(a) The activation barrier ΔW_a for excitation transition into cb* should be such as to ensure a rate of jumping of $\sim 10^{-12}-10^{-13}$ sec. At an oscillation frequency of a free electron of $10^{15}-10^{17}$ sec^{-1} and a temperature of $300°$, we can derive from Boltzman's formula $\Delta W_a \sim 0.3$ eV. This means that molecules of the form λ_m adjacent to TC should be in intimate contact with molecules C and B.

(b) Accordingly, the dissipation of energy ΔW_{tr} must proceed within $10^{-12}-10^{-13}$ sec. As was stated by M.V. Fock and V.I. Godik (unpublished), this condition is easy to meet if the portion of dissipated energy is so small that it can be assimilated by one oscillation, since excitation of many oscillations within $10^{-12}-10^{-13}$ sec is hardly probable from the point of view of physics.

On the other hand, ΔW_{tr} cannot be too small as the back process of feeding energy to antenna will be too effective. Therefore we obtain finally

$$\Delta W_{tr} \simeq (0.05 - 0.1)eV$$

Values close to this (0.12 eV) have been reported by Shuvalov and Klimov (1976) for RC of photosynthesizing bacteria *Rh. viridis*. While in more recent work by Dr. Godik values even smaller than this has been obtained for *Rh. rubrum* chromatophores ($\Delta W_{tr} = 0.03-0.06$ eV).

The P^F state of primary charge separation has recently been discovered in reaction centres of purple bacterium *Rh. sphaeroides* R-26 (see Chapter 3 of this book and also Parson et al., 1975, Rockley et al., 1975). Many features of that state are similar to those discussed above in this section, although some basic questions still need to be answered before we can succeed in making an artificial TC capable of trapping efficiently excitations from extended molecular antenna.

1.7. CONCLUDING REMARKS

Of course, there is a long way to go from the systems discussed above to a working photoelectric model. The most serious difficulties will apparently be encountered in the transition from the microscopic level of biomembranes (dimensions $\sim 10^{-6}$ cm) to elements of "macroscopic" dimensions (~ 1 cm).

But we already have an encouraging example (Drachev et al., 1978) where natural complexes of rhodopsin have been successfully incorporated into thick (~ 1 mm in diameter) artificial membranes. Consequently, the various obstacles mentioned in the above sections may not be insurmountable.

In my opinion, this area of research has quite definite advantages.

(1) Since a crystalline structure is not indispensable, there is hope that film photobatteries will be used which could be glued on to any surface.

(2) Dyes have narrow absorption bands which means that specific regions of the visible spectrum can be utilized. For example, a convenient region would be the near infrared (750—900 nm), since transmitted light could still be used by plants and algae which absorb at shorter wavelengths. In this way artificial and natural photochemical storage processes could operate together.

It is this type of reasoning which makes the author optimistic towards the significance and the perspectives of studies in the primary energy-conversion processes of photosynthesis and of attempts at their modelling.

REFERENCES

Barsky, E.L. and Borisov, A.Yu. (1971) J. Bioenerg., 2, 275—281.

Barsky, E.L., Borisov, A.Yu., Godik, V.I. and Il'ina, M.D. (1974) Mol. Biol. (SSSR), 8, 739—745 (Engl. transl.).

Barsky, E.L., Borisov, A.Yu., Fetisova, Z.G., Il'ina, M.D. and Samuilov, V.D. (1975) Mol. Biol. (SSSR), 9, 221—226 (Engl. transl.).

Beddard, G.S. and Porter, G. (1976) Nature, 260, N 5549, 366—367.

Bolton, J.R. Science (in press).

Borisoc, A.Yu. (1973) Sovremennie Problemi Photosynthesa (Current Problems of Photosynthesis), pp. 161—174. Moscow University Edition, Moscow.

Borisov, A.Yu. (1976) Brookhaven Symp. Biol., N 26, 130—131.

Borisov, A.Yu. and Godik, V.I. (1972) J. Bioenerg., 3, 211—220.

Borisov, A.Yu. and Il'ina, M.D. (1973) Biochim. Biophys. Acta, 325, 240—246.

Borisov, A.Yu., Fetisova, Z.G. and Godik, V.I. (1977) Biochim. Biophys. Acta, 461, 500—509.

Calvin, M. (1976) Photochem. Photobiol., 23, 425—444.

Campillo, A.J., Huyer, R.C., Monger, T.G., Parson, W.W. and Shapiro, S.L. (1977) Proc. Natl. Acad. Sci. U.S.A., 74, 1997—2001.

Clayton, R.K. (1962) Photochem. Photobiol., 1, 313—321.

Clayton, R.K. (1967) Photochem. Photobiol., 5, 807—821.

Drachev, L.A', Kaulen, A.D. and Skulachev, V.P. (1978) FEBS Lett., 87, 161—167.

Förster, Th. (1960) Comparative Effects of Radiation, pp. 300—341. Wiley and Sons Inc., New York and London.

Fuller, R.C., Boyce, C.O. and Oyewole, S.H. (1976) Int. Conf. Primary Electron Transport and Energy Transduction in Photosynthetic Bacteria, p. MB 7.

Godik, V.I. and Borisov, A.Yu. (1977) FEBS Lett., 82, 355—358.

Hall, D.O. (1976) FEBS Lett. 64, 6—16.

Heathcote, P. and Clayton, R.K. (1977) Biochim. Biophys. Acta, 459, 506—515.

Kenkre, V.M. and Knox, R.S. (1974) Phys. Rev. Lett., 33, N 14, 803—806.

Knox, R.S. (1968) J. Theor. Biol., 21, 244—256.

Knox, R.S. (1975) Bioenergetics of Photosynthesis, pp. 183—224. Academic Press, New York, San Francisco and London.

Kok, B., Rurainsky, H. Harmon, E. (1964) Plant Physiol., 39, 513—522.

Kok, B., Rurainsky, H. Owens, O. (1965) Biochim. Biophys. Acta, 109, 347—356.

Leontowitch, M.A. (1974) Usp. Fiz. Nauk (Soviet Advances of Physics Sciences), 114, N 3, 555—558.

Norris, J.R., Scheer, H. and Katz, J. J. (1975) Ann. N.Y. Acad. Sci., 244, 260—280.

Olson, J.M., Prince, R.C. and Brune, D.C. (1976) Brookhaven Symp. Biol., N 28, pp. 238—246.

Parson, W.W., Clayton, R.K. and Cogdell, R.J. (1975) Biochim. Biophys. Acta, 387, 265—278.

Paschenko, V.Z., Kononenko, A.A., Protasov, S.P., Rubin, A.B., Rubin, L.B. and Uspenskaja, N.Y. (1977) Biochim. Biophys. Acta, 461, 403—414.

Rockley, M.G., Windzor, M.W., Cogdell, R.J. and Parson, W.W. (1975) Proc. Natl. Acad. Sci. U.S.A., 72, 2251—2255.

Ross, R.T. and Hsiao, T.L. (1977) J. Appl. Phys., 48, 4783—4785.

26

Searle, G.F., Barber, J., Harris, L., Porter, G. and Tredwell, C.J. (1977) Biochim. Biophys. Acta, 459, 390—401.
Shuvalov, V.A. and Klimov, V.V. (1976) Biochim. Biophys. Acta, 440, 587—599.
Sybesma, C. and Beugeling, T. (1967) Biochim. Biophys. Acta, 131, 357—361.
Sybesma, C. and Vredenberg, W.J. (1963) Biochim. Biophys. Acta, 75, 439—441.
Vredenberg, W.J. and Duysens, L.N.M. (1963) Nature, 193, 355—357.
Vredenberg, W.J. and Duysens, L.N.M. (1965) Biochim. Biophys. Acta, 94, 355—360.
Vredenberg, W.J. and Slooten, L. (1967) Biochim. Biophys. Acta, 143, 583—594.
Wraight, C.A. and Clayton, R.K. (1974) Biochim. Biophys. Acta, 333, 246—260.

Photosynthesis in relation to model systems, edited by J. Barber
© Elsevier/North-Holland Biomedical Press 1979

Chapter 2

Photosynthetic Pigments and Models for their Organization in vivo

J. PHILIP THORNBER * and J. BARBER **

* *Department of Biology, University of California, Los Angeles, California, 90024, U.S.A.*
and ** *Department of Botany, Imperial College of Science and Technology, London,*
SW7 2BB, England

CONTENTS

Abbreviations

ATP, Adenosine triphosphate; CD, circular dichroism; LDAO: lauryl dimethyl amine oxide; NAD^+, nicotinamide adenine dinucleotide; $NADP^+$, nicotinamide adenine dinucleotide phosphate; PSI, Photosystem I; PSII, Photosystem II; SDS: sodium dodecyl sulphate.

2.1. INTRODUCTION

The main purpose of this chapter and the following chapter is to summarise the present state of our knowledge of how pigments are arranged in photosynthetic lipoprotein membranes of plants and bacteria so that they trap solar radiation and convert it efficiently into chemical energy. Current models for the arrangement of pigments within the building blocks of the photosynthetic apparatus as well as of the apparatus itself are presented in this chapter. Equipped with such information, the reader can compare what occurs in vivo with the observations made on in vitro systems that have used some of the components of the photosynthetic machinery in attempts to mimic portions of the in vivo process. Such studies have already been touched upon in chapter 1 and form the essence of this volume.

The major photosynthetic pigment in all organisms is chlorophyll. It has been realized for some time (see section 2.3) that photosynthetic pigments which include carotenoids and bilipigments as well as chlorophyll can be classified into two functionally distinct types: the bulk, termed antenna or light-harvesting pigments, functions to absorb photons and to funnel the resulting electronic excitation to a few specialized chlorophyll molecules which, together with an electron acceptor, constitute the energy-converting site termed the photochemical reaction centre (see chapter 3).

2.2. THE PHOTOSYNTHETIC PIGMENTS

There are a variety of chlorophylls, carotenoids and bilipigments that constitute the antenna pigments of plants and photosynthetic bacteria. On the other hand, little variation occurs in the nature of the primary electron donor in the reaction centre: it is almost certainly composed of a special pair of chlorophyll molecules in all organisms — chlorophyll a in plants and bacteriochlorophyll a, or in some species bacteriochlorophyll b, in photosynthetic bacteria. The distribution of chlorophylls, carotenoids and bilipigments among the various classes of photosynthetic organisms is summarized in Table 2.1. The chemical structure of some of the more commonly encountered pigments is presented in Fig. 2.1.

The exact absorption maximum of these pigments in vivo depends upon the immediate environment of the chromophore (see e.g. Katz et al., 1977; Sauer, 1975; Thornber et al., 1978). It has been determined that most, and perhaps all, of the photosynthetic pigments in vivo occur in a variety of pigment—protein complexes, each complex having <10 pigment molecules within its protein container (Section 2.4.1). Thus pigment—pigment (Katz et al. 1977), pigment—protein and possibly pigment—lipid interactions (Thornber et al. 1978a) give rise to the absorption maximum of a pigment observed in vivo. Since the pigments are not arranged in a regular lattice within the protein but in what appears to be a haphazard array, the interac-

TABLE 2.1A

OCCURRENCE OF PIGMENTS IN PHOTOSYNTHETIC ORGANISMS. DISTRIBUTIONS OF CHLOROPHYLLS AND BILIPROTEINS IN PLANTS AND BACTERIA [a]

	Higher plants	Chlorophyceae	Conjugatophyceae	Charophyceae	Prasinophyceae	Euglenophyceae	Xanthophyceae	Chloromonadophyceae	Eustigmatophyceae	Haptophyceae	Chrysophyceae	Phaeophyceae	Bacillariophyceae	Dinophyceae	Cryptophyceae	Rhodophyceae	Cyanophyceae	Chlorobacteriaceae	Thiorhodaceae	Athiorhodaceae
Chlorophylls																				
Chlorophyll a	+	+	+	+	+	+	+	+	+	+	+	+	+	+	+	+	+			
Chlorophyll b	+	+	+	+	+	+														
Chlorophyll c_1							+			+	+	+	+							
Chlorophyll c_2							+	(+)		+	+	+	+	+	+					
Bacteriochlorophyll a																		+	+	+
Bacteriochlorophyll b																				+
Bacteriochlorophyll c																		+		
Bacteriochlorophyll d																		+		
Bacteriochlorophyll e																		[b]		
Biliproteins																				
C-phycocyanin																+	+			
R-phycocyanin																+	+			
Allo-phycocyanin																+	+			
Allo-phycocyanin B																	+			
B-phycoerythrin																+				
C-phycoerythrin																	+			
R-phycoerythrin																+				
Others [c]															+					

[a] Adapted from Ragan and Chapman (1978) by permission.
[b] In "brown-coloured" *Chlorobium phaeobacterioides* and *C. phaeovibrioides.*
[c] Phycocyanins and phycoerythrins differing in spectral properties from those in Cyano- and Rhodophyceae.

TABLE 2.1B

OCCURRENCE OF PIGMENTS IN PHOTOSYNTHETIC ORGANISMS. DISTRIBUTION OF CAROTENOIDS IN PLANTS AND BACTERIA [d]

	Higher plants	Chlorophyceae	Conjugatophyceae	Charophyceae	Prasinophyceae	Euglenophyceae	Xanthophyceae	Chloromonadophyceae	Eustigmatophyceae	Haptophyceae	Chrysophyceae	Phaeophyceae	Bacillariophyceae	Dinophyceae	Cryptophyceae	Rhodophyceae	Cyanophyceae	Chlorobacteriaceae	Thiorhodaceae	Athiorhodaceae
Olefines																				
Bicyclic hydrocarbons (e.g. α, β, γ-carotene)	+	+	+	+	+	+	+	+	+	+	+	+	+	+	+	+	+			
Bicyclic xanthophylls (e.g. lutein, echinonone)	+	+	+	+	+	+			+		+	+				+	+			
Bicyclic xanthophyll epoxides (e.g. violaxanthin)	+	+	+	+	+	+			+		+	+				*				
Bicyclic xanthophylls (C-8 keto) (e.g. siphonoxanthin)		+			+											(occasionally)				
Monocyclic xanthophyll glycosides (e.g. myxoxanthophyll)																	+			

Acyclic hydrocarbon and xanthophylls (e.g. spirilloxanthin series and spheroidenone series)

Acyclic xanthophyll glycosides (e.g. oscilloxanthin)

Aryls

Monocyclic 1,2,5-trimethyl hydrocarbon (e.g. chlorobactene)

Monocyclic 1,2,3-trimethyl xanthophyll (e.g. okenone, warmingone)

Allenes

Bicyclic xanthophyll epoxides (e.g. neoxanthin, vaucheraxanthin, fucoxanthin, peridinin)

Acetylenes

Bicyclic monoacetylenes (e.g. diatoxanthin)

Bicyclic monoacetylene epoxides (e.g. diadinoxanthin)

Bicyclic diacetylenes (e.g. alloxanthin)

---

d Biosynthetic intermediates which may occur in traces in organisms are not listed; the carotenoid examples given are not necessarily present in all species (The authors are most grateful for the advice of Dr. D.J. Chapman in the preparation of this table).

tions for each pigment molecule in a pigment—protein complex can be quite different (see Fenna and Matthews, 1977). This can, and does, result in any one type of pigment molecule existing in different spectral forms in any one particular organism.

Different spectral forms are generally readily observed in absorption spectra of most photosynthetic bacteria. Monomeric bacteriochlorophyll a in organic solvents absorbs close to 760 nm, but in vivo there exist forms absorbing at 800, 820, 850 and 870—890 nm. The terms B800, B820, B850,

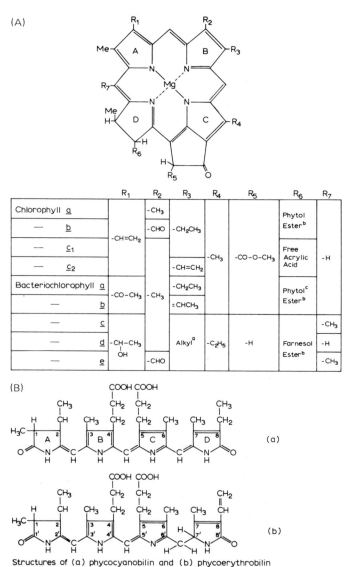

(A)

	R_1	R_2	R_3	R_4	R_5	R_6	R_7
Chlorophyll a			-CH$_3$			Phytol Ester[b]	
— b			-CHO	-CH$_2$CH$_3$			
	-CH=CH$_2$						
— c_1				-CH$_3$	-CO-O-CH$_3$	Free Acrylic Acid	-H
— c_2			-CH=CH$_2$				
Bacteriochlorophyll a		-CO-CH$_3$	-CH$_3$	-CH$_2$CH$_3$		Phytol[c] Ester[b]	
— b				=CHCH$_3$			
— c							-CH$_3$
— d	-CH-CH$_3$ OH		Alkyl[a]	-C$_2$H$_5$	-H	Farnesol Ester[b]	-H
— e			-CHO				-CH$_3$

(B)

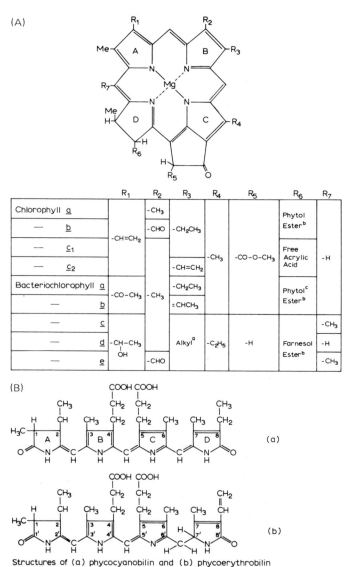

(a)

(b)

Structures of (a) phycocyanobilin and (b) phycoerythrobilin

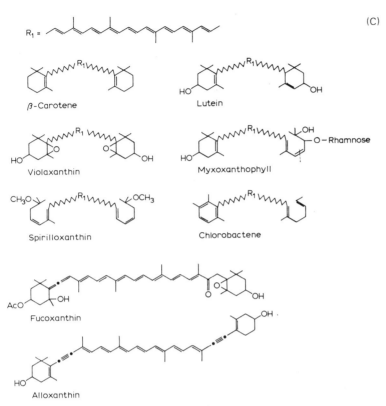

(C)

Fig. 2.1. (A) Structure of chlorophylls occurring in photosynthetic organisms. ([a]), alkyl side chain substituents are ethyl, propyl and isobutyl; ([b]), phytol and farnesol esterified to the propionic acid substituent -CH_2-CH_2-COO-alcohol; ([c]), in some members of the Athiorhodaceae the esterifying alcohol may be geranylgeraniol. Bacteriochlorophyll a and b lack the double bond between R_2 and R_3; two additional hydrogen atoms are present in bacteriochlorophyll a while one additional hydrogen atom at R_2 occurs in bacteriochlorophyll b. Figure adapted from Ragan and Chapman (1978) by permission. (B) Structure of (a) phycocyanobilin and (b) phycoerythrobilin. (C) Structure of some carotenoids typically found in plants and bacteria.

B890 are used to refer to these near IR-absorbing form (Vredenberg and Amesz, 1967). Monomeric chlorophyll a absorbs maximally in the red somewhere between 660 and 665 nm depending on the solubilizing organic solvent; however, in situ chlorophyll a absorbs at 663, 669, 677 nm, etc. (termed Ca 663, Ca 669, Ca 677, etc., see French et al., 1972). Since these forms are less separated spectrally than those observed in bacteria, computer deconvolution or fourth derivatives of low temperature spectra are needed to distinguish the various chlorophyll a forms. All photosynthetic pigment molecules absorb at lower energy wavelengths in vivo than they do in vitro. Sometimes the in vivo to in vitro shift can be quite dramatic. For example, antenna bacteriochlorophyll *b* in photosynthetic organisms (Eimjellen et al., 1963) absorbs at ~1020 nm compared to ~790 nm in organic solvents. The existence of different spectra forms of the same pigment, or of several types of pigment with different wavelength maxima, in an organism is advantageous. For example, almost invariably the special pair of chlorophyll molecules in the reaction centre absorbs at longer wavelengths than any pigments in their associated antenna; this enables the pair to act as a sink for excitation energy. Moreover, as emphasised by Borisov in chapter 1, the wavelengths of light which can be strongly absorbed by the organism, and which therefore can indirectly drive its photochemistry, are extended. Such breadth of absorbance is almost certainly essential for the survival of organisms under light-limited conditions; e.g. an organism growing in the shade of other plants or bacteria. Red algae demonstrate this point: they can frequently be found growing at considerable depths in the ocean under a filter of chlorophyll *a*-, *b*- and/or *c*-containing organisms; they survive photosynthetically because the biliproteins (phycocyanins and phycoerythrins) in their antenna system absorb strongly the only visible wavelengths of light (green and yellow) that are weakly absorbed by the surface organisms and transmitted through the water to the depths where red algae are to be found.

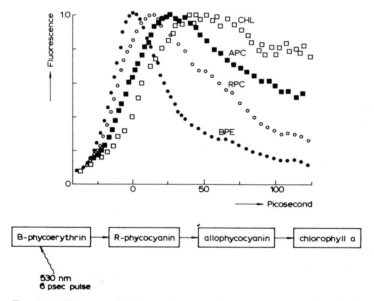

Fig. 2.2. Time resolved energy transfer sequence to chlorophyll *a* via the accessory pigments of the red alga *Porphyridium cruentum*. The measurements were made by selectively exciting B-phycoerythrin using a 6 ps pulse generated by a mode-locked Nd-glass laser. The emission from the pigments was selected out by means of interference filters and time resolved by means of a streak camera (from Porter et al., 1977).

If the spectral forms of the different pigments in the antenna are arranged so that those of lowest energy are closest to the reaction centre chlorophylls, then absorbed light energy will flow in the required direction, i.e. towards the photochemically active site. Evidence for such an arrangement occurring in vivo has been obtained for a red alga (see section 2.4 and also Gantt et al., 1976) and recently demonstrated by picosecond fluorescence spectroscopy (Porter et al., 1978), where it has been possible to time-resolve the energy transfer sequence involved (see Fig. 2.2).

It is striking that evolution has resulted in a collection of photosynthetic organisms whose absorption covers the entire range of photosynthetically useful wavelengths emitted by the sun (see chapter 1 and Clayton, 1971).

2.3. PHOTOSYNTHETIC ELECTRON FLOW

As already indicated above the majority of the pigment in photosynthetic tissue acts only as a light-harvesting system and as such transfers excitation energy to specific reaction centres where energy conversion and storage can occur. In higher plants and algae, that is in those organisms capable of oxidising water to molecular oxygen, there are two different types of reaction centres. They both contain chlorophyll a, probably in the form of a dimer, often denoted as P680 and P700 since when they undergo oxidation after excitation an absorption decrease is seen at about 680 nm and 700 nm, respectively (see Witt, 1971). There is a great deal of experimental evidence which shows that these two reaction centres operate in series so that two photons are absorbed for every electron extracted from water and transferred to a redox level capable of reducing carbon dioxide. Some of the most important experimental observations leading to this conclusion have been summarised by Williams in chapter 3 of Volume 2 of this series. Even without experimental evidence simple thermodynamic arguments also indicated that two photons of red light (i.e. corresponding to the red absorption band of chlorophyll) would be required to span the 1.63 eV of free energy necessary to drive an electron through the maximum potential difference involved in green plant photosynthesis. If the full photon energy was available, i.e. about 1.83 eV, then a one electron/one photon event would be possible. However, as explained by Knox (1977) only 1.33 eV of the photon energy can be supplied as free energy thus indicating the need for the collaboration of at least two photons.

The function of the reaction centre chlorophyll P680 is to "catalyse" light-induced charge separation so as to form a strong oxidant Z^+ ($E_0' \sim$ +0.8 V) and weak reductant Q^- ($E_0' \sim 0.0$ V). The chemical nature of Z^+ is not known but it is sufficiently oxidised to extract electrons from water. In practise, however, the oxidation of water requires the accumulation of four positive charges since the process giving rise to the evolution of molecular

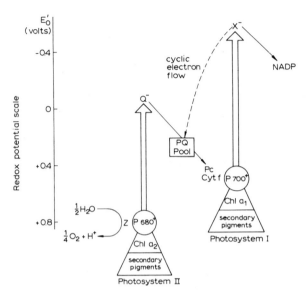

Fig. 2.3. A "Z scheme" showing how two light reactions are in series to drive an electron from water to nicotinamide adenine dinucleotide phosphate (NADP). PQ, plastoquinone; Pc, plastocyanin; cyt f, cytochrome f; $P680^+$ and $P700^+$, the oxidised forms of the reaction centre chlorophyll and Q^- and X^-, the reduced primary acceptors. The potential energy scale is expressed as redox potentials.

oxygen must involve a four electron event (see chapter 8 of this book). The primary reductant Q^- interacts, via a pool of plastoquinone (PQ) and other electron carriers (including cytochrome f and plastocyanin), with a weak oxidant generated by the reaction centre containing P700. This redox interaction gives rise to a supply of free energy which is used to synthesise adenosine triphosphate (ATP), a compound necessary for driving the biochemical cycle responsible for photosynthetic CO_2 fixation (see Walker, 1976). The primary electron acceptor to P700 has a redox potential (E_0') of about -0.6 V which is sufficient to reduce nicotinamide adenine dinucleotide phosphate ($NADP^+$), a compound also necessary to drive the CO_2 fixing reactions. A simple scheme of light-driven electron flow from water to $NADP^+$ is shown in Fig. 2.3 (for more details see various chapters in Volumes 1 and 2 of this series and also Govindjee and Govindjee, 1975). Note that the part of the scheme associated with P700 is called Photosystem one (PSI) and with P680 in Photosystem two (PSII). Also there is evidence that electrons can be cycled around as shown in Fig. 2.3 with the result that ATP is synthesised without the reduction of $NADP^+$. For obvious reasons the scheme is often called the "Z-scheme".

Fig. 2.3 indicates that each photosystem is served by its own light-harvesting pigment array. As will be explained in this chapter, this is an over-

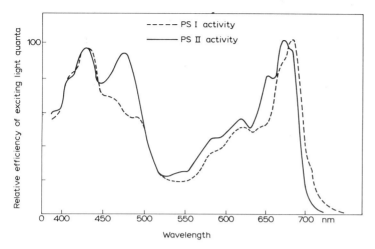

Fig. 2.4. Action spectra for Photosystem I (PSI) and Photosystem II (PSII). Note the inability of PSII to utilise light above 700 nm, and that below about 680 nm light is absorbed by both photosystems with some preference for PSII (data from Ried, 1972). - - - - - -, PSI activity; ————, PSII activity.

simplified picture since there is a certain amount of sharing of the pigment systems between PSI and PSII and moreover the degree of sharing can be varied. Nevertheless, the fact that some pigments are more closely associated with one or other photosystems is clearly seen when excitation (action) spectra for PSII and PSI are measured. Fig. 2.4 shows such spectra where it can be clearly seen that beyond 700 nm only PSI absorbs light energy, a fact which gives rise to the "red-drop" and "enhancement" phenomena (see Williams, 1977). These types of action spectra have been measured for different organisms and indicate that not only are there special forms of chlorophyll a associated with each photosystem but that chlorophyll b and the biliproteins preferentially transfer their excitation energy to PSII. On the other hand the carotenoids can function as accessory pigments to both photosystems. At room temperature mainly the chlorophyll a molecules associated with energy transfer to the PSII reaction centre are fluorescent and the maximum for the emission is at 685 nm. However, at liquid nitrogen temperature (77°K) and below some of the chlorophyll a molecules of PSI fluoresce at 735 nm while emission peaks at 685 nm and 695 nm are due to two different forms of chlorophyll a more closely associated with PSII.

Photosynthetic bacteria differ from green plants and algae in that they cannot oxidise water to molecular oxygen. Instead they extract electrons from more reduced substances such as H_2S. Although the photon energy of excited bacteriochlorophyll (~ 1.4 eV) is less than that of chlorophyll, these organisms have only one type of reaction centre which on oxidation gives rise to a maximum absorption change within the range 870—960 nm,

38

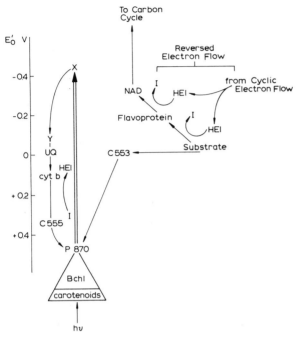

Fig. 2.5. A scheme for electron flow in bacterial photosynthesis. NAD$^+$ reduction probably occurs by reversed electron flow and not by direct reduction. P870, reaction centre bacteriochlorophyll; BChl, antenna bacteriochlorophyll; X, primary electron acceptor; Y, secondary electron acceptor; UQ, ubiquinone; Cyt b, C555, C553, cytochromes; HEI, high energy intermediate (e.g. ATP).

depending on the particular organism, e.g. P870 (see section 2.4 and chapter 3). In some ways the light-induced electron flow process of bacterial photosynthesis can be compared with those of PSI in higher plants. However the cyclic flow of electrons and associated synthesis of ATP seems to be more important in the bacterial system. Apparently these organisms use ATP to reduce NAD$^+$ by reverse electron flow coupled to substrate oxidation although it has been argued that some direct reduction of NAD$^+$ may also occur. A hypothetical scheme for electron flow in bacteria is shown in Fig. 2.5 and is based on a model presented by Govindjee and Govindjee (1975). It can be seen that the light-induced flow of electrons is thought to involve several cytochromes as well as ubiquinone.

2.4. ORGANIZATION OF PIGMENTS IN THE PHOTOSYNTHETIC UNIT

Photosynthesis researchers have long been intrigued by how chlorophylls are arranged in photosynthetic membranes (Kupke and French, 1960;

Thornber, 1975). Many years ago it was reasoned, quite correctly as it turns out, that chlorophyll was conjugated with protein in vivo. This early view was a seemingly logical extension of the discovery that other tetrapyroles occurred in protein complexes (e.g. hemoglobin, cytochromes, biliproteins, etc.). However, in the case of chlorophyll (and carotenoid) its occurrence in a pigment—protein complex(es) was difficult to substantiate, because in almost every photosynthetic organism these pigments are located in the cell in water-insoluble membranous structures which are difficult to dissolve while maintaining the specific interactions of the pigments. Furthermore, the photosynthetic pigments are not covalently bonded to protein.

The advent of synthetic detergents greatly aided an elucidation of the problem. Beginning primarily with Smith (1941), many investigators have used detergents or organic solvents to dissolve photosynthetic membranes, and then applied standard protein fractionation techniques to purify pigment—protein complexes from the resulting solution. Such research has not been devoid of controversy — several scientists believed that the complexes obtained were artifacts produced by the association of solubilized chlorophyll with membrane proteins; and indeed there are some examples of such in the literature (cf. Thornber, 1975). There were also groups that believed that chlorophyll, being a lipid, would be located in the lipid phase of the membrane and therefore would not require protein for its organization. Models for the structure of the photosynthetic membrane based on this notion appeared some years ago and have worked their way into many textbooks, but the gradual and general acceptance of the occurrence of chlorophyll within protein molecules and the appearance of a fluid mosaic model for membrane structure (Singer and Nicholson, 1972) have changed such views (see section *2.4.3* and *2.4.5*). Antenna chlorophylls must be organized in some manner so that the molecules are >10 Å apart or otherwise excitation energy quenching will occur among them (cf. Beddard and Porter, 1976) resulting in lower efficiencies of energy transfer to the reaction centres than observed. Again, lipids (particularly glycolipids which are prevalent in chloroplast membranes) have been proposed to act as spacers between adjacent chlorophyll molecules (cf. Beddard and Porter, 1976); however, the recent acceptance of detergent-isolated, essentially lipid-free chlorophyll—protein complexes as reflecting the organization of chlorophyll in vivo has now turned the attention of workers in the field almost entirely towards protein as being the organizing agent for chlorophyll.

What has been particularly convincing about the reality of chlorophyll—protein complexes has been that the photochemical reaction centre of several purple bacteria has been isolated, following detergent treatment of membranes, in a photochemically active pigment—protein complex that fully reflects the many characteristics deduced for the reaction centre from studies on the intact membrane (see chapter 3). Furthermore, denaturation of the protein moiety eliminates these characteristics. Likewise antenna complexes of plants and bacteria have been similarly isolated and found to have spectral properties (absorption, emission, circular dichroism, etc.) characteristic of pigments in whole membranes. Also persuasive are the observations that plant and bacterial mutants exist which lack one of several chlorophyll—protein complexes observed in the wild type organism, and in each case the effect of the mutation on the photosynthetic activity of the organism is exactly that expected from the function determined for that chlorophyll—protein. There is now little doubt that the majority of chlorophyll in any photosynthetic organism is conjugated with specific proteins (see section 2.4.1), the sole function of which is to organize the chlorophyll molecules so that excitation energy quenching does not occur among the antenna pigments but migrates with high efficiency to the reaction centre. It may ultimately prove to be true that all photosynthetic pigments in any organism are organized by protein.

2.4.1. Pigment—protein complexes of plants and bacteria

Several reviews of this topic giving a more detailed description than space permits here have appeared recently. For example, coverage of plant pigment—proteins can be found in Anderson (1975), Boardman et al. (1978), Thornber (1975) and Thornber et al. (1977); biliproteins have been reviewed by Bogorad (1975), Gantt (1977) and Glazer (1976). For photosynthetic bacteria the reaction centre has been described in several articles in Clayton and Sistrom's (1978) book and in Olson and Thornber (1978) while information on the antenna components can be found in Thornber et al. (1978a) for purple bacteria and in Fenna and Matthews (1977) for green bacteria. A particularly useful single reference covering all the topics in this chapter is Brookhaven Symposium of Biology, Volume 28 (1977). The field is, however, expanding daily and so these reviews are not completely up-to-date; it is nevertheless appropriate to give here a summary of the essence of what has already been discovered.

2.4.1.1. Plant chlorophyll—proteins

There are two reasonably well characterized complexes that have been obtained from a variety of plants; a P700—chlorophyll a—protein and a light-harvesting chlorophyll a/b—protein. The former has been found to be ubiquitous in photosynthetic plants (Brown et al., 1974) while the latter occurs only in those plants that contain chlorophyll b (Table 2.1). The photochemically active P700-containing complex represents the heart of Photosystem I, but it is not a reaction centre preparation per se because it contains twenty or more antenna chlorophylls in addition to the photochemically active components. This complex must be present for photosynthetic competence of any plant, whereas the same is not true of the chlorophyll a/b—protein. Several groups (e.g. Thornber and Highkin, 1974; Anderson and Levine, 1974; Miller et al., 1976) have shown the absence of this complex in a photosynthetically grown chlorophyll b-less mutant of barley. Thus, the chlorophyll a/b—protein must function as antenna component (mainly for PSII (cf. sections 2.3 and 2.4.2)) and not be part of the electron transport chain. It has been measured to represent some 50—55% of the total chlorophyll and to contain all of the chlorophyll b in several different green plants. Thus it is the major pigment—protein complex of such plants. The two complexes can be isolated by chromatography or electrophoresis of detergent extracts (cf. Thornber et al., 1977).

The light-harvesting chlorophyll a/b—protein is obtained as a pigmented complex of approx. 30 000 daltons. It has been determined that each isolated pigment—protein molecule contains six chlorophyll molecules (3 chlorophyll a and 3 chlorophyll b), about one carotenoid molecule, but not of a specific type, and a particularly hydrophobic (cf. Table 2.2) polypeptide(s) of about 21 000—29 000 daltons (Thornber, 1975, Thornber et al., 1977). Some of its threonine residues become phosphorylated when chloroplasts are illuminated (Bennett, 1977). The precise size and exact number (perhaps as many as three) of polypeptide(s) present in a chlorophyll a/b—protein preparation are equivocal (see Apel, 1977, Bar-Nun et al., 1977; Henriques and Park, 1977, Thornber, 1975). In vivo some oligomeric form(s) of the ~30 000 dalton complex must exist; Thornber et al. (1977) calculated that thirty to forty monomers must be associated with each photosynthetic unit to account for all the chlorophyll b present. Van Metter (1976) has shown that energy transfer from chlorophyll b to chlorophyll a is highly efficient in the isolated component. He has proposed a model (Fig. 2.6) for the complex based on its optical properties (fluorescence polarization and circular dichroism). As can be seen it shows a centrally located trimer of chlorophyll b molecules, suggested by the strong exciton coupling of this pigment in CD spectra, surrounded by three chlorophyll a molecules located near the periphery of the whole molecule so that they are not strongly coupled.

The P700—chlorophyll a—protein has not yet been so exactly described. Furthermore, it can be isolated with various ratios of antenna chlorophyll molecules to P700, and containing minor but varying amounts of β-carotene. Most frequently the complex is purified as a component with a chlorophyll/ P700 ratio of 40/1; one to two molecules of β-carotene are present (Thornber et al., 1977). However, lower ratios ($\leqslant$20/1) as well as higher ratios ($\leqslant$120/1) can be obtained (cf. Brown, 1976) in chromatographically isolated material; the higher ratios are not due to destruction of P700. The ~20/1 ratio complex is certainly more difficult to obtain, but it does provide one of the best preparations available at present for studies of the primary event of PSI. Solvent extraction of PSI particles (Ikegami and Katoh, 1975) or the use of a mixture of detergents as the membrane-solubilizing agent for

TABLE 2.2

AMINO ACID COMPOSITION (IN MOL%) OF SOME ISOLATED PHOTOSYNTHETIC PIGMENT–PROTEIN COMPLEXES

	P700–Chl–P [a]	Chl a/b–P [b]	Peridinin– chl-P [c]	Phyco- erythrin [d]		Phyco- cyanin [e]		Allo-phyco- cyanin [f]		Reaction centre subunits [g]			Bchl a–P [h]
				α	β	α	β	α	β	21 KD*	24 KD*	28 KD*	
Asp	8.4	8.8	9.9	9.4	9.8	11.4	12.4	9.6	9.1	6.3	6.4	8.2	8.4
Thr	5.9	5.0	3.6	5.5	5.2	6.3	5.3	4.2	7.3	5.2	4.4	4.7	5.2
Ser	5.9	3.6	10.6	6.3	11.7	7.6	6.5	7.2	7.3	4.4	5.6	5.0	5.5
Glu	6.5	9.1	8.3	7.0	6.8	6.9	6.5	10.2	7.3	4.8	6.5	8.2	7.8
Pro	4.7	7.2	2.6	4.3	2.6	3.8	2.4	3.0	2.4	5.9	4.7	7.8	4.9
Gly	10.0	12.5	10.2	10.2	6.7	8.2	7.7	9.0	7.3	12.0	11.7	10.3	11.0
Ala	10.0	11.2	16.8	19.4	19.1	15.8	18.8	13.3	14.6	9.7	9.9	10.5	9.7
Val	6.2	4.4	6.3	8.1	11.4	5.0	7.7	8.4	7.3	6.0	5.5	7.8	6.2
Cys	0.9	0.6	0	—	—	—	—	0.6	1.2	1.3	0	0.9	0.3
Met	1.9	1.7	3.6	1.4	2.9	0.3	1.8	1.8	2.4	1.6	3.1	2.0	1.3
Ile	6.2	4.4	4.0	5.3	3.7	4.4	5.9	7.8	6.7	6.2	5.1	5.4	6.5
Leu	11.2	11.5	5.3	6.2	7.2	10.1	8.2	7.8	9.1	11.2	11.9	9.9	11.4
Tyr	3.1	3.3	3.3	4.2	2.8	6.3	2.9	4.8	6.7	4.3	3.3	2.5	2.6
Phe	6.2	6.8	3.3	3.0	1.3	3.8	3.5	1.8	1.2	7.5	7.7	4.2	6.8
Lys	3.4	5.3	8.9	3.7	2.1	4.4	2.9	4.2	4.9	2.4	0.9	5.1	2.9
His	4.0	1.5	2.0	0.6	0.5	0.6	0	0	0	2.3	2.1	2.3	4.9
Arg	3.4	3.2	0.7	5.7	6.3	4.4	7.7	6.0	4.9	3.3	4.0	4.4	3.6
Try	1.9	—	0.7	—	—	0.6	0	0	0	5.7	7.3	0.8	1.0

[a] From *Phormidium luridum* (Thornber, 1969);
[b] From *Chlamydomonas reinhardii* (Kan and Thornber, 1975);
[c] From *Glenodinium* sp., pI 7.4 (Prézelin and Haxo, 1976);
[d] From *Fremyella diplosiphon* (Takemoto and Bogorad, 1975);
[e] From *Syneschococcus* sp., strain 6301 (cf. Glazer, 1976);
[f] *Allo*-phycocyanin II from *Mastigocladus laminosum* (Gysi and Zuber, 1974);
[g] From *Rps. sphaeroides* (Steiner et al., 1974);
[h] From green bacterium (*Prostecochloris aestuarii*) (Thornber and Olson, , 1968).
* KD, kilodalton.

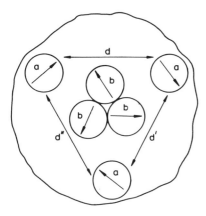

Fig. 2.6. Hypothetical model of light-harvesting chlorophyll a/b—protein. The transition moments of the three chlorophyll b and three chlorophyll a molecules are shown by the short arrows. The distances between the chlorophyll a molecules (d,d'd") are not known. Figure from Van Metter (1976).

a few specific plant materials (cf. Alberte and Thornber, 1978a) seem to be required to obtain the $\leqslant 20/1$ ratio. Olson and Thornber (1978) review what has been learned of the photochemical event in PSI using such materials; however, for our purposes here a biochemical description of the complex is more pertinent. The purified material has an electrophoretic mobility equivalent to that of a denatured protein marker of 100 000—130 000 daltons on SDS-polyacrylamide gel electrophoresis (cf. Chua et al., 1975); this size is almost certainly an underestimate because the pigment—protein is electrophoresed in its native form against markers in their denatured state (cf. Thornber et al., 1977). The amino acid composition of the complex shows a very high proportion of hydrophobic amino acids (Table 2.2). A polypeptide trimer model has been postulated for the blue-green algal component (Fig. 2.7) which explains most of the analytical data available so far (cf. Thornber et al., 1977). Each monomer in the trimer is composed of a ~50 000 dalton polypeptide containing seven chlorophyll a molecules. One monomer of the three is proposed to be composed of a slightly smaller polypeptide containing P700 in addition to five other chlorophyll a molecules. Such a trimer would have a chlorophyll/P700 ratio of close to 20/1, and it may be descriptive of material isolated with that ratio. If another trimer, containing monomers of identical composition to the type described above that does not contain P700, were present in a 1/1 ratio with the P700-containing trimer then such material would have a chlorophyll/P700 ratio of 40/1 — the ratio most frequently obtained in isolated P700—chlorophyll a—proteins (Bengis and Nelson, 1977; Malkin, 1975; Thornber et al., 1977). The model is tentative. It does not explain how chlorophylls in excess of 40 per P700 might be accommodated: Do they occur in additional trimers, or

44

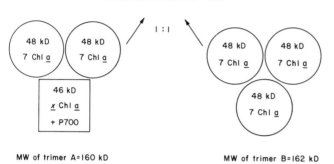

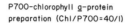

Fig. 2.7. Proposed subunit structure for the two kinds of trimer thought to be present in the P700—chlorophyll a—protein (from Thornber et al., 1977).

are they just loosely associated with the two proposed trimers, or are they located in a biochemically different polypeptide? Much more analytical data are needed to establish an unequivocal model for this important plant constituent. As in the case of the chlorophyll a/b—protein the size (50 000—70 000 daltons) and number of subunits composing a molecule of the isolated complex is debatable (cf. Bengis and Nelson, 1977; Chua et al., 1975; Thornber et al., 1977). The 40-chlorophylls-per-P700 complex represents some 8—15% of the total chlorophyll in eukaryotic organisms; the value in prokaryotes is higher since a considerable portion of the total photosynthetic pigment in these organisms is not chlorophyll but a bilipigment (Thornber, 1975).

Recently, improved SDS-polyacrylamide gel electrophoretic systems have been devised (e.g. Hayden and Hopkins, 1977; Markwell et al., 1978) that reveal more pigmented bands, and permit more of the chlorophyll to remain associated with protein during electrophoresis than occurs with former systems. In the Markwell et al. (1978) system, the percentage of the total chlorophyll associated with the P700—chlorophyll a—protein zone is more than doubled to about 25—30%. Thus the two complexes described in this section are now found to contribute some 75—85% of the total chlorophyll in higher plants and green algae. Little is known about the remaining 15—25% or of the bulk of the antenna pigments in other eukaryotes (however, see section 2.4.1.2). Such pigment generally electrophoreses on SDS-polyacrylamide gels in the free pigment zone (cf. Thornber, 1975); it is thought that the free pigment has most likely been dislodged by the solubilizing detergent from known and from as-yet-unidentified pigment—protein complexes occurring in vivo in these organisms. It is therefore of considerable interest that additional pigment—protein complexes are now being observed on polyacrylamide gel electrophoresis. For example, there are several recent reports of a chlorophyll—protein zone that is thought to represent the reaction centre of Photosystem II (e.g. Hayden and Hopkins, 1977; Miller et al., 1976; Wessels and Borchert, 1978) and other pigmented zones remain to be explained (e.g. Remy et al., 1977; Markwell et al., 1978). It is anticipated that ultimately all the chlorophyll in any plant will be described in terms of chlorophyll—proteins of known structure. Similarly, it is hoped that this can be done for the carotenoids, but at present little of the total carotenoid in plants is contained in the known complexes, and no unequivocal caroteno—protein complexes have been obtained from plants except in one instance, a peridinin—chlorophyll a—protein (see section 2.4.1.2). The majority is converted into free pigment immediately the photosynthetic membranes are solubilized even in the new electrophoretic system of Markwell et al. (1978).

2.4.1.2. Other pigment—protein complexes of plants

The tertiary structure of a pigment—protein complex maintains the pigment molecules within it in a precise orientation and spacing with respect to each other. The tertiary structure must be preserved during isolation if subsequent investigations of the complex are going to be of much value. The proteins of the complexes just described are particularly hydrophobic (see amino acid compositions in Table 2.2), and require detergents to extract them from their location in the photosynthetic membranes and to maintain them in solution during their purification. Since detergents are protein denaturants, the whole study of detergent-soluble pigment—protein complexes is therefore fraught with difficulty. Thus it is not surprising to find that some of the presently better characterized complexes do not require detergents for their removal from the photosynthetic apparatus but are extracted by aqueous buffers in which they are soluble.

Biliproteins have been extensively investigated over the past 30 years (for reviews see Bogorad, 1975; Gantt, 1977; Gantt et al., 1976; and Glazer, 1976). The biliproteins contain the principal photosynthetic antenna pigment(s) in blue-green, red and cryptomonad algae. Study of these pigment—proteins is enhanced not only by their solubility in water but also by the fact that their chromophores are covalently bonded to their associated protein. They can be categorized into three classes: the blue phycocyanins ($\lambda_{max} \sim 615$ nm) and allophycocyanins ($\lambda_{max} \sim 650$ nm), and the red phycoerythrins λ_{max} 540—570 nm). Some algae contain only the blue biliproteins while others contain all three. The ratio of the different types of biliprotein can vary in an organism depending on the color of light under which they are grown (see section 2.4.4). Classical (Duysens, 1952) and modern experiments (Porter et al., 1978) show that excitation energy absorbed among the biliproteins is transferred: phycoerythrin → phycocyanin → allophycocyanin → chlorophyll (see Fig. 2.2); this chlorophyll is predominantly that composing the reaction centre complex of PS II (Ley and Butler, 1977; Wang et al., 1977).

All phycoerythrins and phycocyanins have been found to be composed of two polypeptide chains (α and β) of very similar size and amino acid composition (Table 2.2); the size of the α chain in different organisms varies between 10 000 and 20 000 daltons and the β chain between 14 000 and 22 000 daltons. The α subunit contains one, and the β two, covalently bound chromophores. Allophycocyanin preparations from some algae have been found to contain α and β chains of very nearly the same size ($\sim$16 000 daltons); in other algae where only one chain has been observed it is probable that the α and β chains are of an identical size. Each allophycocyanin subunit contains one chromophore. The hexamer form ($\alpha_6\beta_6$) of phycocyanin, and probably of phycoerythrin, are intermediates in the assembly of the biliproteins into their in vivo structure, the phycobilisome. Phycobilisomes have been observed in electron micrographs of some algae as 32—35 nm diameter particles. A model for a red algal phycobilisome (Fig. 2.8), proposed by Gantt et al. (1976; 1977), is composed of an allophycocyanin core, surrounded by a hemispherical layer of phycocyanin with phycoerythrin on the periphery (see Fig. 2.8). Such a unit would contain over 100 chromophore molecules.

Another water-soluble pigment—protein complex that has been purified and characterized from several, but not all, dinoflagellates is a peridinin—chlorophyll a—protein complex (Prézelin, 1976; Siegelman et al., 1977). The complex is isolated by Sephadex and ion-exchange chromatography of clarified extracts of broken cells. The caroteno—chlorophyll—protein obtained from *Glenodinium* sp. account for most of the peridinin and 20—40% of the chlorophyll a in the organism; cultivation of the organism under lower light intensities increases the cellular content of the peridinin—chlorophyll a—protein (Prézelin, 1976; Prézelin et al., 1976). The

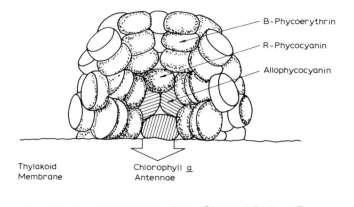

Fig. 2.8. Model of a phycobilisome as envisioned to exist in the red alga, *Porphyridium cruentum*. Redrawn by C.J. Tredwell from Gantt et al. (1976).

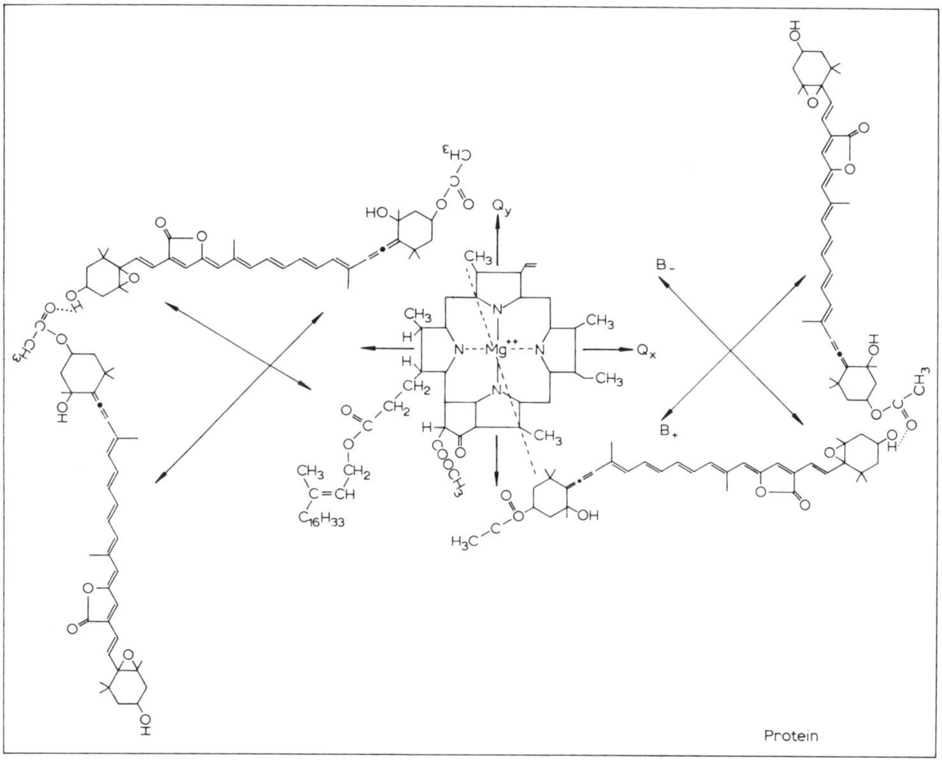

Fig. 2.9. A probable molecular arrangement of chlorophyll *a* and peridinins in the peridinin—chlorophyll *a*—protein of *Glenodinium* sp. From Song et al. (1976).

Glenodinium sp. complex is the most studied. It has been estimated that four peridinin and one chlorophyll *a* molecules are associated with protein in a particle of 35 500 daltons (Prézelin and Haxo, 1976). The amino acid composition is shown in Table 2.2. Detailed spectral analysis (fluorescence polarization and circular dichroism) has given rise to a model for the complex (Song et al., 1976; Koka and Song, 1977) in which a centrally located chlorophyll molecule separates the peridinin molecules into two sets of dimeric excitons at either end of the molecule (Fig. 2.9). Energy transfer from peridinin to chlorophyll *a* occurs with 100% efficiency in this complex (Song et al., 1976). It is not clear whether the *Amphidinium carterae* complex is homologous to that of *Glenodinium* since it has one, rather than two, subunits, and contains nine peridinin and two chlorophyll molecules in its basic unit (39 000 daltons); however, the two protein moieties do have a similar amino acid composition (Haxo et al., 1976; Siegelman et al., 1977).

Finally, evidence is beginning to accumulate to indicate something of the organization of chlorophyll *c* in plants (see Table 2.1). Barrett and Anderson

(1977) have obtained from a brown alga three fractions containing chlorophylls *a* and *c*; one of these fractions was also enriched in fucoxanthin. Light energy absorbed by fucoxanthin and chlorophyll *c* was efficiently transferred to chlorophyll *a* in the complexes. It is postulated that the fucoxanthin—chlorophyll *a/c* fraction has an analogous function to the green plant chlorophyll *a/b*—protein.

2.4.1.3. Pigment—protein complexes of photosynthetic bacteria

Reed and Clayton (1968) were the first to fractionate the reaction centre from any photosynthetic organism. Using Triton X-100 as the membrane-solubilizing agent and a carotenoidless mutant of a purple bacterium, *Rhodopseudomonas sphaeroides*, the reaction centre pigments and some proteins were fractionated from the antenna chlorophylls and other proteins. Since that time reaction centres from other purple bacteria have been isolated. Absence of carotenoids is no longer a prerequisite for a successful isolation of the photochemically active pigments although it does make the task easier. Furthermore, the weakly zwitterionic detergent, lauryl dimethyl amine oxide (LDAO), appears to be the most desirable detergent (compared to SDS and Triton X-100) to use for obtaining a pure product that has undergone the least alteration during purification. Both bacteriochlorophyll *a*- and bacteriochlorophyll *b*-containing purple bacteria but neither green bacteria nor plants, have provided such reaction centre preparations (cf. Olson and Thornber, 1978).

The bacterial reaction centre preparations are composed of pigments and protein. Six chromophore molecules are located in each reaction centre entity, four bacteriochlorophylls and two bacteriopheophytin molecules. In some preparations one molecule of a carotenoid is also present. Two of the four chlorophyll molecules form a special photochemically active pair (see Katz et al., 1977 for review of structure of the special pair) termed the primary electron donor, P870 or P960 in bacteriochlorophyll *a*- and *b*-containing organisms, respectively. One bacteriopheophytin molecule (or maybe both) acts as an intermediate electron carrier between the special pair and the primary electron acceptor, an iron—quinone complex (Feher and Okamura, 1977; Dutton et al., 1978), which is also contained in LDAO-isolated reaction centre preparations (for more details see chapter 3). The protein moiety has been well characterized for the *Rps. sphaeroides* complex (Feher and Okamura, 1977). It is composed of three polypeptides of known amino acid compositions (Table 2.2) and of sizes 21 000, 24 000 and 27 000 daltons which occur in a 1 : 1 : 1 ratio. Interestingly, the amino acid compositions are very similar to that of the P700—chlorophyll *a*—protein (Table 2.2). The 27 000 dalton component can be extracted from the preparation with no effect on the activity or pigment composition of the remaining complex, i.e. the 27 000 dalton component is not essential for the primary event to occur, however it does seem to confer some stability on the complex. A trimer of polypeptides close to the sizes given above occurs in many (Clayton and Haselkorn, 1972; Gingras, 1978) but not in all (Clayton, 1978a) purple bacteria. Nevertheless it does appear from all available data that the reaction centre of all bacteriochlorophyll a- or b-containing purple bacteria is constructed in essentially the same manner (cf. Thornber et al., 1978b). No models have yet been advanced for how the six (or seven when carotenoid is present) chromophores are arranged with respect to the protein or to each other (however, see Sauer, 1975 and Cogdell, 1978). Ongoing studies of the linear dichroism of oriented reaction centres will undoubtedly aid in a solution of this problem. For more detailed coverage of the isolation and characterization of the bacterial reaction centre the reader is directed to articles by Feher (1977, 1978), Gingras (1978), Clayton (1978a, b), Olson and Thornber (1978), Thornber et al. (1978b) and to chapter 3 of this book.

The antenna components of purple bacteria have only recently received any substantial examination. Thornber et al. (1978a) reviewed the information available up to 1976. For reasons not yet understood antenna complexes of purple bacteria appear to be more resilient to denaturation by detergents than chlorophyll a-containing complexes. Thus it has been shown that almost all the bacteriochlorophyll and carotenoid in one bacterium, *Chromatium vinosum*, can be accounted for in three different caroteno—chlorophyll—protein complexes (Thornber, 1970; Cogdell and Thornber, 1978), i.e. after solubilization of the chromatophores, very little pigment occurs as detergent-complexed free pigment (cf. section 2.4.1). Other recent work indicates the same to be true for other purple bacteria (Sauer and Austin, 1978; Cogdell and Thornber, 1978). It is now becoming increasingly clear that the different spectral forms of bacteriochlorophyll (B800, B820, B850, B890) in purple bacteria are associated in vivo with different proteins and not one protein for each form as originally suggested by Wassink et al. (1939). It seems that B890 is associated with one specific protein, B800 and B850 with another, and the six chromophores of the reaction centre are in a third complex. The B800 + B850 component is capable of existing in a slightly different configuration so that it then has chlorophylls giving rise to the B800 and B820 forms. This spectrally different complex occurs together with the B800 + B850— and the B890—proteins in some bacteria (Thornber, 1970; Cogdell and Thornber, 1978). Furthermore, it would appear that the B800 + B850 and the B890 complexes form two classes of homologous proteins (cf. Cogdell and Thornber, 1978; Thornber et al., 1978a). In those species of purple bacteria exhibiting only one long wavelength maximum for the antenna bacteriochlorophyll (e.g. *Rhodospirillum rubrum*) only the B890 complex is present, while in all others both antenna complexes occur.

Recent studies have attempted to describe the basic unit from which each of these two complexes is assembled in vivo (Cogdell and Crofts, 1978; Cogdell and Thornber, 1978; Feick and Drews, 1978; Firsow and Drews, 1977; Sauer and Austin, 1978; Tonn et al., 1977). It seems likely that the

B890 complex is composed of units of approx. 20 000 daltons that contain a pair of excitonically interacting bacteriochlorophyll molecules associated with one carotenoid molecule. The carotenoid type may influence the exact near IR wavelength maximum of the complex (cf. Thornber et al., 1978a). Whereas the B800 + B850—protein's basic unit of a similar size is composed of four chromophores (three bacteriochlorophylls and one carotenoid) in conjugation with two different polypeptides of about 9000 daltons; two of three bacteriochlorophylls are in an exciton interaction and give rise to the 850-nm absorption band, while the other monomeric chlorophyll molecule absorbs at 800 nm. Cogdell and Crofts (1978), working with carotenoid mutants of *Rps. sphaeroides*, found no alteration in the spectral forms of bacteriochlorophyll present when the identity of the carotenoid in the complex was changed by mutation. This probably eliminates carotenoid—chlorophyll interactions as being responsible for these two spectral forms in this bacterium (cf. Thornber et al., 1978a). No models exist showing the spatial arrangement of the chromophores within either protein; however,

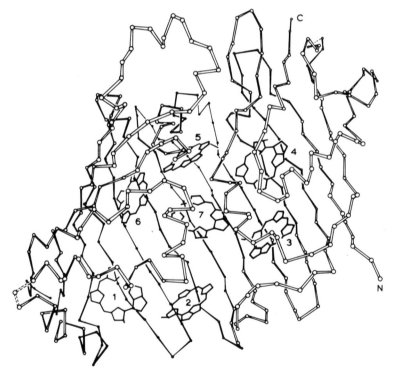

Fig. 2.10. Schematic diagram showing the polypeptide backbone and the seven chlorophyll molecules in one subunit of the trimeric bacteriochlorophyll *a*—protein from a green bacterium, *Prosthecochloris aestuarii* (from Fenna and Matthews, 1977).

since the two complexes are probably the most simple in composition of those in any photosynthetic organism, it should not be long before such models appear.

The best and most exactly described of all photosynthetic pigment—protein complexes is the bacteriochlorophyll *a*—protein complex of green bacteria. This ~140 000 dalton, water-soluble component is a trimer, each monomer (45 000 daltons) of which contains seven chlorophyll molecules. The complex was first isolated 15 years ago by Olson (see Olson (1978) for a review of all aspects of this chlorophyll—protein). Its amino acid composition is given in Table 2.2. The three-dimensional structure of the complex (Fig. 2.10) has been determined by X-ray crystallography (Fenna and Matthews, 1977). As Fig. 2.10 shows each monomer resembles a hollow cylinder that contains seven chlorophyll molecules. The outer wall of the cylinder in the intact trimer is composed almost entirely of fifteen strands of β-pleated sheet. Each porphyrin ring of the central chlorophyll core is spaced at least 12 Å center-to-center from its nearest neighbour as required to prevent concentration quenching of excitation energy (see section 2.4). In the trimer all the chlorophyll molecules are completely surrounded by protein, and hence protected from the solvent. The phytyl chains lie close together within the polypeptide chain.

2.4.2. Models for the energy-trapping portion of the photosynthetic unit

The pigment—protein complexes described above are the building blocks from which the photosynthetic apparatus of the various organisms is largely constructed. In this section models for this apparatus are considered. Most are composed only of those components that are involved in energy capture and in the primary photochemical event; few models exist that incorporate all components of the apparatus (i.e. those showing cytochromes, quinones, iron-sulphur centres, coupling factor, etc.).

The models that have appeared have attempted to account for all phenomena known about the photosynthetic apparatus. Firstly, they take into account that the antenna pigments must be arranged so that they can feed preferentially to one of the two traps thereby enabling the two photosystems to have the different absorption spectra observed (see section 2.3 and Fig. 2.4). Secondly, the models frequently explain how the arrival of excitons at the PSI and PSII traps can be proportioned to optimize electron flow through the Z-scheme (see Barber, 1976; Williams, 1977); theoretically, if the quantum yield of each trap is one, then incoming photons should be divided on a 1 : 1 basis for each photosystem so that maximum rates of electron transport can occur. Murata (1969) and Bonaventura and Myers (1969) have demonstrated that plants do have a mechanism for optimizing electron flow, the so-called state 1-state 2 transition (Wang and Myers, 1974). Although it is almost certain that mono- and divalent cations and the antenna pigments are involved in this mechanism, a precise description of the cation's effect on the photosynthetic apparatus at the molecular level has not been unequivocally determined. Current views (Arntzen et al., 1977; Davis and Gross, 1975; Barber, 1978) are that cations cause conformational changes in the membrane structure, possibly by surface electrostatic interactions involving the chlorophyll a/b—protein, and thus changing the interaction of this major antenna component with the traps (Barber, 1978). Moreover any model must consider whether or not each trap with its closely associated antenna pigments occurs in complete isolation of the other. Separate package is the term used to describe the situation when the photosystems exist as separate entities, continuous array when they do not (cf. Boardman, 1970; Myers, 1971; Seely, 1973). Many recent models of the plant apparatus are of the continuous array type because this concept makes it easier to explain how excitons can be evenly distributed to each trap (see below), and how the considerable changes in the composition of the apparatus that occur in a plant grown under different conditions or in mutants of a plant can be accommodated without losing photosynthetic efficiency (see section 2.4.4). On the other hand, the separate package model simplifies explanation of the different action spectra for the two plant photoacts. Finally any models constructed must take into account the high efficiency observed for energy transfer through the antenna to a reaction centre. Transfer from antenna pigment to antenna pigment molecule and ultimately to the reaction centre must be very rapid -- measurements indicate this occurs within a few picoseconds — to compete successfully against the unwanted processes of fluorescence emission and non-radiative decay that occur within nanoseconds of pigment excitation. Models of the apparatus therefore try to arrange the pigments so that the number of time-consuming interpigment transfers that have to be made between light absorption and arrival of the exciton at the reaction centre is minimized (cf. Seely, 1973). Current views on excitation energy movement through the antenna invoke two mechanisms: one, the diffusion of an exciton wave through closely spaced, strongly interacting chlorophyll molecules such as occurs among the pigments within a chlorophyll—protein complex, and two, a Förster resonance transfer mechanism involving the hopping of excitation energy between weakly interacting pigment molecules such as occurs between pigments in different chlorophyll—protein molecules (Junge, 1977; Knox, 1975, 1977; Sauer, 1975). Since resonance transfer between pigments is more rapid the more strongly the pigments interact, it would seem preferable to invoke solely the first mechanism in models of the apparatus. However, current indications are that a combination of the two mechanisms occurs in vivo, which will certainly enable the exciton to arrive at the reaction centre in a much shorter time than if it had to perform a random walk through many antenna pigments using a Förster mechanism (cf. Beddard and Porter, 1978).

The term, photosynthetic unit, is frequently defined as that morphological entity (cf. sections 2.3 and 2.4.4) which contains all the components of a complete unit of the electron transport chain depicted by the Z-scheme (Fig. 2.3), i.e. one P700, one cytochrome f, two cytochrome b_6, etc. Over the years it has been of interest to measure how many chlorophylls such a unit may contain (e.g. Kelly and Sauer, 1968; Vernon et al., 1971; Thornber et al., 1977; Williams, 1977). A value for plants of 350—450 has often been determined, wheras values of 50—200 for purple bacteria and 1000 for green bacteria are probable. However, it is extremely unlikely that any one photosynthetic unit exists in complete isolation from all others at all times (cf. Williams, 1977). Some components, particularly the PSII and bacterial antenna pigments, are found to feed excitation energy to any one of several reaction centres within their domain (cf. Monger and Parson, 1977; Williams, 1977). And, of course, it is quite possible that such a structurally defined unit does not exist. Nevertheless, in the absence of overwhelming evidence to the contrary, it is still practical to think in terms of the existence of photo-synthetic units for the present. Knox (1968; 1975; 1977) developed an appropriate terminology to describe the organization of pigments in the photosynthetic apparatus: "Lake" is used for an organization in which the antenna pigments form an extended array supplying excitons to many differ-ent reaction centres, and "puddle" for one in which independent photo-synthetic units occur each with their own reaction centre (bacteria) or centres (plants) and antenna pigments. As indicated above, "lakes" occur in both plants and bacteria. However, "puddles" are easier to depict diagrammatically and so most models of the photosynthetic apparatus are shown as such (viz. textbooks covering photosynthesis). These can be readily converted into "lakes" if the shared entity(ies) is specified. It would appear that the continuum might be the chlorophyll a/b—protein in green plants, the B800 + B850—protein and the B890—protein in purple bacteria (cf. Monger and Parson, 1977), the *chlorobium* vesicle in green bacteria (cf. Olson et al., 1977) and the biliproteins in red and blue-green algae. The remaining antenna pigments in these organisms probably occur with the reaction centre(s) and their closely associated electron transport components as "islands" in the "lake".

2.4.2.1. Models for the green plant unit

Some of the first models to appear (e.g. Weier and Benson, 1967; Kreutz, 1970) are now out of contention because they were conjectured before much was known of the occurrence of chlorophyll in protein complexes in vivo. Seely (1973) produced a model and speculated that certain portions of it would ultimately be found to represent some of the plant chlorophyll—protein complexes that were then being studied; this turned out to be true (see below). He used a computer to aid in arranging some 344 chlorophyll a and b molecules in a planar array (Fig. 2.11). The molecules are arranged in sets (1—31) based upon the number of molecules of chlorophyll b and of

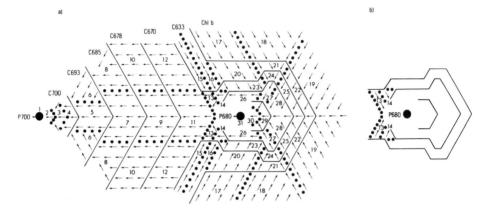

Fig. 2.11(a). Model of the photosynthetic unit proposed by Seely (1973). P700 and P680 designate the reaction centres of PSI and PSII, respectively. Each arrow indicates the transition moment vector of one chlorophyll molecule. The numbers designate sets of molecules of the same spectral type; the solid lines separate regions of different spectral types. (b) Shows the reorientation of the transition moment vectors of the six chlorophylls in sets 13 and 14 which would correspond to an increase in energy transfer of 16% to P680 at the expense of a decrease in transfer to P700.

each chlorophyll a spectral forms present per P700 (or P680) in the photosynthetic apparatus (cf. French et al., 1972). The direction of exciton migration through the array is governed by the molecular orientation of the pigments as shown by arrows (transition moment vectors). The antenna is arranged with respect to the reaction centres so that those chlorophylls feeding excitation energy to P700 are of the longer wavelength type (Photosystem I) while most of those chlorophyll a molecules of shorter wavelength and all of the chlorophyll b would feed to P680 (Photosystem II); this satisfied what was then known about the spectral composition of each plant photosystem (section 2.3). Six strategically located chlorophyll a (C670) molecules (sets 13 and 14) can reorient as shown in Fig. 2.11b so as to deliver excitons to P680 rather than to Photosystem I. When all these six are reoriented, 16% more of the incoming photons are channelled to P680 (Seely, 1973). This arrangement gives a high probability of trapping excitons at the reaction centres with a minimum number of jumps between pigment molecules. It also explains how excitation energy arriving at the two traps could be regulated to optimize electron transport; presumably in this model mono- and divalent cations would in someway influence the orientation of the six chlorophyll molecules. Fluorescence polarization studies do not support such an ordered array of chlorophyll molecules as shown in Fig. 2.11, but nevertheless Seely's model has served as an important and useful forerunner for later models.

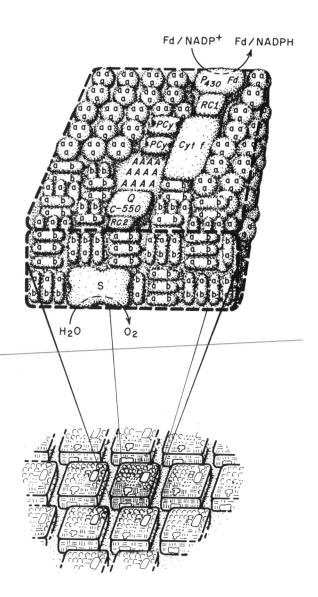

Fig. 2.12. The pebble mosaic model for the photosynthetic unit (Sauer, 1975). At the top is a view of a single unit, consisting of an integrated array of different pebbles (electron transport components and chlorophyll—proteins). Carotenes and uncharged lipids may be interspersed between components acting as a "glue"; charged lipids may be located at the periphery of the unit in contact with the aqueous surroundings. At the bottom is pictured a portion of a single thylakoid membrane, which consists of a two-dimensional mosaic of the individual units. The arrays are not always so regularly arranged as shown; variations in the orderliness may control the extent of electronic excitation energy transfer from one unit to its neighbours.

Based on the increasing knowledge of the substructure of the photosynthetic apparatus, on electron microscopic observation of repeating units in photosynthetic lamellae thought to reflect photosynthetic units (Branton and Park, 1967) and on the fluid mosaic model of Singer and Nicholson (1972), Sauer (1975) proposed a pebble mosaic model (Fig. 2.12). Each pebble consists of a defined set of components of the photosynthetic apparatus and is separated from other pebbles by a lipid matrix. The pebbles are envisaged to occur in regular and irregular arrays, the latter reflecting the fluidity of the membrane (cf. Singer and Nicholson, 1972) as the arrays break up and reform (Sauer, 1978). Sauer's (1975) model includes the secondary electron transport components, and was the first model to include what was then known about the occurrence of chlorophyll in chlorophyll—protein complexes.

The Seely and Sauer models as well as the studies being performed by Butler's (e.g. Kitajima and Butler, 1975) and Vernon's (Vernon et al., 1971) groups greatly influenced a photosynthetic unit model (Fig. 2.13) described quantitatively in terms of chlorophyll—proteins and proposed by one of the authors (Thornber, 1975; Thornber et al., 1977; Alberte and Thornber, 1978b). This model combined aspects of the continuous array and pebble mosaic models but unlike Sauer's was not meant to imply a direct correlation with morphological observations. Thornber (1975) pointed out the excellent agreement between parts of the Seely model and the spectral forms

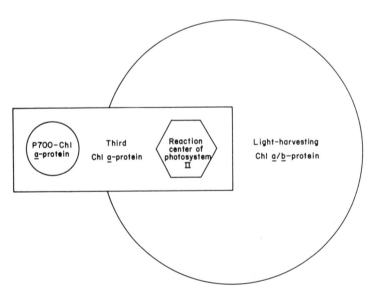

Fig. 2.13. Proposed arrangement of chlorophyll—proteins in the photosynthetic unit of green plants (Thornber et al., 1977). See text for details, and the original publication for the number of chlorophyll molecules in each component of the model.

and proportions of them present in isolated chlorophyll—proteins; sets 13—28 with the chlorophyll a/b—protein, and sets 1—5, 7, 9 and 11 with the P700—chlorophyll a—protein. It is likely that sets 8, 10 and 12 also occur in the P700—chlorophyll a—protein in view of the increased percentage of chlorophyll now obtained in that zone in recently devised electrophoretic systems (section 2.4.1.1); i.e. the postulation (Thornber et al., 1977) that the third chlorophyll a—protein (Fig. 2.13) is more of the P700-less trimer (cf. Fig. 2.7) now seems quite probable. The quantitative nature of the model was provided by determinating the proportions of chlorophyll—proteins and the chlorophyll a/b and chlorophyll/P700 ratios in a variety of plant materials and mutants thereof as well as in plants grown under different environmental conditions (see section 2.4.3 and Alberte and Thornber, 1978b). In particular the model permits one to explain how the considerable variations observed to occur in the composition of a plant's photosynthetic unit can be rationalized. This model is proposed to function in the same manner as Seely's.

During the last four years Butler and his colleagues have been examining the fluorescence characteristics in the presence or absence of Mg^{2+} of plant material containing or lacking the chlorophyll a/b—protein (see Butler (1977) for review). A model (Fig. 2.14) that describes the relative arrangement of the major components of the photosynthetic unit (chl a_1, chl a_2 and chl

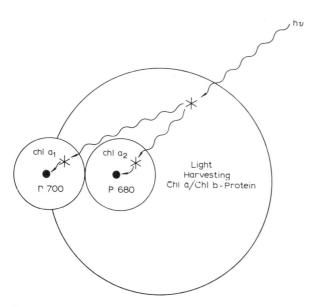

Fig. 2.14. Tripartite model for the arrangement of the three major pigmented components in the photosynthetic apparatus (Butler, 1978). The relationship between the pigmented components and those used in Fig. 2.12, 2.13 and 2.14 is discussed in the text.

a/b—protein) and quantitatively describes spillover of excitons from PSII to PSI has been derived. The components of this tripartite model (Fig. 2.14) probably correspond to those in the Thornber et al. (1977) model as follows: chl a_1 to a P700—chlorophyll a—protein with a chlorophyll/P700 ratio of $\sim 100/1$; chl a_2 to the trap of PSII and its closely associated antenna chlorophylls; light-harvesting chl a/b—protein to some oligomeric form of the chl a/b—protein. Measurements have been made of α and β, the fractions of incoming photons that are absorbed in the three parts and are deliverd to P700 and P680, respectively. A means has been found for determining the fraction of β ($\beta\phi_{T(II \to I)}$) that is spilled over from PSII to PSI. Calculations showed that in the presence of Mg^{2+} α was 0.27 and $\phi_{T(II \to I)}$ was 7% when all PSII traps were open, and 23% when they were closed; in the absence of Mg^{2+} $\alpha = 0.32$ and $\phi_{T(II \to I)}$ varies between 12 and 28%. The distribution of quanta to PSI and PSII can then be calculated to show that only 40% of all absorbed quanta, rather than 50% as expected (see Wang and Myers, 1976), are delivered to PSI under conditions (minus Mg^{2+}) in which spillover is maximal. However, it is possible that the measured values of α are low (Butler, 1977).

All these models come to a similar conclusion about the organization of the energy-trapping components in the photosynthetic unit. Boardman et al. (1978) have proposed a slightly different model (Fig. 2.15) which is essentially a separate package model; however, some interaction between the two photosystems is permitted to enable excitation transfer to occur between the two systems. It is based on arguments similar to those outlined above and also on the observation that PSI and PSII can be readily separated from each other by detergent (particularly by digitonin) treatment (Boardman, 1970; Brown, 1973). Explanation of this phenomenon is largely neglected in those models described above (however, see Thornber et al., 1977). In the Boardman et al. (1978) model PSI is proposed to be an entity composed of the P700—chlorophyll a—protein, and either a third chlorophyll a—protein together with a small proportion of the total chlorophyll a/b—protein (Model A, Fig. 2.15) or a Photosystem I-specific pigment--protein (chlorophyll a/b = 4—5/1) (Model B, Fig. 2.15). The bulk (Model A) or the whole (Model B) of the light-harvesting chlorophyll a/b—protein together with chlorophyll a_{II} constitutes PSII. The number of pigments and the cross-section area for absorption in each photosystem are adjustable to maintain even distribution of excitation energy to each reaction centre. It is, however, somewhat difficult to envisage the control mechanism that could do this when the composition of the apparatus, particularly the content of the chlorophyll a/b—protein, can vary so greatly (see section 2.4.3).

Currently these models are being tested in a variety of ways, and ultimately a sufficiently detailed model will emerge that explains all observed phenomena, all composition data and all electron microscopic observations of the photosynthetic membrane.

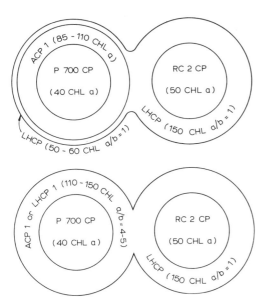

Fig. 2.15. Two models for the organization of pigment—protein complexes in the photo-synthetic unit (Boardman et al., 1978). P700 CP, P700—chlorophyll a—protein; RC 2 CP, reaction centre of Photosystem II complex; LHCP, chlorophyll a/b—protein; ACP 1 or LHCP 1 is a postulated additional chlorophyll—protein having analogy to the third chlorophyll a—protein of Fig. 2.13. The figures in parentheses give the number of chlorophyll molecules and the chlorophyll a/b ratio in each component of the model.

2.4.2.2. Models of the photosynthetic apparatus in other organisms

The photosynthetic unit in other classes of organisms has not received as much attention as that of green plants. Thornber et al. (1977) have proposed that their model can be extended to all plant classes by having the rectangular portion in Fig. 2.13 containing what is essentially chl a_I and chl a_{II}, remain constant during evolution. They hypothesize that it is the nature of the major light-harvesting component (the chlorophyll a/b—protein in green plants) that changes between the various photosynthetic classes. In blue-green and red algae biliproteins, probably occurring in the form of phyco-bilisomes (Gantt, 1977), are thought to occupy this position, and experimental evidence (Ley and Butler, 1977; Wang et al., 1977) supports this notion. Furthermore the chlorophyll/P700 ratio in these organisms is precisely that predicted by the Thornber et al. (1977) model if the chlorophyll a/b—protein has replaced the biliproteins during evolution. In other algal classes a chlorophyll a/c—fucoxanthin—protein and a peridinin—chlorophyll a—protein probably function in this role in brown and dino-flagellate algae, respectively (Barrett and Anderson, 1977; Prézelin et al.,

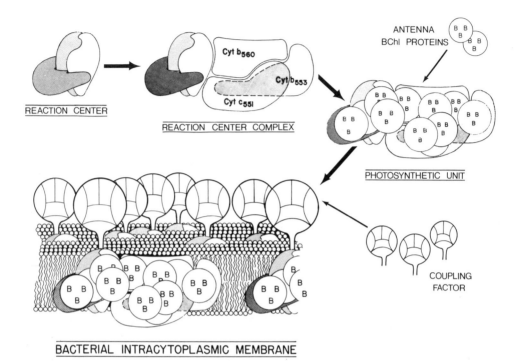

Fig. 2.16. Model for the arrangement of pigment—proteins and secondary electron transport components in the photosynthetic membranes of a purple bacterium (Sauer, 1978). Reaction centres (left) form complexes with cytochromes and, with 12—15 molecules of antenna bacteriochlorophyll—proteins, make up the basic photosynthetic unit of the membrane. The units are assembled with lipids in a two-dimensional sheet that possesses the asymmetry required for ion pumping and electric-field formation.

1976). Euglenoid organisms and yellow-green algae are hypothesized to contain large amounts of a chlorophyll a—protein to fill the role (cf. Thornber et al., 1977).

For purple bacteria, models similar to those described above have been proposed (Thornber, 1970; Monger and Parson, 1977; Sauer, 1978). These models envisage a set of building blocks of a reaction centre, secondary electron donors and acceptors, and one or more antenna bacteriochlorophyll—caroteno–-proteins (e.g. Thornber, 1970). The reaction centre is closely surrounded by molecules of B890—protein (section 2.4.1.3) and secondary electron transport components; a continuum of such assemblies forms an "island" in a "lake" composed of the B800 + B850—protein. Experimental support for such an organization has been obtained by Campillo et al. (1977) and Monger and Parson (1977). A pebble mosaic model (Fig. 2.16) for one of these organisms has recently been presented by Sauer (1978). He

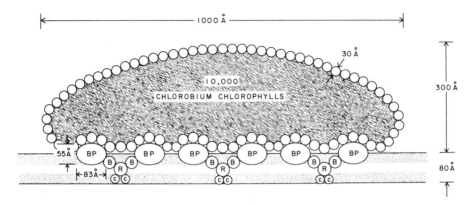

Fig. 2.17. Model for a *Chlorobium* vesicle and associated cytoplasmic membrane in a green bacterium (Olson et al., 1977). For every 10 000 molecules of *Chlorobium* chlorophyll there are ~800 bacteriochlorophyll *a* molecules (B and BP), ~10 reaction centres (R), and ~20 cytochrome *c* molecules (C). One-half of the bacteriochlorophyll *a* molecules are located inside ~ 20 bacteriochlorophyll *a*—protein trimers (BP) (cf. Fig. 2.10). Membrane components are shown in two dimensions in the figure, but are distributed in three dimensions in the membrane.

considers the chromatophore membrane to consist of a two-dimensional array of individual photosynthetic units, each consisting of a set of components (antenna proteins, cytochromes, quinones, reaction centre complex) occurring in their known stoichiometric ratios. Individual units are calculated to have a size of 750 000 daltons which agrees extremely well with the value estimated (650 000—800 000 daltons) from electron micrographs (Reed and Raveed, 1972).

Olson et al. (1977) have presented a model for the photosynthetic membranes of green bacteria, based on composition and electron microscope data. The majority of the pigment present is bacteriochlorophyll *c* and is contained (Fig. 2.17) in a morphologically distinct structure, the *Chlorobium* vesicle. Energy is fed from this to bacteriochlorophyll *a* which occurs in a lipoprotcin cytoplasmic membrane; some bacteriochlorophyll *a* is present in the well characterized water-soluble bacteriochlorophyll *a*—protein, while some occurs in the reaction centre and the remainder in a membrane-bound component (Fig. 2.17).

2.4.3. Variation in the composition of the photosynthetic unit

There is a tendency to think of the photosynthetic unit as a rather static entity, never altering in composition once it is formed. Furthermore it is frequently assumed that all units in a plant or bacterium are of an identical composition. Recent measurements (Aagard and Sistrom, 1972; Arntzen et al., 1977; Cogdell and Schmidt, in preparation; Lien et al., 1973; Thornber et al., 1977; see also Alberte and Thornber, 1978b) of plants and bacteria grown under different environmental conditions and of mutants indicate that the composition and morphology of a unit can change continually during the life cycle of the organism, and can alter radically without affecting photosynthetic activity (see Alberte and Thornber (1978b) and Thornber et al. (1978a) for reviews of relevant plant and bacterial mutants, respectively). The plastic component of the photosynthetic apparatus of green plants seems to be the chlorophyll a/b—protein and its hypothesized replacement in other algal classes (Prézelin et al., 1976; Thornber et al., 1977), whereas in purple bacteria it is the B800 + B850—protein (Aagard and Sistrom, 1972; Cogdell and Schmidt, in preparation; Lien et al., 1973). Thus growth of plants under water, nutrient or temperature stress as well as many virescent mutants show a decreased content of the major antenna components and smaller photosynthetic unit sizes. Furthermore, plants adapt to changes in light intensity by altering their content of this component, increasing it under low light intensities and decreasing it under high light intensities (Alberte and Thornber, 1978b; Boardman, 1977; Prézelin et al., 1976). Biliprotein-containing algae adapt complementarily to the colour of light in which they are grown (Bogorad, 1975): algae illuminated with red light synthesize more of the blue biliprotein, phycocyanin ($\lambda_{max} \sim 615$ nm) and less of the red phycoerythrin (λ_{max} 540—570 nm); green illumination causes an increased synthesis of phycoerythrin and decreased synthesis of phycocyanin.

How changes in chlorophyll content can be incorporated into models of the unit without altering the even distribution of excitation energy to the two traps can be understood in continuous array models if the size of the major light-harvesting component (the chlorophyll a/b—protein array) is increased or decreased and if a decrease or increase, respectively, occurs in the fraction ($\beta\phi_{T(II \to I)}$) of excitation energy transferred from PSII to PSI. In the separate package model equal amounts of the chlorophyll a/b—protein have to be removed from or added to each photosystem, an occurrence that would appear to be difficult for the plant to control effectively. It is also more difficult in a separate package model to understand how those green plants that have no chlorophyll a/b—protein (e.g. flashed bean leaves or the chlorophyll b-less barley mutant (Arntzen et al., 1977, and Strasser and Butler, 1976) could be as photosynthetically competent as they are if nearly all the photosynthetic pigments present feed excitons to the PSI trap. However, one even has to go out on a limb to explain how the continuous array can function efficiently under such circumstances. In both cases it is necessary to propose that either more chlorophylls are present in chl a_{II} than the few that are normally present (cf. Thornber et al., 1977) or that reverse spillover from PSI antenna to the PSII trap occurs in these plants. Interestingly, such a direction for spillover, opposite to that usually hypothesized (Myers, 1963), has been claimed to occur in normal plants by Sun and Sauer (1972).

2.4.4. Pigment organization in photosynthetic membranes as revealed by electron microscopy

The ultrastructure of the thylakoid membrane has been examined over the past 30 years by thin-sectioning, negative-staining and freeze-etching techniques. The first two techniques revealed much of how the photosynthetic membranous system in plants was arranged into stacked and unstacked regions (see Coombs and Greenwood, 1976); however, little information about membrane substructure was obtained until the use of freeze-etching techniques. The use of this technique has revealed the presence of particles (probably proteinaceous) some of which are imbedded in, and some of which are located on, a bilayer lipid membrane (Branton and Park, 1967; Park and Pfeifhofer, 1969; Miller et al., 1976; Muhlethaler, 1976; Sane, 1976; Arntzen et al., 1977; Staehelin et al., 1977). Evidence has accumulated over the years indicating that green plant photosynthetic membranes contain largely two types of intramembranous particles differentiated by size and location: The smaller particle of ~80 Å diameter is located on the PF face and a larger one of two distinct sizes (105 and 164 Å) on the EF face (see Staehelin et al. (1977) for explanation of nomenclature). The larger particles occur only in the stacked membrane region (grana) and give rise to a dense population of clearly spaced particles. The 164 Å particles are occasionally observed to occur in crystalline arrays; it is believed that the quantasome particles, first observed by Park and Biggins (1964), correspond to such particles. There is some evidence that indicates that the smaller particle is a representation of PSI although there are other possibilities (see Fig. 2.18A) while the larger is associated with PSII activity (see Anderson, 1975; Arntzen et al., 1977; Miller et al., 1976; Staehelin et al., 1977; Boardman et al., 1978). The EF particle is hypothesized to contain an 80 Å core (the PSII reaction centre complex) surrounded by varying amounts of the chlorophyll a/b—protein (Fig. 2.18). This notion would explain why varying sizes of the EF particle are observed in higher plants or mutants thereof that contain varying amounts of the complex (see section 2.4.3). It is thought that PSII antenna components bound at one time to a given PSII reaction centre complex (80 Å core) could dissociate from it and migrate to a new binding site within the membrane (Arntzen et al., 1977). Fig. 2.18 gives the current chloroplast membrane models of Arntzen and Staehelin; each model is depicted in terms of chlorophyll—proteins and colorless proteins.

Staehelin et al. (1977) suggested that the formation of grana stacks seems to be a mechanism for increasing the interactions between Photosystem I and II complexes by promoting energy transfer between complexes in

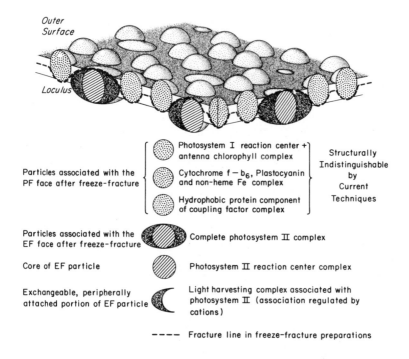

Outer Surface

Loculus

Particles associated with the PF face after freeze-fracture

⬭ Photosystem I reaction center + antenna chlorophyll complex

⬭ Cytochrome f − b₆, Plastocyanin and non-heme Fe complex

⬭ Hydrophobic protein component of coupling factor complex

Structurally Indistinguishable by Current Techniques

Particles associated with the EF face after freeze-fracture — Complete photosystem II complex

Core of EF particle — Photosystem II reaction center complex

Exchangeable, peripherally attached portion of EF particle — Light harvesting complex associated with photosystem II (association regulated by cations)

- - - - Fracture line in freeze-fracture preparations

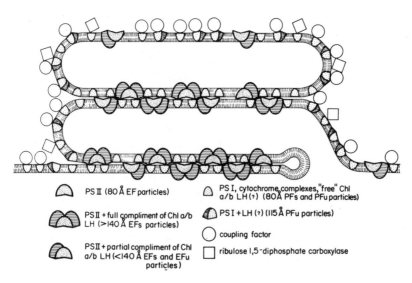

PS II (80 Å EF particles)

PS II + full compliment of Chl a/b LH (>140 Å EFs particles)

PS II + partial compliment of Chl a/b LH (<140 Å EFs and EFu particles)

PS I, cytochrome complexes, "free" Chl a/b LH (?) (80 Å PFs and PFu particles)

PS I + LH (?) (115 Å PFu particles)

coupling factor

ribulose 1,5-diphosphate carboxylase

Fig. 2.18. Models for photosynthetic membranes of higher plant chloroplasts based on freeze-etched electron microscopy and biochemical observations. (A) Model of Arntzen (1978); (B) Model of Staehelin et al. (1977).

adjacent membranes and thereby adding to the lateral interactions that occur between the complexes within an individual membrane. Interaction of the chlorophyll a/b—protein with Mg^{2+} is thought to form the link. It is therefore pertinent that Arntzen et al. (1977) have reported that there is a parallel appearance of grana stacks and Mg^{2+} regulation of energy distribution to the two traps. However it may be fortuitous that these two phenomena (see Telfer et al., 1976; Barber, 1976) occur together and in any case to explain the Mg^{2+} effect does not necessarily require the involvement of chemical linkage. The properties of both the cation-induced structural and energy transfer changes can be accounted for by changes in electrostatic screening of surface changes (Barber et al., 1977; Mills and Barber, 1978). The interaction of thylakoid membranes to form stacks and the associated reorganisation of the membrane particles observed by the freeze-fracture technique can be explained as being due to changes in positive space charge density within a few Ångstroms of the thylakoid surface (Barber, 1978). Staehelin et al. (1977) have pointed out that the membrane model (Fig. 2.18) is most striking in that "all types of particles are asymmetrically organized with respect to the bilayer, but not randomly arranged in the plane of the membrane." There is a rather concise spatial organization in the stacked region (Fig. 2.18).

In other plant species (red and blue-green algae) the two types of intramembranous particles are also observed. The EF particles occur randomly in some species and in rows in others. Interestingly, phycobilisomes (see section 2.4.1.2) also occur in some organisms in rows (Gantt et al., 1977). Since the pigments in phycobilisomes feed preferentially to PSII, these data help subtantiate the notion that the EF particles contain the PSII reaction centre complex; furthermore, the EF particles in these algae (~ 100 Å) are smaller than in green plants as would be expected if the EF particles contain all the chlorophyll a/b—protein in green plants and if pigments in phycobilisomes perform the same function in these algae.

Similar particle-containing membranes have also been observed in photosynthetic purple bacteria (Reed et al., 1970; Slooten, 1972). Positive identification of the origin of the particles observed in the photosynthetic membranes of all organisms awaits a determination of the in vivo supramolecular structure (i.e. the aggregated or oligomeric form) of the pigment—proteins described in section 2.4.1. Such an approach has been invaluable in the elucidation of the phycobilisome structure (cf. Bogorad, 1975; Gantt et al., 1977).

2.5. CONCLUDING REMARKS

This article has attempted to summarize our current knowledge of the occurrence and the in vivo molecular organization of pigments in the photosynthetic apparatus of all classes of plants and bacteria. We have come a long way in our knowledge of this facet of the photosynthetic process in recent years. Not too many years ago the dogma was that chlorophylls were organized by the lipids of the membranes, but now there is general agreement that particularly hydrophobic proteins are the primary organizing agent. We still have a long way to go before we can consider our knowledge of the molecular organization of the photosynthetic apparatus in the membranes to be reasonably complete. Only in a few organisms (some purple bacteria) do we know the gross, but not the detailed, organization of all the pigments. We have virtually no knowledge about the pigments in the Photosystem II trap; we know nothing of the arrangement of carotenoids in most organisms, or indeed of the majority of the pigments in any class of photosynthetic organism except in green plants, dinoflagellates, red and blue-green algae and purple bacteria; finally, details of the molecular arrangement of pigments within any chlorophyll—protein are complete and unequivocal only in one case, the bacteriochlorophyll a—protein of a green bacterium, and even in this case we still await a determination of the amino acid sequence which, together with spectral studies of bacteriochlorophyll a in vitro, should aid immensely in obtaining a molecular explanation of the different spectral forms of chlorophyll in vivo. The precise arrangement of pigments in the bacteriochlorophyll a—protein was delineated because the complex was water-soluble and crystallizable. Apart from this complex, the biliproteins and the peridinin—chlorophyll a—protein, all other major photosynthetic pigment—proteins are soluble only in the presence of detergents. It should not be too long before X-ray crystallographers determine unequivocally the structure of the water-soluble entities; however, before the other complexes can be examined in a similar manner we must await a major scientific breakthrough that permits detergent-soluble proteins to be crystallized. In the interim spectroscopic data (fluorescence polarization, linear and circular dichroism, etc.) and biochemical investigations must be used to provide us with likely models for the pigment organization. More detailed knowledge of the mechanism of energy capture and transfer in photosynthesis and of the structure of the photosynthetic apparatus should come from improvements in electron microscopy techniques, the use of antibodies to specific membrane components and biochemical studies of the supramolecular structure of the pigment—proteins. Finally, it is anticipated that further picosecond spectroscopy experiments will be extremely valuable in contributing to our knowledge of pigment organization in both the building blocks of the photosynthetic apparatus as well as in the apparatus as a whole. As Knox (1977) stated in the last volume of this series: "It is clear that when a whole generation of experiments is repeated with picosecond excitation, the kinetic and dynamic structure of the photosynthetic unit will reveal itself by inference in almost as much detail as one might expect of the molecular structure from X-ray analysis."

ACKNOWLEDGMENTS

The authors are most grateful to Drs. J.M. Anderson, D.J. Chapman, R.J. Cogdell, P.L. Dutton, E. Gantt, G. Gingras, R. Knox, K. Sauer, and R. van Metter who supplied preprints of articles. J.P.T. was supported in part during the preparation of this chapter by the Guggenheim Memorial Foundation; J.P.T.'s research referred to in this article has been largely supported by the National Science Foundation.

REFERENCES

Aagard, J. and Sistrom, W.R. (1972) Photochem. Photobiol., 15, 209—225.

Alberte, R.S. and Thornber, J.P. (1978a) FEBS Lett. 91, 126—130.

Alberte, R.S. and Thornber, J.P. (1978b) Science (in preparation).

Anderson, J.M. (1975) Biochim. Biophys. Acta, 416, 191—235.

Anderson, J.M. and Levine, R.P. (1974) Biochim. Biophys. Acta, 333, 378—387.

Apel, K. (1977) Biochim. Biophys. Acta, 462, 390—402.

Arntzen, C.J., Armond, P., Briantais, J.M., Burke, J.J. and Novitzky, W.P. (1977) Brookhaven Symp. Biol., 28, 316—337.

Arntzen, C.J. (1978) in Current Topics in Bioenergetics (D.R. Sanadi, ed.), Vol. 8. Academic Press, New York pp. 111—160.

Barber, J. (1976) in Topics in Photosynthesis, Vol. 1: The Intact Chloroplast (J. Barber, ed.), pp. 89—134, Elsevier, Amsterdam.

Barber, J. and Mills, J. (1976) FEBS Lett., 68, 288—292.

Barber, J., Mills, J. and Love, A. (1977) FEBS Lett., 74, 174—181.

Barber, J. (1979) CIBA Foundation Meeting, Symp. 61, (new series), pp. 203—209, Elsevier, Amsterdam.

Bar-Nun, S., Schantz, R. and Ohad, I. (1977) Biochim. Biophys. Acta, 459, 451—467

Barrett, J. and Anderson, J.M. (1977) Plant Sci. Lett., 9, 275—283.

Beddard, G. and Porter, G. (1976) Nature, 260, 360—367.

Beddard, G. and Porter, G. (1979) CIBA Foundation Symp. 61, (new series), Elsevier, Amsterdam.

Bengis, C. and Nelson, N. (1977) J. Biol. Chem., 252, 4564—4569.

Bennett, J. (1977) Nature, 269, 344—346.

Boardman, N.K. (1970) Annu. Rev. Plant Physiol., 21, 115—140.

Boardman, N.K. (1977) Annu. Rev. Plant Physiol., 28, 355—377.

Boardman, N.K , Anderson, J.M. and Goodchild, D.J. (1978) in Current Topics in Bioenergetics (D.R. Sanadi, L.P. Vernon, eds.), Vol. 7. (in press).

Bogorad, L. (1975) Annu. Rev. Plant Physiol., 26, 369—401.

Bonaventura, C. and Myers, J. (1969) Biochim. Biophys. Acta, 189, 366—383.

Branton, D. and Park, R.B. (1967) J. Ultrastruct. Res., 19, 283—303.

Brown, J.S. (1973) Photophysiology, 8, 97—112.

Brown, J.S. (1976) Carnegie Inst. Washington, Yearb., 75, 460—465.

Brown, J.S., Alberte, R.S. and Thornber, J.P. (1974) in 3rd Int. Congr. Photosynth. Res., pp. 1951—1962, Elsevier, Amsterdam.

Butler, W.L. (1977) Brookhaven Symp. Biol., 28, 338—346.

Butler, W.L. (1979) CIBA Foundation Symp. 61, (new series), Elsevier, Amsterdam.

Campillo, A.J., Hyer, R.C., Monger, T.G., Parson, W.W. and Shapiro, S.L. (1977) Proc. Natl. Acad. Sci. U.S.A., 74, 1997—2001.

68

Chua, N.-H., Matlin, K. and Bennoun, P. (1975) J. Cell. Biol., 67, 361—377.
Clayton, R.K. (1971) Light and Living Matter, Vol. 2, p. 34. McGraw-Hill, New York.
Clayton, R.K. Rafferty, C.N., Clayton, B.J. and Barouch, Y. (1978a) in Proc. 4th Int. Congr. Photosynth. Res., pp. 45—54, Reading, England.
Clayton, R.K. (1978b) in The Photosynthetic Bacteria (R.K. Clayton, W.R. Sistrom, eds.), Plenum Press, New York (in press).
Clayton, R.K. and Haselkorn, R. (1972) J. Mol. Biol., 68, 97—105.
Clayton, R.K. and Sistrom, W.R. (1978) in The Photosynthetic Bacteria (R.K. Clayton, W.R. Sistrom, eds.). Plenum Press, New York (in press).
Cogdell, R.J. and Crofts, A.R. (1978) Biochim. Biophys. Acta (in press).
Cogdell, R.J. and Thornber, J.P. (1979) CIBA Foundation, Sym. 61, (new series), Elsevier, Amsterdam.
Cogdell, R.J. (1978) Proc. R. Soc. London, Ser. B (in press).
Coombs, J. and Greenwood, A.D. (1976) in Topics in Photosynthesis, Vol. I: The Intact Chloroplast (J. Barber, ed.), pp. 1—52. Elsevier, Amsterdam.
Davis, D.J. and Gross, E.L. (1975) Biochim. Biophys. Acta, 387, 557—567.
Dutton, P.L., Prince, R.C. and Tiede, D.M. (1978) Photochem. Photobiol. (in press).
Duysens, L.N.M. (1952) Ph.D. Thesis. State University, Utrecht, The Netherlands.
Eimjellen, R.E., Aasmundrud, O. and Jensen, A. (1963) Biochem. Biophys. Res. Commun., 10, 232.
Feher, G. (1978) in The Photosynthetic Bacteria (R.K. Clayton, W.R. Sistrom, eds.). Plenum Press, New York (in press).
Feher, G. and Okamura, M.Y. (1977) Brookhaven Symp. Biol., 28, 183—194.
Feick, R. and Drews, G. (1978) Biochim. Biophys. Acta, 501, 499—513.
Fenna, R. and Matthews, B.W. (1977) Brookhaven Symp. Biol., 28, 170—182.
Firsow, N.J. and Drews, G. (1977) Arch. Microbiol., 115, 299—306.
French, C.S., Brown, J.S. and Lawrence, M.C. (1972) Plant Physiol., 49, 421—429.
Gantt, E. (1977) Photochem. Photobiol., 26, 685—689.
Gantt, E., Lipschultz, C. and Zilinskas, B. (1976) Biochim. Biophys. Acta, 430, 375—388.
Gingras, G. (1978) in The Photosynthetic Bacteria (R.K. Clayton, W.R. Sistrom, eds.). Plenum Press, New York (in press).
Glazer, A.N. (1976) in Photochemical and Photobiological Reviews (K.C. Smith, ed.), Vol. I, pp. 71—115, Plenum Press, New York.
Govindjee and Govindjee, R. (1975) in Bioenergetics of Photosynthesis (Govindjee, ed.), pp. 1—50, Academic Press, New York and London.
Gysi, Y. and Zuber, H. (1974) FEBS Lett., 48, 209—213.
Haxo, F.T., Kycia, J.H., Somers, G.F., Bennett, A. and Siegelman, H.W. (1976) Plant Physiol., 59, 297—303.
Hayden, D.B. and Hopkins, W.G. (1977) Can. J. Bot., 55, 2525—2529.
Henriques, F. and Park, R.B. (1977) Plant Physiol., 60, 64—68.
Ikegami, I. and Katoh, S. (1975) Biochim. Biophys. Acta, 376, 588—592.
Junge, W. (1977) Encycl. Plant Physiol., New series. 5, 55—93.
Kan, K-S. and Thornber, J.P. (1975) Plant Physiol., 57, 47—52.
Katz, J.J., Norris, J.R. and Shipman, L.L. (1977) Brookhaven Symp. Biol., 28, 16—55.
Kelly, J. and Sauer, K. (1968) Biochemistry, 7, 882—890.
Kitajima, M. and Butler, W.L. (1975) Biochim. Biophys. Acta, 408, 297—305.
Knox, R.S. (1968) J. Theor. Biol., 21, 244—259.
Knox, R.S. (1975) in Bioenergetics of Photosynthesis (Govindjee, ed.), pp. 183—221, Academic Press, New York.
Knox, R.S. (1977) in Topics in Photosynthesis, Vol. II: Primary Processes in Photosynthesis (J. Barber, ed.), pp. 55—98, Elsevier, Amsterdam.

Koka, P. and Song, P-S. (1977) Biochim. Biophys. Acta 495, 220—231.

Kreutz, W. (1970) Adv. Bot. Res., 3, 53—169.

Kupke, D.W. and French, C.S. (1960) Encycl. Plant Physiol., 1, 298—322.

Ley, A.C. and Butler, W.L. (1977) Proc. Natl. Acad. Sci. U.S.A., 73, 3957—3960.

Lien, S., Gest, H. and San Pietro, A. (1973) J. Bioenerg., 4, 428—434.

Malkin, R. (1975) Arch. Biochem. Biophys., 169, 77—83.

Markwell, J.P., Reinman, S. and Thornber, J.P. (1978) Arch. Biochem. Biophys. 190, 136—141.

Miller, K.R., Miller, G.J. and McIntyre, K.R. (1976) J. Cell Biol., 71, 624—638.

Mills, J. and Barber, J. (1978) Biophys. J., 21, 257—272.

Monger, T.G. and Parson, W.W. (1977) Biochim. Biophys. Acta, 460, 393—407.

Muhlethaler, K. (1976) Encycl. Plant Physiol., New series, 5, 503—521.

Murata, N. (1969) Biochim. Biophys. Acta, 172, 242—251.

Myers, J. (1963) in Photosynthetic Mechanisms of Green Plants. N.A.S.—N.R.C., Publ. 1145, 301—317.

Myers, J. (1971) Annu. Rev. Plant Physiol., 22, 289—312.

Olson, J.M. (1978) in The Photosynthetic Bacteria (R.K. Clayton, W.R. Sistrom, eds.), Plenum Press, New York (in press).

Olson, J.M. and Thornber, J.P. (1979) in Membrane Proteins in Energy Transduction (R.A. Capaldi, ed.). Marcel Dekker, New York (in press).

Olson, J.M., Prince, R.C. and Brune, D.C. (1977) Brookhaven Symp. Biol., 28, 238—246.

Park, R.B. and Biggins, J. (1964) Science, 144, 1009—1011.

Park, R.B. and Pfeifhofer, A.A. (1969) J. Cell Sci., 5, 313—319.

Porter, G., Tedwell, C.J., Searle, G.F.W. and Barber, J. (1978) Biochim. Biophys. Acta, 501, 232—245.

Prézelin, B. (1976) Planta, 130, 225—233.

Prézelin, B. and Haxo, F.T. (1976) Planta, 128, 133—141.

Prézelin, B., Ley, A.C. and Haxo, F.T. (1976) Planta, 130, 251—256.

Ragan, M.A. and Chapman, D.J. (1978) A Biochemical Phylogeny of the Protists, Academic Press, New York.

Reed, D.W. and Clayton, R.K. (1968) Biochim. Biophys. Res. Commun., 30, 471—475.

Reed, D.W. and Raveed, D. (1972) Biochim. Biophys. Acta, 283, 79—91.

Reed, D.W., Raveed, D. and Israel, H.W. (1970) Biochim. Biophys. Acta, 223, 281—291.

Remy, R., Hoarau, J. and LeClerc, J.C. (1977) Photochem. Photobiol., 26, 151—158.

Ried, A. (1972) in Proc. 2nd Int. Congr. Photosynth. Res. (G. Forti, M. Avron, A. Melandri, eds.), pp. 763—772. Junk, The Hague.

Sane, P.V. (1976) Encycl. Plant Physiol. New series, 5, 522—542.

Sauer, K. (1975) in Bioenergetics of Photosynthesis (Govindjee, ed.), pp. 118—181. Academic Press, New York.

Sauer, K (1978) Acc. Chem. Res. 11, 257—264.

Sauer, K. and Austin, L. (1978) Biochemistry, 17, 2011—2019.

Seely, G.R. (1973) J. Theor. Biol., 40, 173—199.

Siegelman, H.W., Kycia, J.H. and Haxo, F.T. (1977) Brookhaven Symp. Biol., 28, 162—169.

Singer, S.J. and Nicholson, G.L. (1972) Science, 175, 720—731.

Slooten, J. (1972) Biochim. Biophys. Acta, 256, 452—466.

Smith, E.L. (1941) J. Gen. Physiol., 24, 565—582.

Song, P-S., Koka, P., Prézelin, B. and Haxo, F.T. (1976) Biochemistry, 15, 4422—4427.

Staehelin, L.A., Armond, P.A. and Miller, K.R. (1977) Brookhaven Symp. Biol., 28, 278—315.

Steiner, L.A., Okamura, M.Y., Lopes, A.D., Moskowitz, E. and Feher, G. (1974) Biochemistry, 13, 1403—1410.

70

Strasser, R.J. and Butler, W.L. (1976) Plant Physiol., 58, 371—376.

Sun, A.S.K. and Sauer, K. (1972) Biochim. Biophys. Acta, 256, 409—427.

Takemoto, J. and Bogorad, L. (1975) Biochemistry, 14, 1211—1216.

Telfer, A., Nicolson, J. and Barber, J. (1976) FEBS Lett., 65, 77—83.

Thornber, J.P. (1969) Biochim. Biophys. Acta, 172, 230—241.

Thornber, J.P. (1970) Biochemistry, 9, 2688—2698.

Thornber, J.P. (1975) Annu. Rev. Plant Physiol., 26, 127—158.

Thornber, J.P. and Highkin, H.R. (1974) Eur. J. Biochem., 41, 109—116.

Thornber, J.P. and Olson, J.M. (1968) Biochemistry, 7, 2242—2249.

Thornber, J.P., Alberte, R.S., Hunter, F.A., Shiozawa, J.A. and Kan, K-S. (1977) Brookhaven Symp. Biol., 28, 132—148.

Thornber, J.P., Trosper, T.L. and Strouse, C.E. (1978a) in The Photosynthetic Bacteria (R.K. Clayton, W.R. Sistrom, eds.). Plenum Press, New York (in press).

Thornber, J.P., Dutton, P.L., Fajer, J., Forman, A., Holten, D., Olson, J.M., Parson, W.W., Prince, R.C., Tiede, D.M. and Windsor, M.W. (1978b) in Proc., 4th Int. Congr. Photosynth. Res., pp. 55—70. Reading, England.

Tonn, S.J., Gogel, G.E. and Loach, P.A. (1977) Biochemistry 16, 877—885.

Van Metter, R. (1976) Thesis. University of Rochester, New York.

Vernon, L.P., Shaw, E.R., Ogawa, T. and Raveed, D. (1971) Photochem. Photobiol., 14, 343—357.

Vredenberg, W.J. and Amesz, J. (1967) Brookhaven Symp. Biol., 19, 49—61.

Walker, D.A. (1976) in Topics in Photosynthesis, Vol. I: The Intact Chloroplast (J. Barber, ed.), pp. 235—278. Elsevier, Amsterdam.

Wang, R.T. and Myers, J. (1974) Biochim. Biophys. Acta, 347, 134—140.

Wang, R.T. and Myers, J. (1976) Photochem. Photobiol., 23, 411—414.

Wang, R.T., Stevens, C.L.R. and Myers, J. (1977) Photochem. Photobiol., 25, 103—108.

Wassink, E.C., Katz, E. and Dorrestein, R. (1939) Enzymologia, 7, 113—129.

Weier, T.E. and Benson, A.A. (1967) Am. J. Bot., 54, 389—402.

Wessels, J.S.C. and Borchert, M. (1978) Biochim. Biophys. Acta, 503, 78—93.

Williams, W.P. (1977) in Topics in Photosynthesis, Vol. II: Primary Processes of Photosynthesis (J. Barber, ed.), pp. 99—147. Elsevier, Amsterdam.

Witt, H.T. (1971) Q. Rev. Biophys., 4, 365—477.

Photosynthesis in relation to model systems, edited by J. Barber
© Elsevier/North-Holland Biomedical Press 1979

Chapter 3

Kinetics and Thermodynamics of Electron Transfer in Bacterial Reaction Centers

ROBERT E. BLANKENSHIP and WILLIAM W. PARSON

Department of Biochemistry SJ-70, University of Washington, Seattle, WA 98195, U.S.A.

CONTENTS

3.1. INTRODUCTION

Chapters 1 and 2 of this volume have introduced the concept of light harvesting pigment complexes in photosynthetic tissue and their ability to transfer energy to specific reaction centers where efficient energy conversion and storage takes place. In this chapter, we shall discuss current views on the initial electron transfer processes that occur in reaction centres. Our discussions will be focused on those reaction centres occurring in photosynthetic bacteria since compared to higher plants and algae far more is known about the chemical nature of the components involved and the kinetics of the reactions that take place. For other recent reviews on the topic see Vol. 2 of this series and Blankenship and Parson (1978); Clayton and Sistrom (1978); Govindjee (1975); Holten and Windsor (1978); Loach (1976, 1977); Ke (1978); Olson and Hind (1977), and Parson and Cogdell (1975).

When the reaction center absorbs a photon or is excited by excitation transfer from the antenna pigments, its bacteriochlorophyll (BChl) complex is promoted to an excited singlet state (P*). By far the most probable fate of P* is to transfer an electron to an initial acceptor (I) that is thought to be a bacteriopheophytin (BPh). This occurs within about 10 ps. I$^-$ then passes an electron to quinone (Q$_A$) in about 200 ps. The oxidized BChl complex (P$^+$) in turn draws an electron from a c-type cytochrome, in a time that varies from about 0.5 μs to 20 μs, depending on the bacterial species. These reactions are summarized in Fig. 3.1.

In the following section, we discuss the kinetics of the electron transfer

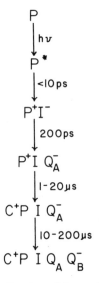

Fig. 3.1. Scheme illustrating electron transfer in photosynthetic bacteria. P, primary electron donor; I, initial acceptor; Q$_A$ and Q$_B$, quinones; C, cytochrome.

reactions that BChl and BPh undergo in vitro. Our aim here is to provide background and perspective for understanding the photochemistry that occurs in vivo. In section 3.3 we summarize the experimental evidence for the kinetic scheme given in Fig. 3.1. Section 3.4 discusses the thermodynamics of photosynthetic reactions. Finally, section 3.5 considers current theories of the kinetics of electron transfer reactions, and tries to apply them to electron transfer in the reaction center.

3.2. PHOTOCHEMICAL REACTIONS OF BChl AND BPh IN VITRO

3.2.1. General considerations

In order to understand the electron transfer reactions that occur in the reaction center complex, it is useful to have some familiarity with the photochemical reactions of BChl, BPh and related pigments in vitro. For additional discussion of work in this area, see Seely's (1977) chapter in Vol. 2 of this series.

The relevant photochemical reactions begin with the excitation of the pigment to its lowest excited singlet state ($^1B^*$). The molecule can decay from the excited state directly back to the ground singlet state, either by emitting fluorescence or by a non-radiative decay; it can undergo intersystem crossing to a triplet state ($^3B^*$); or it can be quenched by reacting with another molecule in one of several different ways. The most important quenching process for our purposes involves electron transfer from the excited pigment to the quencher. The photosynthetic pigments also can accept electrons from appropriate donors, but less is known about these reactions. In the absence of quenchers, fluorescence and intersystem crossing probably account for most of the decay of the excited singlet states of BChl and BPh. However, the actual yield of fluorescence has not been measured directly for either molecule. The triplet yield is about 50% in the case of BPh (Holten et al., 1976, 1978a). The lifetime of the excited singlet state of BChl is about 5 ns in vitro (Tumerman and Rubin, 1962); that for BPh is about 2 ns (Holten et al., 1976, 1978a).

If it is formed, the excited triplet $^3B^*$ has a similar set of possible fates. It can decay to the ground state directly, either radiatively or non-radiatively by intersystem crossing. Phosphorescence near 970 nm has been measured from the triplet states of chlorophyll a and b (Krasnovskii et al., 1974), but it has never been detected from BChl or BPh. (This may be simply a problem of inadequate instrumentation; the phosphorescence probably occurs near 1.2 μm, where present photomultipliers are insensitive.) in the presence of an appropriate acceptor or donor, electron transfer also can occur from the triplet state.

Fig. 3.2 elaborates on the steps that are involved in electron transfer from a molecule in an excited single state, $^1B^*$, to an electron acceptor, A. $^1B^*$

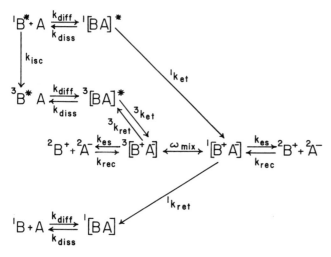

Fig. 3.2. Scheme illustrating photochemical reactions of photosynthetic pigments in vitro. In this diagram the radical-pair state is represented as lying below the level of the excited triplet. It may also lie above the triplet, depending on the redox potential of the acceptor molecule. In the latter case electron transfer from the triplet would not be observed. The singlet and triplet character radical pair states are shown as degenerate, but they may be very slightly split if electronic exchange coupling between the radicals is significant. Coulombic effects may similarly affect the relative energies of the radical pair vs. doublet species. Superscripts 1, 2 and 3 refer to singlet, doublet and triplet. Subscripts are: diff, diffusion; diss, dissociation; et, electron transfer; ret, reverse electron transfer; es, escape; and rec, recombination. ω_{mix} refers to the coherent mixing of singlet and triplet character in the radical pair.

and A first must diffuse together (with rate constant k_{diff}) to form a loose encountercomplex, $^1[B*A]$. The complex can dissociate into $^1B*$ and A again (k_{diss}) or it can undergo electron transfer ($^1k_{et}$) to give a radical pair that has singlet character, $^1[B^+A^-]$. "Singlet character" means that the correlation of the spins of the unpaired electrons on B^+ and A^- is the same as it would be if the two electrons were on the same molecule in a singlet state, although B^+ and A^- in the radical pair may be far enough apart so that the exchange interaction between the two electrons is very weak. The radical pair can dissociate (k_{es}), yielding the free doublet ions $^2B^+$ and $^2A^-$, or it can decay to the ground state by reverse electron transfer (geminate recombination). Electron transfer starting with an excited triplet state is similar, but the $^3[B^+A^-]$ radical pair that is generated has triplet character. In addition, $^1B*$ generally is a much stronger reductant than $^3B*$ is. For example, BPh is capable of reducing methyl viologen from the excited singlet state, but not from the triplet state (see below).

Radical pairs can interconvert between singlet and triplet character by a rephasing of the spins on the two radicals (Brocklehurst, 1976; Atkins and Lambert, 1975). This is a coherent effect that is driven principally by hyperfine coupling with nuclear spins on the two molecules, and it is therefore not describable by an ordinary rate constant. It generally occurs on the time scale of nanoseconds. External magnetic fields can influence both the rate and extent of interconversion (see section 3.3.5). Spin-lattice relaxation also interconverts singlet and triplet radical pairs, but this gives a stochastic relaxation rather than a coherent effect. It typically is considerably slower than the coherent mixing.

If the doublet radicals $^2B^+$ and $^2A^-$ succeed in escaping from the radical pair, they generally will recombine with counter-radicals at some later time (k_{rec}). In this case, the spins on the radicals are uncorrelated with respect to each other. Statistically, three times as many triplet as singlet radical pairs will be formed by this process.

The lowest excited singlet state of BChl lies approximately 1.55 eV in energy above the ground state. This is determined from the absorption and fluorescence emission spectra. The energy gap for the triplet state is not well known, but it probably is about 1.0 eV (Gouterman and Holten, 1977). For the excited singlet and triplet states of BPh, the corresponding energy gaps are 1.6 and (probably) about 1.0 eV (Gouterman and Holten, 1977). The estimates of the triplet energies come from a combination of theoretical considerations, comparisons with chlorophyll a and chlorophyll b and studies of the quenching of the triplet state by excitation transfer to β-carotene.

The thermodynamics of photochemical electron transfer reactions are discussed in section 3.4, but it may be helpful to point out one simple relationship here. To calculate the free energy change associated with the oxidation of BChl in an excited state, one adds the free energy decrease that accompanies the decay of the excited state and the free energy change that is associated with oxidation in the ground state. BChl in vitro can be oxidized electrochemically with a midpoint redox potential (E_m) of approximately +0.5 V relative to the standard H_2 electrode (Fuhrhop and Mauzerall, 1969; Fajer et al., 1974). Combining this with the excitation energy gives an E_m of about −10 V for oxidation in the excited singlet state, and about −0.5 V for the triplet state. The excited molecules thus are extremely strong reductants. One way to view this is that the electron that is elevated to an upper orbital is only loosely bound and can be removed from the molecule relatively easily.

BChl also can be reduced electrochemically, with an E_m of about −0.8 V (Fajer et al., 1973). Again, excitation facilitates this process, making the excited molecule also a strong oxidant. An electron can be added relatively easily to the hole left in a lower molecular orbital in the excited state. The E_m for reduction in the excited singlet state is about +0.7 V, and that for the triplet state is about +0.2 V. Bacteriopheophytin is harder to oxidize than BChl is, but easier to reduce. The pyrole nitrogens appear to bind the two H atoms more covalently than the central Mg is bound in BChl (Fuhrhop and

Mauzerall, 1969). This makes the electron density in the macrocyclic π system lower in BPh, so that an additional electron can be accomodated more readily. The E_m values for oxidation and reduction of BPh in the ground state are both about 0.3 V more positive than the corresponding values for BChl (Fajer et al., 1974, 1975). This may explain why BPh rather than BChl acts as the initial electron-acceptor in the bacterial reaction center (section 3.3).

3.2.2. Kinetics of electron transfer in collision complexes; comparisons with electron transfer in vivo

Let us now return to a more quantitative discussion of the kinetics of electron transfer in the excited singlet and triplet states. The quenching of ^{1}BChl* or ^{1}BPh* by electron acceptors in vitro is manifested in a decrease in the quantum yield of fluorescence, a similar decrease in the yield of triplet states, and an accelerated decay of optical absorbance signals associated with the excited singlet species (Seely, 1969, Holten et al., 1976, 1978a). If every collision between the excited molecule and the acceptor results in quenching, the second-order rate constant for the quenching will be limited by the rate of diffusion at a maximum of about 2×10^{10} M^{-1} s^{-1}. The rate constant attains this value in some cases, but it generally is smaller, presumably because the encounter-complexes formed by collisions frequently dissociate before electron transfer can occur (Holten et al., 1976, 1978a). From statistical mechanical principles or diffusion theory, one can calculate that the lifetimes of such complexes are on the order of 10 ps (Gouterman and Holten, 1977). The probability that electron transfer will occur during this lifetime appears to depend strongly on the free energy change accompanying the electron transfer. The more exothermic the reaction, the more likely it is to occur. This was shown by Seely (1969) in an extensive study of the quenching of fluorescence of pyrochlorophyll by nitrobenzene derivatives. Similar results have been obtained for the quenching of excited singlet states of other aromatic molecules such as pyrene (Rehm and Weller, 1970) and similar, but much less extensive results for the quenching of ^{1}BPh* (Holten et al., 1976, 1978a). The same relationship between quenching and the decrease in free energy has been found for reactions of BChl in the excited triplet state (Connolly et al., 1973). Theoretically, one might expect that the probability of electron transfer would decrease again when the free energy change for the reaction becomes very strongly negative (section 3.5), but there is no evidence that this occurs.

Knowing the lifetime of the collision-complexes formed from ^{1}BChl* or ^{1}BPh* and an electron acceptor, and knowing the fraction of the collision-complexes that enter into the electron transfer reaction and result in quenching, one can calculate the first-order rate constant for the electron transfer. The rate constants for electron transfer from ^{1}BPh* to m-dinitrobenzene (m-DNB) or methyl viologen (MV^{++}) are on the order of 10^{11} s^{-1}

(Holten et al., 1978a). These reactions are exothermic by about 0.4 eV (including estimates of coulombic interactions between the radicals in the radical pair products). The rate constants for these reactions in vitro are comparable to the constant for the initial electron transfer in the bacterial reaction center, which probably involves a comparable decrease in free energy (sections 3.3 and 3.4). The high quantum yield of P^+I^- in the reaction center therefore does not appear to require an unusually high rate constant for electron transfer. However the high yield of P^+I^- does depend on the fact that P and I are held close together in the reaction center so that the excited complex ($P*I$) cannot dissociate.

The high quantum yield of products in vivo depends also on another factor that is not so well understood. In reaction centers, the P^+I^- state lasts for about 200 ps before it passes an electron to a quinone. The second step affords $P^+Q_A^-$ in nearly 100% yield. In other words, virtually no back reactions occur during the 200-ps lifetime of P^+I^-. Even if the transfer of an electron to Q is blocked, P^+I^- survives for a time on the order of 10 ns. In vitro, the situation is dramatically reversed. Despite the fact that the radical pair complexes can dissociate, nearly all of the radical pairs usually decay by reverse electron transfer. Picosecond studies of electron transfer from $^1BPh^*$ and $^1Chl\ a^*$ to m-DNB or p-benzoquinone indicate that the rate constants for reverse electron transfer in the singlet radical pairs are greater than $10^{11}\ s^{-1}$ (Huppert et al., 1976; Holten et al., 1976, 1978a). The formation of free doublet products from the excited singlet state is not observed at all in these systems. The formation of doublets can be detected when MV^{++} is used as an electron-acceptor, presumably because coulombic repulsion destabilizes the radical pair complex in this case and the ions tend to fly apart (Holten et al., 1978a). However, even with this coulombic effect, about 75% of the singlet radical pairs decay by geminate recombination and only 25% escape to form doublet products. Judging from the yield of doublets, and an estimate of the time required for dissociation of the radical pair (2 ps), the rate constant for reverse electron transfer in the $[BPh^+\ MV^+]$ radical pair must be on the order of $2 \times 10^{12}\ s^{-1}$ (Holten et al., 1978a). This is about 10^4 times faster than the rate constant for back reactions in P^+I^-, although the free energy changes are similar in the two systems. We shall discuss some possible explanations for the extremely low back reaction rate in reacton centers in section 3.5.8.

Back reactions appear to be much slower if the electron transfer occurs from an excited triplet state, because doublet products can be observed readily in such cases (Lamola et al., 1975; Tollin, 1976; Holten et al., 1976). Reverse electron transfer in a triplet radical pair would have to regenerate the excited triplet species rather than the ground state. This will be thermodynamically unfavorable, if the triplet state lies above the radical pair in energy, as shown in Fig. 3.2. The interconversion between triplet and singlet radical pairs evidently is slow enough so that the ions can escape before singlet character develops to any significant extent.

The photo-oxidation of chlorophyll derivatives generating stable doublet products in vitro thus usually occurs by way of an excited triplet state. This contrasts with the photo-oxidation of BChl in vivo, which almost certainly occurs from the excited singlet state.

3.3. ELECTRON TRANSFER IN BACTERIAL REACTION CENTERS

3.3.1. The primary electron donor (P)

The primary electron donor in bacterial photosynthesis is frequently called "P870" after the wavelength of its absorption maximum in the near IR. In species of bacteria that contain BChl *b* rather than BChl *a*, the major absorption maximum is at 960 nm, and the electron donor has been called "P960". (See chapter 2.) We shall use the abbreviation "P" in a general sense, because the two classes of reaction centers appear to be functionally very similar (Prince et al., 1977; Holten et al., 1978b). Fig. 3.3 summarizes

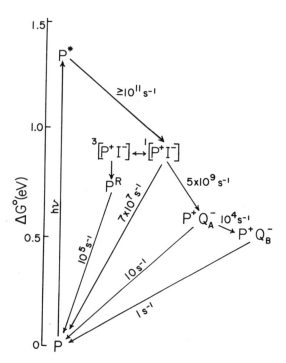

Fig. 3.3. Energy-kinetic diagram summarizing important reactions in bacterial reaction centers at room temperature. The thermodynamic evidence used to place components vertically is given in section 3.4.1. Experimental evidence for the first-order rate constants shown in the diagram is summarized in section 3.3.

current thinking on the kinetic pathways for electron flow in bacterial reaction centers.

When P is oxidized by illumination or by chemical oxidants, the absorption band at 870 or 960 nm bleaches and several new bands appear. Picosecond kinetic measurements have shown that these absorbance changes are complete within 10 ps after the excitation of reaction centers with a short flash (Kaufmann et al., 1975, 1976; Rockley et al., 1975; Dutton et al., 1975a; Netzel et al., 1977; Holten et al., 1978b; Moscowitz and Malley, 1978). From a deconvolution of the rise time of the absorbance changes and the shape of the excitation pulse, Moscowitz and Malley (1978) calculated that the rate constant for the electron transfer is $\geqslant 2 \times 10^{11}$ s^{-1}. The fluorescence lifetime of P* has been measured by picosecond pulse techniques and reported to be 15 ± 8 ps (Kononenko et al., 1976; Paschenko et al., 1977). This seems slightly too long to be consistent with the picosecond absorbance measurements, but the discrepancy probably is within the experimental uncertainty of the methods. A fluorescence lifetime of 7 to 10 ps has been calculated from the fluorescence yield and the natural radiative lifetime (Zankel et al., 1968; Slooten, 1972a) *.

A structureless ESR signal is associated with the oxidized BChl complex, P$^+$. In reaction centers that contain BChl a, the ESR signal has a "g" factor of 2.0025, a Gaussian lineshape, and a 1st-derivative linewidth of 9.5 G. The ESR spectrum is similar to that of the π-radical cation of BChl in vitro, but is narrower by a factor of about $\sqrt{2}$ (McElroy et al., 1972). The electron nuclear double resonance (ENDOR) spectrum of P$^+$ shows hyperfine splittings that are half as large as those seen with BChl in vitro (Norris et al., 1974; Feher et al., 1975). These facts have been interpreted to mean that the unpaired electron in P$^+$ is delocalized equally over two of the four BChl molecules in the reaction center. However, the delocalization appears not to be so extensive in reaction centers from species that contain BChl b (McElroy et al., 1972; Fajer et al., 1977). This may indicate that the environments of the two BChl b molecules are not strictly equivalent.

Very little information is available on the arrangement of the special pair of BChl molecules in the reaction center, or on how these molecules interact with the two other BChl molecules and the two BPh molecules in the complex. For further discussion of this point, see Blankenship and Parson (1978).

As already mentioned in chapter 1 the quantum yield of photo-oxidation of P has been measured to be 1.02 ± 0.04 in isolated reaction centers at room temperature (Wraight and Clayton, 1973). Virtually every quantum that is absorbed results in electron transfer. The quantum yield does not

---

* See Note 1 added in proof on page 114.

change significantly if the temperature is lowered to $5°K$ (Clayton and Yamamoto, 1976). This indicates that the rate constant for the electron transfer reaction is not strongly dependent on temperature. However, the quantum yield would not be very sensitive to changes in the rate constant, as long as electron transfer remained fast relative to the other routes for the decay of P*. The fluorescence yield from P* increases rather abruptly by a factor of about 2 as one lowers the temperature through the region between $180°$ and $220°K$ (Clayton, 1977). This could mean that the rate of electron transfer decreased by 50% in this range. (It is also possible that a small fraction of the reaction center population simply becomes completely inactive at this point.) At lower temperatures, the fluorescence yield seems not to change further.

3.3.2. Initial electron acceptor (I)

In reaction centers that contain BChl a, the reduction of the initial electron-acceptor (I) causes the bleaching of absorption bands at 540 and 760 nm, and the appearance of a broad band near 650 nm. Similar absorbance changes occur at somewhat longer wavelengths in reaction centers that contain BChl b. In both cases, the absorbance changes occur within 10 ps after excitation of the reaction center, along with the absorbance changes associated with the oxidation of P (Kaufmann et al., 1975, 1976; Rockley et al., 1975; Dutton et al., 1975a; Netzel et al., 1977; Holten et al., 1978b). However they can also be generated by continued illumination of reaction centers at low redox potentials, under conditions that keep P fully reduced (Shuvalov et al., 1976; Shuvavlov and Klimov, 1976; Tiede et al., 1976a, 1976b; Van Grondelle et al., 1976; Prince et al., 1977; Okamura et al., 1977; Trosper et al., 1977; Holten et al., 1978b). The same changes occur on excitation of reaction centers that have been depleted of the quinone that acts as the next electron acceptor (Kaufmann et al., 1976). The spectrum of the absorbance changes supports the idea that I is one of the two BPh molecules in the reaction center (Fajer et al., 1975), but the spectrum also includes complex changes in some of the absorption bands of the BChl molecules of the reaction center. These probably are the two "extra" BChl molecules, rather than the two that constitute P.

The ESR and ENDOR spectra of I⁻ have been obtained by illuminating reaction centers at low potentials and then lowering the temperature to trap I in the reduced state (Shuvalov and Klimov, 1976; Tiede et al., 1976b; Prince et al., 1977; Feher et al., 1977; Fajer et al., 1977). The spectra are consistent with the identification of I⁻ as an anion radical of either BPh or BChl. In this case, the unpaired spin appears not to be delocalized over more than one molecule. Under some conditions, however, the ESR spectrum indicates strong spin—spin coupling between I⁻ and another radical, which probably is the quinone—Fe complex.

3.3.3. The quinone electron acceptor (Q_A)

The involvement of ubiquinone as an early electron-acceptor has been known for many years, and the quinone frequently has been called the "primary" acceptor. In the light of the evidence presented above, this designation no longer seems appropriate. We shall use the term quinone acceptor (Q_A).

Reduction of Q_A in isolated reaction centers causes absorbance changes in the near UV and visible regions that are very similar to those accompanying the formation of the anionic radical (semiquinone) of ubiquinone in vitro (Clayton and Straley, 1970, 1972; Slooten, 1972b; Vermeglio, 1977; Vermeglio and Clayton, 1977; Wraight, 1977). The ESR spectrum of Q_A^-, however, is dramatically different from the narrow spectrum near g = 2.00 that one expects for a free semiquinone. It has a principal g factor of 1.8, and a linewidth of about 600 G (Feher, 1971; Feher et al., 1974; Leigh and Dutton, 1972; Dutton et al., 1973a; Okamura et al., 1975). The broad ESR spectrum initially was attributed to the nonheme Fe of the reaction center, and seemed to indicate that the Fe was the "primary" electron-acceptor. However, the Mössbauer spectrum of the Fe indicated that the redox state of the Fe did not change when the acceptor was reduced; the Fe was Fe^{II} before and after the reduction (Debrunner et al., 1975). In addition, reaction center preparations that were depleted of Fe were found to exhibit normal photochemical activity (Loach and Hall, 1972; Feher et al., 1972; Blankenship and Parson, 1977). In the depleted reaction centers, the reduced electron-acceptor had an ESR signal that was characteristic of ubisemiquinone. When ubiquinone was extracted from reaction centers, on the other hand, normal photochemical activity and the g = 1.8 ESR signal both were lost (Cogdell et al., 1974; Okamura et al., 1975). Readdition of ubiquinone restored both the activity and the ESR signal. The current view is that ubiquinone accepts electrons from I^-, and that the Fe is close enough nearby so that it severely perturbs the ESR spectrum of the semiquinone. In some species of bacteria, Q_A appears to be a menaquinone rather than ubiquinone (Feher and Okamura, 1977; Peucheu et al., 1976; Romijn and Amesz, 1977).

The kinetics of the electron transfer between I^- and Q_A have been measured with picosecond absorbance techniques, using the decay of the absorption bands associated with I^-. The transfer has a rate constant of approximately 5×10^9 s^{-1} at room temperature (Kaufmann et al., 1975; Rockley et al., 1975; Holten et al., 1978b). The rate increases slightly if a second quinone in the reaction center (Q_B) is reduced (Pellin et al., 1978) *.

A second quinone (Q_B) acts as a secondary acceptor, removing electrons from Q_A (Clayton and Yau, 1972; Halsey and Parson, 1974; Parson, 1978a). Recent work indicates that the Fe is required for this step (Blankenship and Parson, 1977). At room temperature, the reaction has a half-time of 10—

---

* See Note 1 added in proof on page 114.

200 μs, depending on the bacterial species and the pH (Parson, 1969; Chamorovski et al., 1976; Vermeglio and Clayton, 1977). Near room temperature, it has a classical activation energy of about 0.35 eV (Parson, 1969; Chamorovsky et al., 1976). Electron transfer between Q_A and Q_B can be blocked by o-phenanthroline, or by extracting Q_B.

3.3.4. Cytochrome oxidation

In whole cells of photosynthetic bacteria, P^+ is reduced by a c-type cytochrome. Many species contain two major types of such cytochromes, one with a midpoint redox potential (E_m) near 0 mV, and one with an E_m near +300 mV. There generally are either two copies of the high-potential cytochrome or two copies of each type of cytochrome present per reaction center (Parson, 1974; Dutton and Prince, 1978). The cytochromes are solubilized when cells of some species are broken, but in others they remain bound to the reaction center and are purified along with it. At room temperature, the rate constants for the transfer of an electron from the low-potential cytochromes to P^+ are typically about 1×10^6 s^{-1}. (Kihara and Chance, 1969; Parson and Case, 1970; Case et al., 1970; Seibert and DeVault, 1970). Photo-oxidation of the low-potential cytochromes frequently occurs even at very low temperatures (DeVault and Chance, 1966; DeVault et al., 1967; Kihara and Chance, 1969; Dutton et al., 1970, 1971; Kihara and McCray, 1973). The rate decreases as one lowers the temperature, but becomes independent of temperature at about 120° K. This behavior has prompted considerable theoretical effort, which will be discussed in section 3.5.6.

Rate constants for photo-oxidation of the high potential cytochromes range from 3.5×10^6 s^{-1} for the tightly bound high-potential cytochrome in *Rhodopseudomonas viridis* (Holten et al., 1978b) and 3.5×10^5 s^{-1} for the similar cytochrome of *Chromatium vinosum* (Parson, 1968; Parson and Case, 1970; Case et al., 1970) down to about 3×10^4 s^{-1} for the loosely bound cytochrome-C_2 of *Rps. sphaeroides* (Dutton et al., 1975b). (The reaction in *Rps. sphaeroides* also has a slower component with a variable rate constant.) These reactions do not occur at low temperatures, but their temperature dependence has not been studied in detail.

3.3.5. Back reactions; the triplet state

The reduction of P^+ by cytochromes will be blocked if the cytochromes are already oxidized, if they have been dissociated from the reaction center, or (in the case of the high potential cytochromes) if the temperature is sufficiently low. Under these conditions, P^+ can be reduced by any of the three acceptors discussed above (see Fig. 3.3).

If both of the quinones are present and functional, the reaction center will be in the state $P^+IQ_AQ_B^-$ shortly after excitation. The reduction of P^+ by Q_B^- is very slow, with a half-time on the order of 1 s (Clayton and Yau, 1972).

If electron transfer from Q_A to Q_B is blocked by adding o-phenanthroline, by extracting Q_B, or by lowering the temperature, the reaction center will be trapped in the state $P^+IQ_A^-$ after excitation. The back reaction from Q_A^- to P^+ occurs with a rate constant of $10-20$ s^{-1} at room temperature (Clayton and Yau, 1972; Clayton et al., 1972; Hsi and Bolton, 1974; Halsey and Parson, 1974; Loach et al., 1975). This reaction has a remarkable temperature dependence, which we shall discuss in more detail in section 3.5.6 (Parson, 1967; Clayton and Yau, 1972; Hsi and Bolton, 1974; McElroy et al., 1974; Loach et al., 1975). The rate *increases* by a factor of about 2.5 as the temperature is lowered to 150°K. It continues to increase at lower temperatures, but more gradually. By 4°K, the rate constant is $30-40$ s^{-1}. In isolated reaction centers, one can change the rate of the back-reaction over about a 4-fold range by substituting other quinones for the ubiquinone that acts as Q_A (Okamura et al., 1975). Removing the Fe or replacing it by Mn has no significant effect on the rate (Loach and Hall, 1972; Feher et al., 1972, 1974).

If Q_A is removed from the reaction center or is chemically reduced, the photochemical apparatus remains in the state P^+I^- after excitation. This state also has been called P^F (Parson et al., 1975; Cogdell et al., 1975; Parson and Monger, 1977). The back reaction from P^+I^- is considerably more rapid than the ones discussed above. Its rate constant is about 7×10^7 s^{-1} at room temperature and about 2×10^7 s^{-1} at 30°K (Parson et al., 1975; Cogdell et al., 1975; Holten et al., 1978b).

The back reaction between P^+ and I^- is more complicated than the reactions involving Q_A and Q_B. Not all of the reaction centers return directly to the ground state, PI. At room temperature, 10 to 20% of the reaction centers first enter what appears to be a triplet state of the BChl, state P^R. At low temperature, virtually all of the decay of P^+I^- proceeds via the triplet state (Wraight et al., 1974; Parson et al., 1975; Clayton and Yamamoto, 1976). The triplet state can be detected optically (by triplet—triplet absorption) or by its characteristic ESR spectrum at low temperatures.

The formation of the triplet state can be explained on the assumption that the radical pair state P^+I^- is created initially in a singlet state, but that rephasing of the spins on P^+ and I^- can occur prior to the back reaction. If the transfer of an electron from I^- to P^+ occurs when the radical pair has singlet character, the result would be to return the system to the ground singlet state, PI. If electron transfer occurs when the radical pair has triplet character, the BChl would be placed in a triplet state.

External magnetic fields influence the yield and the properties of P^R in a way that is consistent with this interpretation. The ESR spectrum exhibits an unusual spin polarization which is not observed with chlorophyll or BChl triplet states in vitro (Dutton et al., 1972; Leigh and Dutton, 1974; Uphaus et al., 1974; Thurnauer et al., 1975; Prince et al., 1976). Transitions from the T_0 spin sublevel to the T_{+1} sublevel show greatly enhanced absorption, and transitions from T_0 to T_{-1} show microwave emission. This means that the triplet state is formed almost exclusively in the T_0 sublevel. This is what one expects, if P^R forms from P^+I^- and if P^+I^- is created initially in a singlet state. In the presence of an external magnetic field (such as the field used for the ESR measurements), the singlet state of a radical pair can mix with the T_0 sublevel but not with $T_{\pm 1}$ (Atkins and Lambert, 1975; Atkins, 1976; Brocklehurst, 1976). In agreement with this, optical measurements have shown that an external field decreases the quantum yield of P^R at room temperature (Blankenship et al., 1977; Hoff et al., 1977). Magnetic fields do not affect the triplet yield significantly at low temperatures, probably because under these conditions the rate constant for electron transfer is much greater in the triplet state than in the singlet state (Wraight et al., 1974; Blankenship et al., 1977). The temperature dependence of the yield can be explained on the assumption that lowering the temperature decreases the rate constant for electron transfer in the singlet state, relative to that in the triplet state (section 3.5.8).

The nature of the interaction that mixes the S and T_0 states of the radical pair is not entirely clear, but hyperfine coupling with nuclei in P^+ and I^- appears to be sufficient. Interactions with Q_A^- or the Fe do not seem to be important here, because the removal of either of these components from the reaction center does not decrease the triplet yield (Blankenship and Parson, 1977).

The T_0 triplet ESR spin polarization and the direction of the magnetic field effect on the triplet yield constitute strong evidence that the initial electron transfer from P^* to I occurs from the excited singlet state of the BChl complex, without requiring intersystem crossing to a triplet state. This conclusion is consistent with the very high rate of the initial electron transfer.

The energy of the triplet state P^R is not well known. If the gap between P^* and P^R is similar to the singlet—triplet splitting that has been estimated for BChl and BPh in vitro (section 3.2.1), P^R would be about 0.75 eV above the ground state. This estimate puts the energy of P^R about 0.15 eV below the apparent free energy of P^+I^- (section 3.4).

In reaction centers that do not contain carotenoids, P^R decays by intersystem crossing, with a rate constant of approximately $10^5 \ s^{-1}$ at room temperature (Parson et al., 1975; Cogdell et al., 1975; Holten et al., 1978b). The decay rate decreases by about a factor of 10 as one lowers the temperature to $120°K$, but then becomes independent of temperature (Parson et al., 1975). We shall discuss this further in section 3.5.8. The decay becomes multiphasic at temperatures below about $10°K$, when spin-lattice relaxation is very slow and the three triplet sublevels of P^R decay independently (Clarke et al., 1976; Hoff, 1976; Parson and Monger, 1977). In reaction centers that contain a carotenoid, P^R decays in about 20 ns by transferring energy to the carotenoid (Cogdell et al., 1975; Parson and Monger, 1977). The energy transfer does not occur at very low temperatures, probably because the carotenoid triplet state lies slightly above P^R in energy.

3.4. THERMODYNAMICS OF THE PRIMARY REACTION

3.4.1. Midpoint redox potentials

The primary electron-donor (P) can be oxidized in the dark by raising the ambient redox potential with chemical oxidants such as $K_3Fe (CN)_6$. The oxidation follows a 1-electron titration, with a midpoint redox potential (E_m) of approximately +480 mV [See Parson and Cogdell (1975), Prince and Dutton (1978), and references listed therein]. The E_m is not strongly dependent on the pH (Case and Parson, 1971; Jackson et al., 1973), but it does depend on the ionic strength (Case and Parson, 1973). The dependence of the E_m on temperature indicates that a substantial increase in entropy (about 32 cal mole^{-1} deg^{-1}) accompanies the oxidation in C. vinosum (Case and Parson, 1971), but apparently not in Rps. viridis (Carithers and Parson, 1975).

The quinone acceptor (Q_A) can be titrated similarly by chemical reduction with $Na_2S_2O_4$. In chromatophores, Q_A exhibits a 1-electron titration with an E_m of −50 to −100 mV at pH 7, depending on the species [See Parson and Cogdell (1975) for tabulation]. In this case, the E_m depends on pH, behaving as though the uptake of a proton accompanies the reduction. The E_m decreases to about −180 mV by pH 10 and then becomes independent of pH, suggesting that the reduced form of the electron acceptor has a pK_a of about this value (Prince and Dutton, 1976; Prince et al., 1976). The temperature dependence of the E_m indicates that a substantial entropy decrease (71 cal mole^{-1} deg^{-1}) occurs when the quinone is reduced in C. vinosum (Case and Parson, 1971).

In preparations of isolated reaction centers from Rps. sphaeroides, the E_m of the quinone does not depend on the pH, but remains about −50 mV between pH 6 and 9 (Reed et al., 1969; Dutton et al., 1973b). The reason for the different behavior in chromatophores and isolated reaction centers is not clear.

The E_m of the initial electron acceptor (I) has been measured in isolated reaction centers from Rps. viridis. Prince et al. (1976, 1977) found that the amount of the spin-polarized triplet state that was generated by excitation with a single flash decreased if the ambient redox potential was made sufficiently low. The decrease followed a 1-electron titration with an E_m of about −400 mV apparently independent of pH. Shuvalov et al. (1976) obtained a more negative value, −620 mV, by measuring the optical absorbance changes that occur when I is reduced during continuous illumination. The reason for the disagreement between the two measurements is not clear.

The free energy change that accompanies the transfer of an electron from P to I is given by

$$\Delta G_{PI \to P^+I^-} = \Delta G^0_{PI \to P^+I^-} + k_B T \ln([P^+I^-]/[PI])$$

$$\approx (e)(E_m^P - E_m^I) + k_B T \ln([P^+I^-]/[PI]) \tag{1}$$

Here the standard free energy change, ΔG^0, is *approximately* equal to the difference between the two midpoint redox potentials, multiplied by the electron's charge (e). [PI] and [P$^+$I$^-$] are the concentrations of reaction centers in the ground and radical-pair states and k_B is Boltzmann's constant. The E_m values quoted above give $\Delta G^0_{PI \to P^+I^-} \sim 0.9$ eV. Similarly, for the formation of $P^+Q_A^-$, $\Delta G^0_{PQ \to P^+Q^-} \sim 0.6$ eV. There are two reasons why the E_m values give one only approximate values for ΔG^0 (Case and Parson, 1971). First when one performs a redox titration of P, I is invariably in an oxidized state, and when one titrates I, P is invariably reduced. Titrating the two components separately in this way neglects any interactions that occur between the components when P is oxidized and I reduced. In other words, the two half reactions that one considers in the separate redox titrations are

$$PI \to P^+I + e^- \text{ and } PI^- \to PI + e^-. \tag{2}$$

Substracting the second of these from the first gives

$$2 PI \to P^+I + PI^-. \tag{3}$$

The photochemical process that actually occurs in the reaction center, however, is

$$PI \to P^+I^-. \tag{4}$$

Equations (3) and (4) differ by the dismutation

$$P^+I + PI^- \to P^+I^- + PI, \tag{5}$$

and the free energy change in the dismutation depends on the interaction between P^+ and I^- in P^+I^-. Unfortunately, we have no information on the magnitude of this interaction.

The second difficulty with using the E_m values that are quoted above is that they necessarily reflect equilibrium states of the system. In performing a redox titration, one adjusts the ambient redox potential and then waits a relatively long time before measuring the redox states of the components. During this time interval, comparatively slow processes such as the binding of protons or other ions or changes in protein conformation have an opportunity to proceed to completion. The final equilibrium state may not accurately reflect the transient states of the system that are important during the electron transfer reactions. This problem is particularly acute in the case of P^+I^-, which normally survives for only about 200 ps. Again, we have insufficient information to evaluate the problem quantitatively. Even the longer-lived $P^+Q_A^-$ may decay rapidly enough so that the uptake of protons by Q_A^- has little or no chance to occur (Prince and Dutton, 1976).

3.4.2. The relationship between free energy capture and light intensity

With the above reservations about the significance of E_m values in mind, let us compare the free energy changes in the electron transfer reactions with the free energy that becomes available upon excitation of the reaction center with light. The free energy change associated with raising the reaction center's BChl complex from the ground state (P) to the lowest excited singlet state (P^*) is

$$\Delta G_{P \to P*} = \Delta G^0_{P \to P*} + k_B T \ln([P^*]/[P]) \tag{6}$$

The standard free energy change $\Delta G^0_{P \to P*}$ is given by

$$\Delta G^0_{P \to P*} = \Delta H^0_{P \to P*} - T \Delta S^0_{P \to P*} \tag{7}$$

Here ΔH^0 and ΔS^0 represent the changes in the standard partial molecular enthalpy and entropy accompanying the excitation, i.e., the standard changes in the enthalpy and entropy of the system for each reaction center that is excited. $\Delta H^0_{P \to P*}$ is approximately equal to the energy difference $h\nu_0$, where ν_0 is the frequency of the 0-0 absorption band and h is Planck's constant. From the absorption maximum at 870 nm and the fluorescence emission maximum at 920 nm, ν_0 is approximately 10 800 cm^{-1}, giving $\Delta H^0_{P \to P*} \sim 1.34$ eV. (In equating enthalpy and energy changes, one makes the safe assumption that the excitation does not involve significant work of expansion or contraction of the reaction center against atmospheric pressure.) The standard partial molecular entropy change is

$$\Delta S^0_{P \to P*} = k_B \ln(n^*/n) \tag{8}$$

where n^* and n are the numbers of degenerate (isoenergetic) substrates in the excited state and in the ground state. Although there is no experimental information on this point, it seems likely that $n^* \sim n$, so that $\Delta S^0_{P \to P*}$ is negligibly small. (If P^* were a triplet state, the change in the number of spin substrates would give $T\Delta S^0_{P \to P*} \sim 0.025$ eV at room temperature.) With these assumptions, we have $\Delta G^0_{P \to P*} \sim 1.34$ eV.

One should distinguish clearly between $\Delta G_{P \to P*}$ and $\Delta G^0_{P \to P*}$ in eq. (6).

The *standard* free energy change $\Delta G^0_{P \to P*}$ is an intrinsic molecular property of the reaction center. It is independent of the light intensity. It can be used to describe a sample consisting of an individual reaction center. The complete free energy change $\Delta G_{P \to P*}$, on the other hand, is meaningful only for systems containing large numbers of molecules. In addition to depending on molecular properties, it depends on the ratio of the numbers of molecules in the ground and excited states (eq. 6). This ratio is a function of the intensity of the light that causes the excitation, and of the kinetics of the reactions that cause the excited state to decay. To a good approximation, $[P*]/[P]$ will be equal to $I\sigma/k_D$, where I is the intensity of light incident on the sample, σ is the absorption cross-section, and k_D is the sum of the rate constants for all the reactions by which P* decays (Parson, 1978b). In the photosynthetic reaction center, the predominant decay path is the electron transfer reaction that generates P^+I^-.

To satisfy oneself with this expression for $[P*]/[P]$, consider the general system illustrated in Fig. 3.4a. Illumination elevates molecules from the ground state (M_1) to the excited state (M_2), from which there is a rapid and efficient transformation to a metastable state (M_3). The k's in Fig. 3.4a are composite rate constants. For example, k_{12} is the sum of the constants for excitation by the actinic light ($I\sigma$), for excitation by the ambient black body radiation from the surroundings, and for spontaneous thermal excitations.

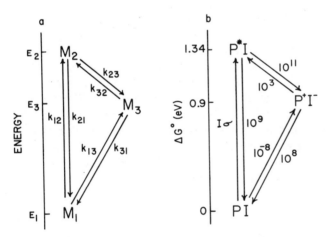

Fig. 3.4. (a) General scheme for a photochemical system consisting of a ground state M_1 (with energy E_1), an excited state M_2 (energy E_2) that is reached by absorption of light, and a metastable state M_3 (energy E_3). The k's are composite rate constants; see text for details. (b) Similar representation of the bacterial photosynthetic apparatus. The rate constants are given in s^{-1}. $I\sigma$ is the rate constant of light excitation, equal to intensity I times the absorption cross-section σ. The vertical scale is partial molecular free energy relative to that of the ground state PI. The 10^9 s^{-1} rate of $P*I \to PI$ was estimated assuming a 99% yield of photochemistry. It could be somewhat smaller.

The first of these predominates if $I\sigma > 10^{-15}$ s^{-1}. k_{21} is the sum of the rate constants for fluorescence and nonradiative decay to the ground state. Fig. 3.4b gives approximate values for the rate constants in the specific case of the bacterial reaction center (see section 3.3). Solving the steady-state rate equations for the general system, one obtains

$$\frac{[M_2]}{[M_1]} = \frac{k_{12}k_{32} + k_{13}k_{32} + k_{31}k_{12}}{k_{21}k_{32} + k_{21}k_{31} + k_{31}k_{23}} \tag{9}$$

In the reaction center, $k_{23} \gg k_{21}$ and $k_{31} \gg k_{32}$. At any light intensity such that $I\sigma$ is greater than about 10^{-13} s^{-1}, the terms $k_{31}k_{12}$ and $k_{31}k_{23}$ predominate heavily and eq. 9 simplifies to

$$[M_2]/[M_1] \sim k_{12}/k_{23} \sim I\sigma/k_{23}, \tag{10}$$

which is the same as the expression given above.

Table 3.1 gives the values of $[P^*]/[P]$ and $\Delta G_{P \to P*}$ calculated from equations (6) and (10) with different values of the light intensity I. For these calculations, we took the excitation wavelength to be 870 nm, where bacterial reaction centers have $\sigma \sim 4.9 \times 10^{-16}$ cm^2 (Straley et al., 1973). (The absorption cross section is $3.817 \times 10^{-21}\epsilon$ cm^2, where ϵ is the molar extinction coefficient.) k_{23} was taken to be 10^{11} s^{-1} (section 3.3.1).

From the values of $\Delta G_{P \to P*}$ listed in Table 3.1, one can calculate the maximum values of the ratio $[P^+I^-]/[PI]$ that can be generated by illumination at

TABLE 3.1

FREE ENERGY CAPTURE IN REACTION CENTERS

Illumination	$I\sigma$ [a] (s^{-1})	$\dfrac{[P^*]}{[P]}$ [b]	$\Delta G_{P \to P*}$ [c] (eV)	$\dfrac{\Delta G_{P \to P*}}{h\nu_0}$
Continuous light $(6 \times 10^{13}$ quanta cm^{-2} s$^{-1})$	2.9×10^{-2}	2.9×10^{-13}	0.61	0.45
Continuous light $(6 \times 10^{14}$ quanta cm^{-2} s$^{-1})$	2.9×10^{-1}	1.9×10^{-12}	0.67	0.50
Q-switched laser flash $(3 \times 10^{22}$ quanta cm^{-2} s$^{-1})$	1.5×10^7	1.5×10^{-4}	1.12	0.84
Mode-locked laser flash [d] $(3 \times 10^{25}$ quanta cm^{-2} s$^{-1})$	1.5×10^{10}	1.5×10^{-1}	1.29	0.96

[a] Calculated for monochromatic excitation at 870 nm and $\sigma = 4.9 \times 10^{-16}$ cm^2. For broadband excitation, $I\sigma$ should be replaced by $\int I(\nu)\sigma(\nu)d\nu$.
[b] Calculated from eq. (10), assuming $k_{23} = 10^{11}$ s^{-1}.
[c] Calculated from eq. (6). with $h\nu_0 = 1.34$ eV.
[d] Eq. (10) will not be strictly applicable at such a high light intensity, because singlet-singlet annihilation and stimulated emission will decrease the lifetime of P^*.

various intensities. These are obtained by supposing that the P*I and P⁺I⁻ states come to equilibrium during the illumination. At equilibrium the free energy change for the electron transfer process $P*I \rightarrow P^+I^-$ is zero, and $\Delta G_{P \rightarrow P*} = \Delta G_{PI \rightarrow P^+I^-}$. This gives a maximum value for the $[P^+I^-]/[PI]$ ratio, because in practice P⁺I⁻ decays rapidly by further electron transfer reactions and the system never really comes to equilibrium. Moderately strong continuous illumination (6×10^{14} quanta cm^{-2} s^{-1}) can keep a maximum of only about 0.01% of the reaction centers in the radical pair state, but a Q-switched laser flash could put essentially all of the reaction centers in this state.

Note that with moderately strong continuous illumination $\Delta G_{P \rightarrow P*}$ is only about 50% of the energy gap $h\nu_0$ (Table 3.1). Each photon that is absorbed increases the free energy of the system, not by $h\nu_0$, but by considerably less. This point has been made frequently by previous authors, who have used more elaborate thermodynamic treatments than the one given above. (See the chapter by Knox (1977) in Vol. 2 of this series.) It seems not to be generally appreciated, however, that it is $\Delta G_{P \rightarrow P*}$, not $\Delta G^0_{P \rightarrow P*}$, that is less than $h\nu_0$. $\Delta G_{P \rightarrow P*}$ is a function of the light intensity, and it goes to zero if one turns off the light. But this does not give any information about the mechanism of the photochemical reaction. The magnitude of $\Delta G_{P \rightarrow P*}$ does not put any limits on $\Delta G^0_{PI \rightarrow P^+I^-}$ or on $(E^P_m - E^I_m)$. The light intensity determines only how frequently the reaction center is excited. The electron transfer reaction is thermodynamically favorable in individual reaction centers because $\Delta G^0_{P \rightarrow P*}$ (the *standard* partial molecular free energy of the excited state relative to that of the ground state) is greater than $\Delta G^0_{PI \rightarrow P^+I^-}$ (the *standard* partial molecular free energy of the radical pair state, relative to the ground state). Since $\Delta G^0_{P \rightarrow P*} \sim 1.34$ eV and $\Delta G^0_{PI \rightarrow P^+I^-} \sim 0.9$ eV, any reaction center that absorbs a photon has more than sufficient free energy to energize the electron transfer reaction. The thermodynamic arguments that Fong (1976) has advanced against the scheme shown in Fig. 3.3 are, therefore, unfounded. For further discussion of this point, see Parson (1978b).

The decrease of about 0.44 eV in the standard partial molecular free energy accompanying the electron transfer reaction may be critical for maximizing the rate of the electron transfer. We shall consider this point in more detail in the following section. In addition, the substantial drop in free energy has the effect of preventing the back reaction that would return the reaction center to the excited state (P*). Referring to Fig. 3.4A, the rate constant for the reaction from state M_3 back to M_2 will be given by

$$k_{32} = k_{23} \exp(-\Delta G^0_{23}/kT) \tag{11}$$

In the bacterial reaction center,

$$k_{32} \sim (10^{11} s^{-1}) \exp(-0.44 \text{ eV}/kT) \sim 10^3 s^{-1} \tag{12}$$

This is very low, compared to the rate constant of about 5×10^9 s^{-1} with which the radical pair state normally decays by transferring an electron from I$^-$ to Q_A. The serious problem remains, however, of preventing the back reaction that returns the system to the ground state (k_{31} in Fig. 3.4a). This process is strongly downhill thermodynamically, and it must be prevented by kinetic barriers. We shall return to this point also in the following section *.

3.5. THEORIES OF ELECTRON TRANSFER

3.5.1. Introduction

In this section, we discuss several theories of electron transfer, with the aim of clarifying the factors that determine the kinetics of the primary photosynthetic reactions. Electron transfer reactions can be characterized as either "outer-sphere" or "inner-sphere" processes, depending on whether they involve only the direct transfer of an electron, or whether the transfer is mediated by a bridging ligand. Because the primary photosynthetic processes probably are outer-sphere processes, we shall limit our discussion to reactions of this type. For an excellent general discussion of electron transfer reactions, see Reynolds and Lumry (1966).

An electron transfer reaction in solution can be visualized as occurring in several steps: (1) The reactants approach each other from an infinite distance. (2) The nuclear geometry of the reactants undergoes distortions, generating an "activated complex". (3) An electron moves from donor to acceptor, generating an activated complex of the products. (4) The geometry of the products relaxes. (5) The products separate to infinite distance. In the photosynthetic apparatus, the primary electron carriers probably are held at essentially fixed distances, so that steps (1) and (5) are not necessary. The important point is that the nuclear configuration of the activated complexes involving the reactants and the products must be identical, because an electron is transferred in a time that is short compared to vibrational periods. This is the Franck-Condon principle. As a consequence, no energy, entropy, or angular momentum (either spin or orbital) can be removed from the system during the actual electron transfer. If the overall electron transfer reaction is exothermic, the extra energy must be stored transiently by placing the products or the solvent in excited vibrational states. In other words, the activated complex must have a strained geometry. Relaxation to the final equilibrium states of the products will occur following the electron transfer.

An electron transfer reaction is called "adiabatic" if distortion of the reactants into the geometry of the activated complex almost always results

---

* See Note 2 added in proof on page 114.

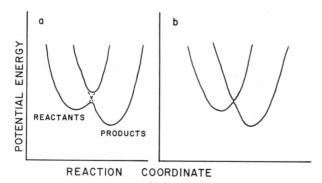

Fig. 3.5. Potential surfaces for electron transfer between reactants and products. (a) Large interaction energy, corresponding to an adiabatic reaction. (b) Small interaction energy, corresponding to a non-adiabatic reaction. (a) might represent a reaction of molecules with a small intermolecular separation, while (b) might represent the same reaction at a larger distance. For an introduction to potential energy surfaces and the concept of the reaction coordinate, see chapter 4 of Reynolds and Lumry (1966).

in electron transfer. It is called "nonadiabatic" if electron transfer in the activated complex is a rare event. The degree to which a reaction is adiabatic increases as a function of the electronic coupling between the reactant and product states of the system. This is illustrated in Fig. 3.5, which shows a classical view of the potential energies of the reactants and products as a function of the geometry of the system. The abscissa in such a diagram, the "reaction coordinate", represents a path through an $(N + 1)$-dimensional space, where N is the number of vibrations in the two molecules (Reynolds and Lumry, 1966). The distance between the electron donor and acceptor is considered to be constant. An activated complex occurs at a point where the geometry and potential energy of the products are the same as those of the products. If the electronic coupling between the reactant and product states of the system is strong, the potential energy surfaces will split in the region of the activated complex (Fig. 3.5a). Reacting molecules then will remain on the lower potential energy surface, and the system will pass smoothly from the reactant to the product states. If the electronic interaction is weak, for example because the electron donor and acceptor are far apart, there will be little or no splitting of the potential energy surfaces. The system then will tend to remain on the reactant surface and will cross through the activated complex region without generating products (Fig. 3.5b).

The strength of the coupling between the reactant and product states depends on the electronic orbital overlap between the wave functions describing these states. It thus can be highly sensitive to the orientations of the electron donor and acceptor with respect to each other, as well as to the distance between the reactants. The splitting of the potential energy surfaces in the activated complex is equal to twice the "interaction matrix element",

T_{da}, which is given by

$$T_{da} = \langle \phi_{d+1} \phi_a | \mathcal{H} | \phi_d \phi_{a+1} \rangle. \tag{13}$$

Here ϕ_d and ϕ_{d+1} are the electronic wave functions for the oxidized and reduced donor, and ϕ_a and ϕ_{a+1} are similar functions for the acceptor. $\mathcal{H}$ is the Hamiltonian operator, and the symbol $\langle \| \rangle$ indicates an integral taken over all space. T_{da} has the units of energy and is analogous to the resonance integral β encountered in Hückel molecular orbital theory. The reactant and product states are isoenergetic in the active complex, and the two states will be in resonance if the orbital overlap is strong. As a rule of thumb, electron transfer should be adiabatic if $T_{da} > 0.01$ eV, and nonadiabatic if $T_{da} <$ 0.001 eV.

It usually is assumed that T_{da} decreases approximately exponentially with increasing distance between the electron donor and acceptor. Electron transfer reactions therefore should become more nonadiabatic as the distance between the reactants increases. Reactions in solution generally are thought to be adiabatic, because the reactants are able to diffuse freely to within 1—2 Å of each other. Rotational diffusion also allows the molecules to sample various orientations with respect to each other. The mean value of T_{da} in the collision complexes formed between the reactants thus is high enough so that the rate of electron transfer becomes relatively insensitive to the exact magnitude of T_{da}. This probably is not true for electron transfer between large biological molecules, where the electron donor and acceptor could be held separated by considerably greater distances. Electron transfer can still occur over distances on the order of 10 Å, but one would expect it to be nonadiabatic (Hopfield, 1974, 1976, 1977; Jortner, 1976). In this case, small changes in the orientation or separation of the reactants could alter the rate of the reaction profoundly.

3.5.2. Classical theories

According to classical absolute rate theory, (Glasstone et al., 1941), the rate constant for a reaction can be written as

$$W = \kappa (k_B T/h) \exp(-\Delta G^{\ddagger}/k_B T). \tag{14}$$

Here κ is the transmission coefficient, or the probability that the activated complex of the reactants will be transformed into the activated complex of products; k_B and h are the Boltzmann and Planck constants; T is the absolute temperature; and $\Delta G^{\ddagger}$ is the free energy of activation (the standard partial molecular free energy change for forming the activated complex of reactants). For adiabatic reactions in solution, $\kappa \sim 1$, so the central theoretical problem is to calculate $\Delta G^{\ddagger}$. Using a classical, macroscopic model of the solvent and the reactants, Marcus (1964) derived the expression

$$\Delta G^{\ddagger} = W_r + (\lambda + \Delta G_r^0)^2/4\lambda. \tag{15}$$

Here W_r is the work that is required to bring the reactants together to a separation distance r, and ΔG_r^0 is the standard partial molecular free energy change for the electron transfer reaction when the reactants are separated by this distance. (ΔG_r^0 includes coulombic interactions between the reactants and between the products; see sections 3.2 and 3.4 for further discussion of this point.) The parameter λ represents the work that must be done to distort the reactants and solvent so as to satisfy the Franck-Condon principle in the activated complex. Marcus (1964) and others (Levich, 1966; Dogonadze, 1971) obtained explicit expressions for λ based on classical concepts. Polarization of the solvent was considered to make a major contribution to λ.

Equations (14) and (15) indicate that the rate of the reaction will be maximal when $\Delta G_r^0 = -\lambda$, and will fall off steeply for values of ΔG_r^0 on either side of this. One can understand this qualitatively by considering the potential energy surfaces illustrated in Fig. 3.6. The potential energy barrier separating the reactants and products becomes very small as ΔG_r^0 approaches $-\lambda$ (Fig. 3.6B). The barrier is greater if ΔG_r^0 is either much less negative than λ (Fig. 3.6A) or much more negative (Fig. 3.6C).

The classical theory outlined above accounts reasonably well for the rates

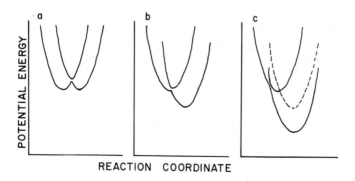

Fig. 3.6. Diagram illustrating the classical dependence of the activation energy barrier on the free energy change associated with an electron-transfer reaction. (a) $\Delta G^0 = 0$; a potential barrier exists between reactants and products and hence a normal activation energy is observed. (b) $\Delta G^0 = -\lambda$; no potential energy barrier exists between reactants and products, so the reaction has maximal rate. (c) $\Delta G^0 < -\lambda$; a potential barrier occurs once again, causing the reaction rate to slow down. The dotted line represents an excited vibrational state of the product which satisfies the energy conservation requirement and therefore allows the reaction to have maximal rate in this region also. A similar figure illustrating this point is given by Efrima and Bixon (1974).

of many electron transfer reactions in solution, particularly if $\Delta G^0 \sim 0$ (Reynolds and Lumry, 1966; Chou et al., 1977). Reactions that are strongly exothermic, however, can have rates that are much higher than predicted (Rehm and Weller, 1970; Van Duyne and Fischer, 1974; Creutz and Sutin, 1977). An explanation of these observations was provided by Van Duyne and Fischer (1974), Efrima and Bixon (1974, 1976), Ulstrup and Jortner (1975), Schmickler (1976) and Fischer and Van Duyne (1977). They pointed out that excited vibrational levels of the products can act in addition to the solvent polarization term to satisfy the energy conservation requirements for strongly exothermic reactions. Qualitatively, this is illustrated by the dotted line in Fig. 3.6C, in which an excited vibrational state of the product intersects the reactant surface near the bottom; reducing the activation energy barrier. We shall discuss this effect, called vibronic coupling, in more detail below.

In addition to failing to account for the rates of very exothermic reactions, eqs. (13) and (14) have other limitations that restrict their usefulness in the problem of photosynthetic electron transfer reactions. Because they are based on classical concepts, they are inapplicable at low temperatures, when the quantization of vibrational energy levels becomes important. The temperature dependence of the photosynthetic electron transfer reactions is one of their most intriguing aspects (section 3.3), and one that certainly needs to be explored theoretically. In addition, eq. (14) is most useful if $\kappa \sim 1$. As we have mentioned above, the electron carriers in the photosynthetic apparatus are held apart at what could be relatively large distances. Their electronic wave functions could overlap so weakly that electron transfer would be nonadiabatic, with κ much closer to zero than to 1. Under these conditions, it is fruitful to view the progression of the system from reactants to products as a tunneling through a barrier of high potential energy.

3.5.3. Franck-Condon factors

The idea of tunneling in photosynthetic electron transfer reactions was first advanced by DeVault and Chance (1966) to account for their observation that the kinetics of oxidation of cytochrome $c552$ in *Chromatium vinosum* became independent of temperature below $120°K$ (Section 3.4). Early discussions of electron tunneling in photosynthetic systems (DeVault and Chance, 1966; McElroy et al., 1969; Kihara and Chance, 1969; Hales, 1976) overestimated the distances over which tunneling can occur. These discussions generally were based on a simple model in which there were few or no restrictions on the energy levels available to the electron on either side of the barrier. In fact, the available energies are quantized, because the electron must move from one bound state to another. The requirements imposed by the Franck-Condon principle are therefore severe (Grigorov and

Chernovskii, 1972; Blumenfeld and Chernovskii, 1973; Hopfield, 1974; Jortner, 1976).

If the electronic interaction between the electron-donor and acceptor is not too strong, one can use time-dependent perturbation theory to express the evolution of the system from reactants into products (Ulstrup and Jortner, 1975; Efrima and Bixon, 1976). Suppose that the reactants are in a particular vibrational state that can be described by a nuclear wave function χ_i. When electron transfer occurs, the products might be generated in any one of many different vibrational states with wave functions χ_f. According to time-dependent perturbation theory, the rate constant for electron transfer starting in χ_i will be

$$W_i = \frac{1}{h} |T_{da}|^2 \sum_f |\langle \chi_i | \chi_f \rangle|^2 \, \delta(E_i - E_f).$$ (16)

In this expression, T_{da} is the electronic interaction matrix element that was discussed above (eq. 13), and E_i and E_f are the total (electronic plus vibrational) energies in the initial and final states. The delta function $\delta(E_i - E_f) = 1$ when $E_i = E_f$ and zero otherwise, in accordance with the Franck-Condon principle. The term $|\langle \chi_i | \chi_f \rangle|^2$ is the square of the overlap integral for the nuclear wave functions in the initial and final states; its magnitude also must be nonzero if electron transfer is to occur, according to the Franck-Condon principle. This term is frequently called a Franck-Condon factor. In general, the reactants may be distributed among several (or even many) initial vibrational states, depending on the temperature. To obtain the total rate of electron transfer, one therefore needs to sum W_i over these different initial states, weighting each state so as to give the proper thermal average. Equation 16 is not readily evaluated, so approximate approaches are required.

3.5.4. Hopfield's theory

Hopfield (1974, 1976, 1977) has suggested a semiclassical solution to the problem, based on an analogy to the Förster description of excitation transfer between molecules. Förster (1946) showed that the rate of excitation transfer is proportional to the square of an interaction matrix element, and to an integral which represents the overlap of the fluorescence emission spectrum of the energy donor with the absorption spectrum of the acceptor. The interaction matrix element contains the dependence on the distance between the donor and acceptor and on the orientations of the molecules, while the overlap integral depends only on the nature of the molecules. Pointing out the similarity between excitation transfer and electron transfer, Hopfield expresses the rate constant for electron transfer as:

$$W = \frac{1}{h} |T_{da}|^2 \int_{-\infty}^{\infty} D_d(E) D_a(E) \, dE.$$ (17)

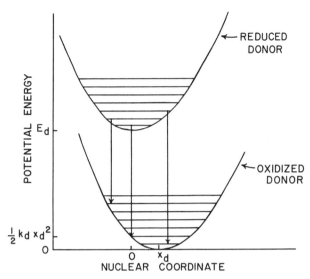

Fig. 3.7. Schematic diagram illustrating the origin of the electron removal spectrum $D_d(E)$. The arrows represent only a few of the many possible transitions between the reduced and oxidized forms of the donor molecule. The two potential curves are offset by a displacement x_d. The peak of the electron removal spectrum will occur at $E = E_d - k_d x_d^2/2$, indicated by the center arrow. The horizontal lines are vibrational levels of the two electronic states. For a similar diagram including the electron insertion spectrum of the acceptor see Potasek and Hopfield (1977).

In eq. (17), the matrix element T_{da} contains the dependence on distance and orientation (see eq. 13). The integral gives the overlap in energy (E) between the "electron removal spectrum" of the donor, $D_d(E)$, and the "electron insertion spectrum" of the acceptor, $D_a(E)$.

The origin of the electron removal spectrum is shown schematically in Fig. 3.7. The upper curve represents the potential energy of the electron donor in its reduced state, and the lower curve the energy in the oxidized state. The abscissa represents the length of a bond in the molecule. (Note that this is not the same as the reaction coordinate that forms the abscissa in Fig. 3.6.) One obtains the electron removal spectrum by summing all possible vertical transitions between the reduced and oxidized states of the molecule, weighting each transition according to its probability. If there are no changes in bond lengths (i.e., no horizontal displacement of the curves) or in the curvature of the vibrational potential energy curves, all vertical transitions would involve the same change in energy. The electron removal spectrum then would be a sharp line at this energy. If there are changes in bond lengths or in curvature, transitions with different energies are possible and the electron removal spectrum is broadened. Similar considerations apply to the electron insertion spectrum for the acceptor. The broadening, or vibronic coupling of the two spectra is critical for electron transfer. Without it, the spectral overlap in eq. (17) would be zero and the Franck-Condon requirements could not be met, unless electron removal and insertion happened to involve identical changes in electronic energy. The horizontal lines in Fig. 3.7 indicate that the vibrational energy levels are quantized. Orchin and Jaffe (1971) give a clear introduction to the role of vibronic coupling in determining the width of electronic absorption bands, a problem similar to the present case of electron transfer. For a comprehensive discussion of the exact meaning of "vibronic coupling", see Azumi and Matsuzaki (1977).

As a simplifying assumption, Hopfield (1974, 1976, 1977) supposed that $D_d(E)$ is a Gaussian function of E:

$$D_d(E) = (2\pi\sigma_d^2)^{-1/2} \exp[-(E-E_d + \tfrac{1}{2}k_d x_d^2)^2/2\sigma_d^2] \tag{18}$$

with a temperature-dependent linewidth

$$\sigma_d^2 = k_d x_d^2 (k_B T_d/2) \coth(T_d/2T) . \tag{19}$$

These expressions suppose that oxidation of the electron donor causes the mean length of a particularly important bond to change by a distance x_d. The force constant for this vibration, k_d, (the curvature of the vibrational potential energy curve) is taken to be the same for the reduced and oxidized forms of the molecule. Since the bond length in the oxidized donor immediately after the electron transfer is the same as it was in the reduced donor, the most probable vibrational energy in the oxidized donor is $k_d x_d^2/2$. E_d is the electronic energy difference between the reduced and oxidized forms of the donor (Fig. 3.7). T_d describes the spacing between the vibrational energy levels, expressed in the form of a characteristic temperature in °K. Fig. 3.8 shows the temperature dependence of σ and $D_d(E)$, according to equations (18) and (19).

Writing a similar expression for $D_a(E)$, and substituting into eq. (17), Hopfield obtained the following equation:

$$W = \frac{1}{h}|T_{da}|^2(2\pi\sigma^2)^{-1/2} \exp[-(\Delta E - \Delta)^2/2\sigma^2] \tag{20}$$

with $\sigma^2 = \sigma_d^2 + \sigma_a^2$, $\Delta = k_d x_d^2/2 + k_a x_a^2/2$, and $\Delta E = E_d - E_a$. (Note that ΔE is defined to be positive for an exothermic reaction.)

Hopfield's treatment agrees with the classical expression that the rate of electron transfer will go through a maximum when $\Delta E = \Delta$. The parameter Δ is analogous to the classical parameter λ, except that Δ describes vibrations in the electron carriers whereas λ was attributed mainly to motion of the solvent. Equation (20) indicates further that W will fall with decreasing

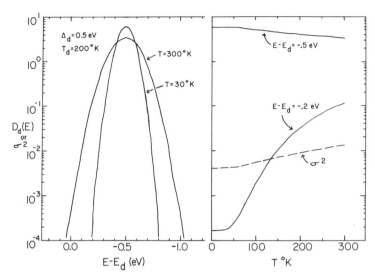

Fig. 3.8. Temperature dependence of the electron removal spectrum. The solid lines are plots of eq. 18 for $D_d(E)$, and the dotted line is a plot of eq. 19 for the width σ_d^2 of the Guassian. For these plots $\Delta_d \equiv k_d x_d^2/2 = 0.5$ eV and $T_d = 200°K$. The left hand side of the figure shows $D_d(E)$ versus $E-E_d$ at $30°K$ and $300°K$. The right hand side shows $D_d(E)$ versus T for $E-E_d = -0.2$ and -0.5 eV. The curves are constant in shape until about $50°K$, then become progressively broader as temperature is increased. The constant width at low temperature results from the different nuclear positions possible in the zero point vibrational mode, while the increase in width at higher temperatures reflects population of excited vibrational levels of the reduced donor molecule (see Fig. 3.7).

temperature when the critical vibrations settle into their zero-point modes. This accounts qualitatively for the temperature dependence that has been found for electron transfer between the low-potential cytochromes and P^+ (section 3.3.4).

Note that Hopfield's theory attributes all of the temperature dependence of W to the effect of temperature on the width of the electron removal and insertion spectra. The distance between the donor and acceptor, which is the main determinant of T_{da}, is assumed to be constant. It is not clear whether this is a realistic assumption. Earlier discussions of electron tunneling frequently invoked the idea that the width of the barrier separating the donor and acceptor was a function of temperature (DeVault and Chance, 1966; Hales, 1976). This could well be the case, in spite of the other limitations of the earlier work (see below).

Although the semiclassical treatment accounts qualitatively for the dependence of electron transfer rates on T_{da}, ΔE, and T, it rests on a number of assumptions and approximations that are not always tolerable. The assumption of Gaussian electron-removal and electron-insertion functions requires that the electron transfer reaction be exothermic, and that the geometrical distortion accompanying the reaction be large (Hopfield, 1974, 1976, 1977). It requires also that transfer of energy to the solvent be insignificant. Motions of the solvent cannot be treated by the Förster approach, because they affect the electron donor and acceptor coherently (Jortner, 1976). Even with these restrictions, the treatment may not be quantitatively accurate at low temperatures (Jortner, 1976).

3.5.5. Jortner's theory

A more rigorous quantum mechanical treatment of electron transfer in biological systems has been developed by Jortner (1976). He expresses the rate as the product of the electronic term $|T_{da}|^2/h$ and a nuclear term that includes both molecular vibrations and solvent vibrations. The molecular vibrations are approximated by an average frequency spacing $\bar{\omega}$ and an effective "coupling strength" $S = \Sigma x_j^2 \mu_j \omega_j/2h$, where x_j is the change that the reaction causes in the mean length of bond j, μ_j is the reduced mass of the vibrating groups in the bond, and ω_j is the frequency of the vibration. S is a dimensionless parameter expressing the change in geometry (of both reaction partners) that is coupled to the electron transfer. The motions of the solvent are taken to have a comparatively low average frequency, ω_s. For the case when the solvent coupling strengths are so small as to be negligible, and when a single mode of molecular vibrations dominates the problem (note again the contrast with the classical theories of electron transfer in solution), Jortner's expression for the rate constant is

$$W = A \exp[-S(2\bar{\nu} + 1)]I_p\{2S[\bar{\nu}(\bar{\nu} + 1)]^{1/2}\}[(\bar{\nu} + 1)/\bar{\nu}]^{p/2} \tag{21}$$

In this expression,

$$A = 2\pi|T_{da}|^2/\hbar^2\omega_s,$$

$$I_p\{z\} = (z/2)^p \sum_{n=0}^{\infty}(z^2/4)^n/n!(n+p)!$$

is a modified Bessel function.

$$p = \Delta E/\hbar\bar{\omega}, \bar{\nu} = [\exp(\hbar\bar{\omega}/k_BT) - 1]^{-1}$$

expresses the thermal excitation of the molecular vibration, and $\hbar = h/2\pi$.

The vibrational energy $\hbar\bar{\omega}$ in the Jortner's theory is analogous to $\frac{1}{2}k_B(T_d + T_a)$ in Hopfield's theory. $Sh\bar{\omega}$ is analogous to Hopfield's Δ. In agreement with the semiclassical theory, eq. (21) predicts that the rate of electron transfer will decrease with falling temperature, leveling off when k_BT becomes $\ll\hbar\bar{\omega}$. In Jortner's view, electron transfer at very low temperatures depends on *nuclear* tunneling from the zero-point vibrational state of the reactants to a nearly isoenergetic vibronic state of the products.

3.5.6. The cytochrome oxidation kinetics

Fig. 3.9 shows plots of the rate predicted by eq. (21), as a function of the temperature. For comparison, the figure includes experimental data on the rates of electron transfer between P^+ and the low-potential c-type cytochrome of *C. vinosum* (see section 3.3). If ΔE for the electron transfer is known, the data can be fit by only a narrow range of the other parameters, but if ΔE is uncertain many different fits are possible. Unfortunately, there is an element of uncertainty about ΔE. The difference between the E_m values of P and the cytochrome is approximately 0.45 V (Case and Parson, 1971; Dutton, 1971). This probably is approximately equal to ΔE, because the net entropy change in the reaction appears is be small (Case, 1972). One must bear in mind that the E_m values can not reflect interactions between P^+ and the reduced cytochrome, and that they may not reflect transient states that are important kinetically (section 3.4.1).

In the absence of definitive information on these points, we set $\Delta E = 0.45$ eV and adjusted the other parameters to fit eq. (21) to the data. The magnitude of $\bar{\omega}$ is fixed fairly tightly by the temperature at which the rate

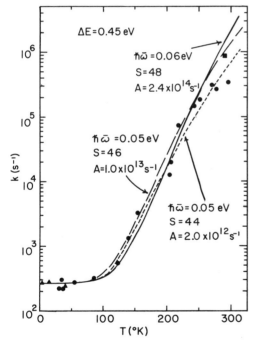

Fig. 3.9. Rate constant for cytochrome c552 oxidation in *C. vinosum* versus temperature. ●, DeVault and Chance (1966); ▲, DeVault et al. (1967); ■, Parson and Case (1970) and Seibert and DeVault (1978). The lines are alternate fits to the data using eq. 21. Fitting parameters are given in the figure.

begins to rise with increasing temperature. S primarily determines the steepness of the curve above this point. A sets the vertical scale for the entire curve. The data are fit well with $\hbar\bar{\omega}$ = 0.05 to 0.06 eV, S = 44 to 48, and A = 10^{12} to 10^{14} s^{-1}.

The value $\hbar\bar{\omega} \sim 0.05$—0.06 eV obtained by the curve-fitting in Fig. 3.9 seems physically reasonable for vibrations in the cytochrome. Transition metal ions such as Fe have metal-ligand vibrations of their first coordination spheres with energy spacings of this magnitude (Kestner et al., 1974). Note that k_BT is smaller than $\hbar\bar{\omega}$ even at room temperature (k_BT = 0.0254 eV at 300°K). "High temperature" approximations of the rate equation therefore are of questionable significance.

Whether the values of S $\sim$ 44—48 obtained by fitting the data are physically reasonable is not so clear. These values imply quite large changes of bond lengths. Outer-sphere electron transfer reactions in solution typically have S values on the order of 1 to 10, although instances of considerably larger values are known (Kestner et al., 1974). A large S value would be expected if the Fe in the cytochrome moved relative to the heme plane upon a change in redox state. Such a movement occurs in hemoglobin, but not in soluble high-potential c-type cytochromes (Salemme, 1977); whether it occurs in the membrane-bound low-potential cytochrome of C. vinosum is problematic.

The value for A $\sim 10^{12}$ to 10^{14} s^{-1} allows one to calculate the magnitude of the electronic interaction matrix element, T_{da}, if one also makes an estimate of ω_s, the average frequency of the motions of the solvent. Taking $\hbar\omega_s \sim 5 \times 10^{-3}$ eV (Jortner, 1976), one gets $T_{da} = 10^{-3}$ to 10^{-2} eV. This is much larger than the estimates given by Jortner (1—3×10^{-5} eV) or Hopfield (4×10^{-4} eV). Our fit to eq. (21) to the data differs from Jortner's in two ways. First, we set ΔE = 0.45 eV (see above), while Jortner used 0.1 eV. Second, we emphasized recent measurements of the cytochrome oxidation kinetics at 295°K, while Jortner used earlier data which apparently underestimated the rate at high temperatures (see Fig. 3.9). The value of T_{da} indicated by our analysis is so large that it calls into question the assumption that the electron transfer is nonadiabatic. The perturbation treatment underlying eq. (16) may not be valid if the electronic interaction is this strong. In any case, the extremely wide range of T_{da} resulting from various fits to the data prevents one from using T_{da} to calculate the distance between the cytochrome and P *.

Equation 21 assumes that motions of the solvent are relatively unimportant in determining the electron transfer rate, compared to the molecular vibrations. Jortner (1976) justifies this assumption by noting that the cytochrome oxidation rate is essentially independent of temperature in the region between 10 and 100°K. If solvent motions were important, some

---

* See Note 3 added in proof on page 114.

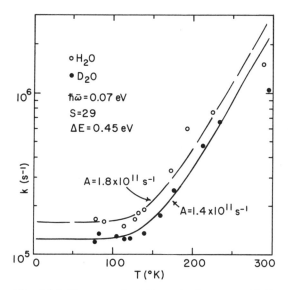

Fig. 3.10. Rate constant for cytochrome oxidation vs. T in *Rhodopseudomonas* sp. N.W. Data points are from Kihara and McCray (1973). ○, H_2O as solvent; ●, D_2O as solvent. The lines are fits to the data using Eq. 21. Fitting parameters are given in the figure.

temperature dependence should be observed in this region, where $\hbar\omega_s$ is likely to be comparable to k_BT.

Fig. 3.10 shows a similar attempt to fit eq. (21) to data on the kinetics of oxidation of the low potential cytochrome in *Rhodopseudomonas* sp. N.W. (Kihara and McCray, 1973). Estimates of ΔE are even less certain for this species than they are for *C. vinosum*, because the E_m values of P and the cytochrome have not been measured. However the E_m values are likely to be close to those of P and the low-potential cytochromes in other related species (see sections 3.3 and 3.4). We therefore set $\Delta E = 0.45$ eV in this case also. The experimental data obtained with H_2O as the solvent can be fit reasonably well, but not perfectly, with $\hbar\bar\omega \sim 0.07$ eV, $S \sim 29$, and $A \sim 1.8 \times 10^{11}$ s^{-1}. The data for D_2O can be fit with the same values of $\hbar\bar\omega$ and S, by decreasing A to 1.4×10^{11} s^{-1}. In both cases, the fit becomes poor at high temperatures; however, it is possible that the data underestimate the rate in this region. The high rate of the reaction makes these measurements difficult technically and subject to error in this direction. The change in rate between high and low temperatures is not nearly as dramatic as it is in *C. vinosum*.

The effect of D_2O could be ascribed to an increase in ω_s or a decrease in T_{da}, since only A appears to change. One might expect deuteration to decrease ω_s instead of increasing it, but the nature of the "solvent" in the chromatophore membrane is not well defined. Kihara and McCray (1973) found that D_2O affected several other oxidation—reduction reactions of cytochromes similarly; the effect apparently does not depend on the nature of the cytochrome or its reaction partner. Deuteration of P and the cytochrome would be expected to increase the rate of the cytochrome oxidation by decreasing $\bar{\omega}$, if C-H vibrations make an important contribution to $\bar{\omega}$. An inverse isotope effect of this nature has been observed in certain reactions in vitro (Razem et al., 1977).

3.5.7. Dependence of the electron transfer rate on ΔE

Like the semiclassical treatment, eq. (21) predicts that the rate of electron transfer will be maximal if $\Delta E = S\hbar\bar{\omega}$. Fig. 3.11 shows the dependence of W on ΔE at 300°K for several different values of S and $\hbar\bar{\omega}$. The decrease of W on either side of the maximum is most abrupt when $S\hbar\omega$ is small.

Fig. 3.12 illustrates how the dependence of W on ΔE is related to the temperature. The decrease of W on either side of the maximum is steeper at lower temperatures. When ΔE is near the optimal value, W is virtually independent of the temperature, actually showing a slight inverse temperature dependence.

The curves shown in Figs. 3.11 and 3.12 are not symmetrical about their maxima. W falls off somewhat more gradually on the right-hand limbs of the curves, when $\Delta E > S\hbar\bar{\omega}$. Efrima and Bixon (1974, 1976) and Ulstrup and

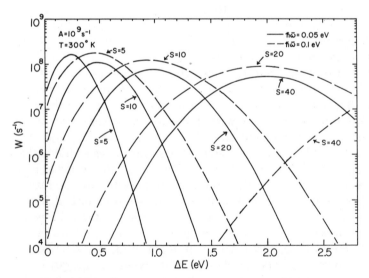

Fig. 3.11. Plot of rate constant (W) vs. ΔE at 300°K, calculated using eq. 21. Solid lines are used for $\hbar\bar{\omega} = 0.05$ eV, while dashed lines are used for $\hbar\bar{\omega} = 0.1$ eV. Plots are shown for S = 5, 10, 20, and 40.

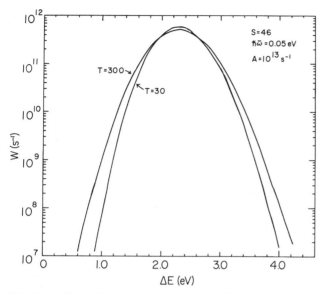

Fig. 3.12. Plot of rate constant (W) vs. ΔE, calculated using eq. 21. Plots are shown for T = 30°K and T = 300°K. Note the inverse temperature dependence at the peak and the increasing temperature dependence as a function of ΔE on either side of the peak. Parameters are listed on the figure.

Jortner (1975) have shown that the asymmetry becomes much more pronounced if a larger number of vibrational modes play a significant role in the electron transfer. If the number of vibrational modes with substantial contributions to S is sufficiently large, W becomes very insensitive to ΔE in the highly exothermic region. Each additional mode provides new opportunities for excited vibrational states of the molecules to satisfy the Franck-Condon requirements in this region.

The fall-off in rate with decreasing ΔE when ΔE is below the optimum has been well documented for electron transfer reactions in solution. The photo-oxidation of BChl and BPh that were discussed in section 2 illustrate this clearly. In contrast, there appears to be no evidence that electron transfer rates also fall off significantly when the reaction becomes highly exothermic. Rehm and Weller's (1970) study of the quenching of fluorescence of aromatic compounds by oxidants and reductants is a case in point here. As we discussed in section 3.2, the second-order rate constant for the fluorescence quenching increases with increasing ΔE, until it reaches the limit set by diffusion. Rehm and Weller (1970) found that the rate did not fall below this limit again, even when ΔE was made as large as 2.6 eV. Unfortunately, the fluorescence quenching constant is relatively insensitive to the actual (first-order) rate constant for electron transfer in the collision complexes between the reactants, as long as electron transfer is fast relative to dissociation of the complexes. Creutz and Sutin (1977) have observed a small (2-fold) decrease in the rate of electron transfer in excited Ru complexes when the reactions become extremely exothermic.

3.5.8. Kinetics of electron transfer between P and I

Let us now return to the electron transfer reactions that occur between P and I. If the photochemical apparatus is to work with the greatest possible efficiency, the rate of electron transfer from P* to I must be maximized, but the rate of the back reaction from I^- to P^+ must be minimized. This is not a trivial problem: in the in vitro photochemical systems that are discussed in section 3.2, the back reactions are faster than the forward reactions. Yet in the bacterial reaction center the rate constant for the forward reaction is more than 10^3-times that for the back reaction (section 3.3). How is this achieved?

In view of the discussion in section 3.5.7, the energy changes accompanying the forward and backward reactions could be a critical factor (Felton, 1978). Unfortunately these are not known with certainty, although the redox titrations of P and I do provide estimates of the free energy changes (section 3.4). The transfer of an electron from P* to I appears to involve a midpoint potential change of about +0.4 eV. A substantially larger ΔE_m of about +0.9 eV appears to accompany the back reaction from I^- that returns P to the ground state. (We shall consider the back reaction that forms the triplet state below.) For the sake of discussion, assume that the ΔE values for the two reactions are close to the ΔE_m values. It is clear from Fig. 3.11 that the rates of the reactions also will depend strongly on the values of S and $\hbar\bar{\omega}$. If $\hbar\bar{\omega} \sim 0.05$ eV (the value obtained by fitting the data on cytochrome oxidation), the ratio of the rates of the forward and backwards reactions would be maximal if S is about 5. The left-most curve in Fig. 3.11 shows that a reaction with $\Delta E = 0.4$ eV then would fall on the peak of the curve relating W to ΔE and would be 10^4-times faster than a reaction with $\Delta E = 0.9$ eV.

A value of 5 for S would be considerably smaller than the values that were obtained from the data on cytochrome oxidation, but it appears to be in line with values obtained from Raman and crystallographic data on porphyrins (R. Felton, personal communication; Spaulding et al., 1974). A more serious reservation, as section 3.5.7 points out, is that there is no experimental evidence that the rates of electron transfer reactions involving BChl and BPh in vitro actually do decrease when the reactions become strongly exothermic.

If the initial reaction between P* and I does fall on the peak of the curve for W as a function of ΔE, the rate of the electron transfer should increase slightly as one lowers the temperature (Fig. 3.12). Section 3.3.1 mentions fluorescence data that suggest that the electron transfer rate decreases slightly at low temperatures, which disagrees with this expectation. Direct measurements of the temperature dependence of the primary reaction are needed.

The ratio of the rate constants for the forward and backward reactions also will depend on the electronic interaction matrix elements. T_{da} could be made greater for the forward reaction than it is for the back reaction if the distance between P^+ and I^- increased immediately after the initial electron transfer. One would imagine that the distance would decrease rather than increase, because of coulombic attraction between P^+ and I^-, but it is possible that P^+ and I^- are pulled apart by interactions with other charged components of the reaction center.

Even with a fixed distance, T_{da} could be different for the forward and backward reactions, because different molecular orbitals on P are involved (Hopfield, 1976; Holten et al., 1978a; Blankenship and Parson, 1978). Electron transfer from P^* to I requires removal of an electron from the orbital that is the lowest unoccupied molecular orbital (LUMO) of P. Electron transfer from I^- to P^+ requires insertion of an electron into a hole in the highest occupied molecular orbital (HOMO) of P. Both reactions involve the LUMO of I. Because the HOMO's and LUMO's have different symmetry, it is possible that P and I could be oriented in such a way that the LUMO of I overlaps well with the LUMO of P, but poorly with the HOMO.

The back reaction from I^- that generates the localized triplet state of P^R raises special problems. Lowering the temperature causes the quantum yield of the triplet state to increase dramatically, although the lifetime of the P^+I^- radical pair seems to change only slightly (section 3.3.5). The quantum yield of the triplet state depends on the frequency at which P^+I^- evolves from singlet to triplet character (and back), and on the rate constants (k_s and k_t) for electron transfer from I^- to the singlet and triplet states of P. The rate constants k_s and k_t will be different for several reasons. First, the electronic interaction matrix elements are not identical, and second, ΔE is almost certainly different for the two back reactions. Whereas the free energy decrease for the decay of the singlet radical pair to the ground state is approximately 0.9 eV, the free energy decrease for the decay of the triplet radical pair to P^R probably is on the order of 0.15 eV (section 3.3.5). This could give the two rate constants a different dependence on temperature, if eq. (21) gives a correct description of the relationship between W and ΔE. If S and $\hbar\bar{\omega}$ are small, k_t could increase slightly with decreasing temperature, while k_s drops precipitously. Because the back reaction to give P^R requires the development of triplet character in P^+I^-, the decay of the radical pair at low temperatures should not begin as soon as P^+I^- is formed. Instead, there should be a lag period, which might be extended by the application of an external magnetic field. More accurate kinetic measurements are needed to test these expectations.

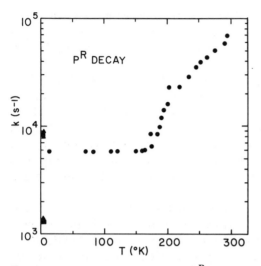

Fig. 3.13. Decay kinetics of state P^R versus temperature. ●, *Rps. sphaeroides* strain R-26 reaction centers (Parson et al., 1975); ■, R-26 reaction centers at 4°K (Parson and Monger, 1976); ▲, R-26 whole cells at 2°K (Hoff, 1976). The decay kinetics are multiphasic at very low temperatures.

Fig. 3.13 shows the temperature dependence of the decay of state P^R. Like the cytochrome oxidations, these decay kinetics exhibit a normal temperature dependence at high temperatures, but become independent of temperature below about 150°K. Jortner (1976) has pointed out that the theoretical considerations that are outlined above apply as well to radiationless processes other than electron transfer. The decay of the triplet state P^R depends on a spin-orbit coupling matrix element, rather than an electronic interaction matrix element, but it also depends on Franck-Condon factors of the type that we have been discussing (Englman and Jortner, 1970; Kleibeuker et al., 1978).

The problem of preventing back-reactions between Q_A^- and P^+ is less difficult than the problem of preventing reactions between I^- and P^+, because Q_A^- probably is farther away from P^+ than I^- is. If the three components are arranged in a row, with I in the middle, the orbital overlap between Q_A^- and P^+ could be very small (See for example Fig. 1.5 in this book). It is therefore not surprising that the back-reaction from Q_A^- is 10^7-times slower than the back-reaction from I^- (section 3.3.5). Other things being equal, electron transfer could be slowed by this amount if the distance from P^+ were greater by 6 to 10 Å.

The rate of the back-reaction from Q_A^- increases as one lowers the temperature (section 3.3). Fig. 3.14 shows data on the temperature dependence of the rate in chromatophores and isolated reaction centers from *Rps. sphaeroides*. The inverse temperature dependence cannot be explained

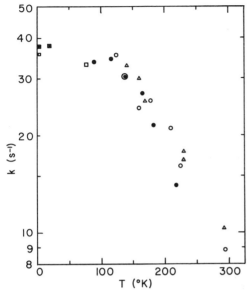

Fig. 3.14. Kinetics of back reaction between P^+ and Q_A^- versus temperature. $\bullet$, *Rps. sphaeroides* strain R-26 reaction centers; $\circ$, R-26 reaction centers +3 mM *o*-phenantrholine; $\triangle$, *Rps. sphaeroides* wild type chromatophores + 3 mM *o*-phenanthroline. All taken from Hsi and Bolton (1974). $\square$,$\blacksquare$, R-26 reaction centers (McElroy et al., 1969, 1974).

adequately by any of the theories discussed above, without additional ad hoc hypotheses. At most, one would expect the rate to increase by about 10%, if ΔE, S, and $\hbar\bar{\omega}$ are such that the back-reaction falls on the peak of the curve relating W to ΔE (Fig. 3.12). Probably the simplest way to explain the increase in the rate is by postulating that P and Q_A move slightly closer together at low temperatures (Parson, 1967; Hsi and Bolton, 1974; Hales, 1976). A movement of 1 Å probably would be more than sufficient. The absorption spectrum of P shifts about 20 nm to longer wavelengths at low temperatures, and this could be another indication of structural changes in the reaction center (Vredenberg and Duysens, 1963; Clayton and Yau, 1972; Kononenko et al., 1976).

3.6. ACKNOWLEDGEMENTS

The preparation of this chapter was supported in part by National Science Foundation grant PCM 77 13290 and by an NSF National Needs Post-doctoral Fellowship to R.E.B. Many of the ideas discussed here have developed as part of our collaboration with D. Holten and M.W. Windsor. We also enjoyed helpful discussions with R.H. Felton, J.J. Hopfield, and R.S. Knox.

110

REFERENCES

Atkins, P.W. (1976) Chem. Br., 12, 214—218.
Atkins, P.W. and Lambert, T.P. (1975) Annu. Rep. Chem. Soc., 72A, 67—88.
Azumi, T. and Matsuzaki, K. (1977) Photochem. Photobiol., 25, 315—326.
Blankenship, R.E. and Parson, W.W. (1977) Abstr. 4th Int. Conf. Photosynth. Reading p. 37.
Blankenship, R.E., Schaafsma, T.J. and Parson, W.W. (1977) Biochim. Biophys. Acta, 461, 297—305.
Blankenship, R.E. and Parson, W.W. (1978) Annu. Rev. Biochem., 47, 635—653.
Blumenfeld, L.A. and Chernavskii, D.S. (1973) J. Theor. Biol., 39, 1—7.
Brocklehurst, B. (1976) J. Chem. Soc., Faraday Trans. II, 72, 1869—1884.
Carithers, R.P. and Parson, W.W. (1975) Biochim. Biophys. Acta, 387, 194—211.
Case, G.D. (1972) Ph.D. Thesis, University of Washington.
Case, G.D., Parson, W.W. and Thornber, J.P. (1970) Biochim. Biophys. Acta, 223, 122—128.
Case, C.D. and Parson, W.W. (1971) Biochim. Biophys. Acta, 253, 187—202.
Case, G.D. and Parson, W.W. (1973) Biochim. Biophys. Acta, 292, 677—684.
Charmorovsky, S.K., Remennikov, S.M., Mononenko, A.A., Venediktov, P.S. and Rubin, A.B. (1976) Biochim. Biophys. Acta, 430, 62—70.
Chou, M., Creutz, C. and Sutin, N. (1977) J. Am. Chem. Soc., 99, 5615—5623.
Clarke, R.H., Connors, R.E. and Frank, H.A. (1976) Biochem. Biophys. Res. Commun., 71, 671—675.
Clayton, R.K. and Yamamoto, T. (1976) Photochem. Photobiol., 26, 67—70.
Clayton, R.K. (1977) in Photosynthetic Organelles: Structure and Function (Miyachi, S., Katoh, S., Fujita, Y., and Shibata, K., eds. Special Issue of Plant Cell Physiol., No. 3, pp. 87—96.
Clayton, R.K. and Sistrom, W.R., eds. (1978) in The Photosynthetic Bacteria. Plenum Press, New York.
Clayton, R.K. and Straley, S.C. (1970) Biochem. Biophys. Res. Commun., 39, 1114—1119.
Clayton, R.K. and Straley, S.C. (1972) Biophys. J., 12, 1221—1234.
Clayton, R.K., Szuts, E.Z. and Fleming, H. (1972) Biophys. J., 12, 64—79.
Clayton, R.K. and Yau, H.F. (1972) Biophys. J., 12, 867—881.
Cogdell, R.J., Brune, D.C. and Clayton, R.K. (1974) FEBS Lett., 45, 344—347.
Cogdell, R.J., Monger, T.G. and Parson, W.W. (1975) Biochim. Biophys. Acta, 408, 189—199.
Connolly, J.S., Gorman, D.S. and Seely, G.R. (1973) Ann. N.Y. Acad. Sci., 206, 649—669.
Creutz, C. and Sutin, N. (1977) J. Am. Chem. Soc., 99, 241—243.
Debrunner, P.G., Schulz, C.E., Feher, G. and Okamura, M.Y. (1975) Biophys. J., 15, 226a (abstr.).
DeVault, D. and Chance, B. (1966) Biophys. J., 6, 825—847.
DeVault, D., Parkes, J.H. and Chance, B. (1967) Nature, 215, 642—644.
Dogonadze, R.R. (1971) in Reactions of Molecules at Electrodes (M.S. Hush, ed.). Wiley-Interscience, New York.
Dutton, P.L. (1971) Biochim. Biophys. Acta, 226, 63—80.
Dutton, P.L., Kihara, T. and Chance, B. (1970) Arch. Biochem. Biophys., 139, 236—240.
Dutton, P.L., Kihara, T., McCray, J.A. and Thornber, J.P. (1971) Biochim. Biophys. Acta, 226, 81—87.
Dutton, P.L., Leigh, J.S. and Seibert, M. (1972) Biochem. Biophys. Res. Commun., 46, 406—413.

Dutton, P.L., Leigh, J.S. and Reed, D.W. (1973a) Biochim. Biophys. Acta, 292, 654—664.

Dutton, P.L., Leigh, J.S. and Wraight, C.A. (1973b) FEBS Lett., 36, 169—173.

Dutton, P.L., Kaufmann, K.J., Chance, B. and Rentzepis, P.M. (1975a) FEBS Lett., 60, 275—280.

Dutton, P.L., Petty, K.M., Bonner, H.S. and Morse, S.D. (1975b) Biochim. Biophys. Acta, 387, 536—556.

Dutton, P.L. and Prince, R.C. (1978) in The Photosynthetic Bacteria (R.K. Clayton, W.R. Sistrom, eds.). Plenum Press, New York (in press).

Efrima, S. and Bixon, M. (1974) Chem. Phys. Lett., 25, 34—37.

Efrima, S. and Bixon, M. (1976) Chem. Phys., 13, 447—460.

Englman, R. and Jortner, J. (1970) Mol. Phys., 18, 145—164.

Fajer, J., Borg, D.C., Forman, A., Dolphin, D. and Felton, R.H. (1973) J. Am. Chem. Soc., 95, 2739—2741.

Fajer, J., Borg, D.C., Forman, A., Felton, R.H., Dolphin, D. and Vegh, H. (1974) Proc. Natl. Acad. Sci. U.S.A., 71, 994—998.

Fajer, J., Brune, D.C., Davis, M.S., Foreman, A. and Spaulding, L.D. (1975) Proc. Natl. Acad. Sci. U.S.A., 72, 4956—4960.

Fajer, J., Davis, M.S., Holten, J.D., Parson, W.W., Thornber, J.P. and Windsor, M.W. (1977) Abstr. 4th Int. Congr. Photosynth., p. 108.

Feher, G. (1971) Photochem. Photobiol., 14, 373—387.

Feher, G., Okamura, M.Y. and McElroy, J.D. (1972) Biochim. Biophys. Acta, 267, 222—226.

Feher, G., Issacson, R.A., McElroy, J.D., Ackerson, L.C. and Okamura, M.Y. (1974) Biochim. Biophys. Acta, 368, 135—139.

Feher, G., Hoff, A.J., Issacson, R.A. and Ackerson, L.C. (1975) Ann. N.Y. Acad. Sci., 244, 239—259.

Feher, G., Issacson, R.A. and Okamura, M.Y. (1977) Biophys. J., 17, 149a (abstr.).

Feher, G. and Okamura, M.Y. (1977) Brookhaven Symp. Biol., 28, 183—194.

Felton, R.H. (1978) in The Porphyrins (D. Dolphin, ed.). Academic Press, New York (in press).

Fischer, S.F. and Van Duyne, R.P. (1977) Chem. Phys., 26, 9—16.

Fong, F. (1976) J. Am. Chem. Soc., 98, 7840—7843.

Förster, T. (1946) Naturwissenschaften, 33, 166—182.

Fuhrhop, J.H. and Mauzerall, D. (1969) J. Am. Chem. Soc., 91, 4174—4181.

Glasstone, S., Laidler, K.J. and Eyring, H. (1941) The Theory of Rate Processes. McGraw-Hill, New York.

Gouterman, M. and Holten, D. (1977) Photochem. Photobiol., 25, 85—92.

Govindjee (ed.) (1975) Bioenergetics of Photosynthesis, Academic Press, New York, 698 pp.

Grigorov, L.N. and Chernavskii, D.S. (1972) Biophysics (USSR), 17, 202—209. (Engl. transl.).

Hales, B.J. (1976) Biophys. J., 16, 471—480.

Halsey, Y.D. and Parson, W.W. (1974) Biochim. Biophys. Acta, 347, 404—416.

Hoff, A.J. (1976) Biochim. Biophys. Acta, 440, 765—771.

Hoff, A.J., Rademaker, H., van Grondelle, R. and Duysens, L.N.M. (1977) Biochim. Biophys. Acta, 460, 547—554.

Holten, D., Gouterman, M., Parson, W.W., Windsor, M. and Rockley, M.G. (1976) Photochem. Photobiol., 23, 415—423.

Holten, D., Windsor, M., Parson, W.W. and Gouterman, M. (1978a) Photochem. Photobiol. (in press).

Holten, D., Windsor, M.W., Parson, W.W. and Thornber, J.P. (1978b) Biochim. Biophys. Acta, 501, 112—126.

Holten, D. and Windsor, M. (1978) Annu. Rev. Biophys. Bioeng., 7, 189—227.
Hopfield, J.J. (1974) Proc. Natl. Acad. Sci. U.S.A., 71, 3640—3644.
Hopfield, J.J. (1976) in Proc. 29th Int. Congr. Societe de Chimie Physique, pp. 471—490. Elsevier, Amsterdam.
Hopfield, J.J. (1977) Biophys. J., 18, 311—321.
Hsi, E.S.P. and Bolton, J.B. (1974) Biochim. Biophys. Acta, 347, 126—153.
Huppert, D., Rentzepis, P.M. and Tollin, G. (1976) Biochim. Biophys. Acta, 440, 356—364.
Jackson, J.B., Cogdell, R.J. and Crofts, A.R. (1973) Biochim. Biophys. Acta, 292, 218—225.
Jortner, J. (1976) J. Chem. Phys., 64, 4860—4867.
Kaufmann, K.J., Dutton, P.L., Netzel, T.L., Leigh, J.S. and Rentzepis, P.M. (1975) Science, 188, 1301—1304.
Kaufmann, K.J., Petty, K.M., Dutton, P.L. and Rentzepis, P.M. (1976) Biochem. Biophys. Res. Commun., 70, 839—845.
Ke, B. (1978) in Current Topics in Bioenergetics, Vol. 7A. Academic Press, New York (75—138).
Kestner, N.R., Logan, J. and Jortner, J. (1974) J. Phys. Chem., 78, 2148—2166.
Kihara, T. and Chance, B. (1969) Biochim. Biophys. Acta, 189, 116—124.
Kihara, T. and McCray, J.A. (1973) Biochim. Biophys. Acta, 292, 297—309.
Kleibeuker, J.F., Platenkamp, R.J. and Schaafsma, T.J. (1978) Chem. Phys., 27, 51—64.
Knox, R.P. (1977) in Topics in Photosynthesis, Vol. 2: Primary Processes of Photosynthesis (J. Barber, ed.), pp. 55—97. Elsevier, Amsterdam.
Kononenko, A.A., Knox, P.P., Admamova, N.P., Paschenko, V.Z., Timofeev, K.N., Rubin, A.B. and Morita, S. (1976) Stud. Biophys., 55, 183—198.
Krasnovskii Jr., A.A., Lebedev, N.N. and Litvin, F.F. (1974) Dokl. Akad. Nauk SSSR, 216, 39—42 (Engl. transl.).
Lamola, A.A., Manion, M.L., Roth, H.D. and Tollin, G. (1975) Proc. Natl. Acad. Sci. U.S.A., 72, 3265—3269.
Leigh, J.S. and Dutton, P.L. (1972) Biochem. Biophys. Res. Commun., 46, 414—421.
Leigh, J.S. and Dutton, P.L. (1974) Biochim. Biophys. Acta, 357, 67—77.
Levich, V.G. (1966) Adv. Electrochem. Electrochem. Eng., 4, 249—371.
Loach, P.A. and Hall, R.L. (1972) Proc. Natl. Acad. Sci. U.S.A., 69, 786—790.
Loach, P.A., Kung, M.C. and Hales, B.J. (1975) Ann. N.Y. Acad. Sci., 244, 297—319.
Loach, P.A. (1976) Prog. Bioorg. Chem., 4, 89—192.
Loach, P.A. (1977) Photochem. Photobiol., 26, 87—94.
Marcus, R.A. (1964) Annu. Rev. Phys. Chem., 15, 155—196.
McElroy, J.D., Feher, G. and Mauzerall, D.C. (1969) Biochim. Biophys. Acta, 172, 180—183.
McElroy, J.D., Feher, G., Mauzerall, D.C. (1972) Biochim. Biophys. Acta, 267, 363—374.
McElroy, J.D., Mauzerall, D.C. and Feher, G. (1974) Biochim. Biophys. Acta, 333, 261—277.
Moscowitz, E. and Malley, M.M. (1978) Photochem. Photobiol., 27, 55—59.
Netzel, T.L., Rentzepis, P.M., Tiede, D.M., Prince, R.C. and Dutton, P.L. (1977) Biochim. Biophys. Acta, 460, 467—478.
Norris, J.R., Scheer, H., Druyan, M.B. and Katz, J.J. (1974) Proc. Natl. Acad. Sci. U.S.A., 71, 4897—4900.
Olson, J.M. and Hind, G. (eds.) (1977) Brookhaven Symposia in Biology, Vol. 28, 385 pp.
Okamura, M.Y., Issacson, R.A. and Feher, G. (1975) Proc. Natl. Acad. Sci. U.S.A., 72, 3491—3495.
Okamura, M.Y., Issacson, R.A. and Feher, G. (1977) Biophys. J., 17, 149a (Abstr.).

Orchin, M. and Jaffe, H.H. (1971) Symmetry, Orbitals and Spectra, pp. 222—229. Wiley-Interscience, New York.

Parson, W.W. (1967) Biochim. Biophys. Acta, 131, 154—172.

Parson, W.W. (1968) Biochim. Biophys. Acta, 153, 248—259.

Parson, W.W. (1969) Biochim. Biophys. Acta, 189, 384—396.

Parson, W.W. (1974) Annu. Rev. Microbiol., 23, 41—59.

Parson, W.W. (1978a) in The Photosynthetic Bacteria (R.K. Clayton, W.R. Sistrom, eds.), Plenum Press, New York.

Parson, W.W. (1978b) Photochem. Photobiol. 28, 389—393.

Parson, W.W. and Case, G.D. (1970) Biochim. Biophys. Acta, 205, 232—245.

Parson, W.W., Clayton, R.K. and Cogdell, R.J. (1975) Biochim. Biophys. Acta, 387, 268—278.

Parson, W.W. and Cogdell, R.J. (1975) Biochim. Biophys. Acta, 416, 105—149.

Parson, W.W. and Monger, T.G. (1977) Brookhaven Symp. Biol., 28, 195—211.

Paschenko, V.Z., Kononenko, A.A., Protasov, S.P., Rubin, A.B., Rubin, L.B. and Uspenskaya, N.Ya. (1977) Biochim. Biophys. Acta, 461, 403—412.

Pellin, M.J., Wraight, C.A. and Kaufmann, K.J. (1978) Biophys. J., in press.

Peucheu, N.L., Kerber, N.L. and Garcia, A. (1976) Arch. Microbiol., 109, 301—305.

Potasek, M.J. and Hopfield, J.J. (1977) Proc. Natl. Acad. Sci. U.S.A., 74, 3817—3820.

Prince, R.C. and Dutton, P.L. (1976) Arch. Biochem. Biophys., 172, 329—334.

Prince, R.C. and Dutton, P.L. (1978) in The Photosynthetic Bacteria (R.K. Clayton, W.R. Sistrom, eds.). Plenum Press, New York (in press).

Prince, R.C., Leigh, J.S. and Dutton, P.L. (1976) Biochim. Biophys. Acta, 440, 622—636.

Prince, R.C., Tiede, D.M., Thornber, J.P. and Dutton, P.L. (1977) Biochim. Biophys. Acta, 462, 467—490.

Ražem, D., Hamilton, W.H. and Funabashi, K. (1977) J. Chem. Phys., 67, 5404—5405.

Reed, D.W., Zankel, K.L. and Clayton, R.K. (1969) Proc. Natl. Acad. Sci. U.S.A., 63, 42—46.

Rehm, D. and Weller, A. (1970) Isr. J. Chem., 8, 259—271.

Reynolds, W.L. and Lumry, R. (1966) Mechanisms of Electron Transfer, 175 pp. Ronald Press, New York.

Rockley, M.G., Windsor, M.W., Cogdell, R.J. and Parson, W.W. (1975) Proc. Natl. Acad. Sci. U.S.A., 72, 2251—2255.

Romijn, J.C. and Amesz, J. (1977) Biochim. Biophys. Acta, 461, 327—338.

Salemme, F.R. (1977) Annu. Rev. Biochem., 46, 299—329.

Schmickler, W. (1976) J. Chem. Soc., Faraday Trans. II, 72, 307—312.

Seely, G.R. (1969) J. Phys. Chem., 73, 125—129.

Seely, G.R. (1977) in Topics in Photosynthesis, Vol. 2: Primary Processes of Photosynthesis (J. Barber, ed.), pp. 1—53. Elsevier, Amsterdam.

Seibert, M. and DeVault, D. (1970) Biochim. Biophys. Acta, 205, 220—231.

Shuvalov, V.A. and Klimov, V.V. (1976) Biochim. Biophys. Acta, 440, 587—599.

Shuvalov, V.A., Krakhmaleva, I.N. and Klimov, V.V. (1976) Biochim. Biophys. Acta, 449, 597—601.

Slooten, L. (1972a) Biochim. Biophys. Acta, 256, 452—466.

Slooten, L. (1972b) Biochim. Biophys. Acta, 275, 208—218.

Spaulding, L.D., Eller, P.G., Bertrand, J.A. and Felton, R.H. (1974) J. Am. Chem. Soc. 96, 982—993.

Straley, S.C., Parson, W.W., Mauzerall, D.C. and Clayton, R.K. (1973) Biochim. Biophys. Acta, 305, 597—609.

Thurnauer, M.C., Katz, J.J. and Norris, J.R. (1975) Proc. Natl. Acad. Sci. U.S.A., 72, 3270—3274.

Tiede, D.M., Prince, R.C., Reed, G.H. and Dutton, P.L. (1976a) FEBS Lett., 65, 301—304.

Tiede, D.M., Prince, R.C. and Dutton, P.L. (1976b) Biochim. Biophys. Acta, 449, 447–469.

Tollin, G. (1976) J. Phys. Chem., 80, 2274–2277.

Trosper, T.L., Benson, D.L. and Thornber, J.P. (1977) Biochim. Biophys. Acta, 460, 318–330.

Tumerman, L.A. and Rubin, A.B. (1962) Dokl. Akad. Nauk SSSR, 145, 202–205.

Ulstrup, J. and Jortner, J. (1975) J. Chem. Phys., 63, 4358–4368.

Uphaus, R.A., Norris, J.R. and Katz, J.J. (1974) Biochem. Biophys. Res. Commun., 61, 1057–1063.

Van Duyne, R.P. and Fischer, S.F. (1974) Chem. Phys., 5, 183–197.

Van Grondelle, R., Romijn, J.C. and Holmes, N.G. (1976) FEBS Lett., 72, 187–192.

Vermeglio, A. (1977) Biochim. Biophys. Acta, 459, 516–524.

Vermeglio, A. and Clayton, R.K. (1977) Biochim. Biophys. Acta, 461, 159–165.

Vredenberg, W.J. and Duysens, L.M.N. (1963) Nature, 197, 355–357.

Wraight, C.A. (1977) Biochim. Biophys. Acta, 459, 525–531.

Wraight, C.A. and Clayton, R.K. (1973) Biochim. Biophys. Acta, 333, 246–260.

Wraight, C.A., Leigh, J.S., Dutton, P.L. and Clayton, R.K. (1974) Biochim. Biophys. Acta, 333, 401–408.

Zankel, K.L., Reed, D.L. and Clayton, R.K. (1968) Proc. Natl. Acad. Sci. U.S.A., 61, 1243–1249.

NOTES ADDED IN PROOF

Note 1

The temperature dependence of the electron transfer from I^- to Q_A has been measured (Peters, K. et al., Biophys. J. (1978) 23, 207–217 and Parson, W. et al. (1978) in "Frontiers of Biological Energetics", Johnson Foundation Symposium, in press); the rate was found to be nearly independent of temperature over a wide range. The latter authors also report a time constant of 3 ps for electron transfer from P* to I. However, Shuvalov, V. et al. (FEBS Lett., (1978) 91, 135–139) present evidence for transient reduction of an intermediate BChl and give 30 ps as the time for reduction of the BPh.

Note 2

Recent fluorescence measurements by Van Grondelle et al. (Biochim. Biophys. Acta (1978) 503, 10–25) suggest that the energy of P^+I^- may be higher than indicated in section 3.4.1; at room temperature they propose that the P^+I^- back reaction proceeds via P*.

Note 3

Teide, D. et al. (Biochim. Biophys. Acta (1978) 503, 524–544) have concluded from the lack of a dipolar broadening by the cytochrome on the P^+ ESR spectrum that the center-to-center distance of the two species is $>25A$. Kuznetsov, A. et al. (Chem. Phys. (1978) 29, 383–390) have presented a theoretical treatment of the cytochrome oxidation kinetics.

Photosynthesis in relation to model systems, edited by J. Barber
© Elsevier/North-Holland Biomedical Press 1979

Chapter 4

Photoeffects in Pigmented Bilayer Lipid Membranes

H. TI TIEN

Biophysics Department, Michigan State University, East Lansing, MI 48824, U.S.A.

CONTENTS

116

4.1. INTRODUCTION

Without any doubt, photosynthesis in green plants is the most crucial reaction on Earth. It is therefore only proper to expend great effort in investigating this vital process so that we may achieve as full an understanding of the mechanism involved as possible. Indeed, the publications of 9 mammoth volumes of the proceedings of 3 international congresses held within the past decade is evidence of this (Metzner, 1969; Forti et al., 1972; Avron, 1975). In spite of extensive efforts, however, we are still far from solving all the major problems of photosynthesis. Among the outstanding problems that remain to be solved at the molecular level are: (i) the primary events of quantum conversion (Barber, 1977), (ii) photophosphorylation and electron transport (Jagendorf, 1977), and (iii) oxygen evolution (Joliot and Kok, 1975; Metzner, 1978). It is generally accepted that all major unsolved problems are closely connected with the lamellar structure of the thylakoid membranes of the chloroplasts of higher plants and green algae (Staehelin et al., 1977; Arntzen et al., 1977).

A closer look at photosynthesis reveals that it is truly a multifarious physico-chemical biological reaction. It is well recognized that a reaction of such complexity and magnitude is most unlikely to be understood by any individual effort or any single approach. Thus, to facilitate the understanding and to maximize the effort, photosynthesis research has been operationally divided into two domains: physico-chemical and bio-chemical (see Vols. 1 and 2 of this series). An overview of photosynthesis by green plants may be seen in the flow diagram below:

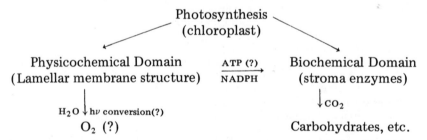

In the biochemical domain, carbohydrate is formed from CO_2 by a series of dark enzymic reactions driven by ATP and NADPH (Calvin, 1976). The process is relatively well understood. In the biophysical domain, the lamellar structure or the thylakoid membrane is strongly implicated in the two different photoreactions (see chapter 2). Therefore, the resolution of the three major problems (identified by question marks in the flow diagram) cannot be achieved without, first of all, having some appreciation of the role of the thylakoid membrane in terms of chemical composition and molecular organization.

The gross composition of thylakoid membranes consists of 47% protein,

45% lipids, and 8% pigments (Anderson, 1975). Structurally, the spatial arrangement of lipids, proteins, and pigments is not known and much of our knowledge about the thylakoid membranes of the chloroplasts comes from electron microscopy (Menke, 1967; Miller, 1976). From the images of electron micrographs, the structure of the thylakoid membrane has been variously interpreted. According to Park and Pon (1961), the membrane consists of an ultrathin lamella with attached globular units, termed quantasomes, which they considered as the morphological expression of the photosynthetic unit. A different interpretation has been given by Howell and Moudrianakis (1967) who suggest that the entire membrane is involved in photoreactions. Benson (1966) has interpreted his images in a somewhat unique fashion, and conjured up a highly intricate picture. On the basis of X-ray diffraction studies, the ultrastructure of chloroplast membranes has been analyzed and reviewed by Kreutz (1970). Using the freeze-etching technique. Muehlethaler et al. (1965) have suggested a picture for the thylakoid membrane based upon the bimolecular leaflet model (see Fig. 4.1A). The salient feature in all these proposed models lies in their oriented lipid core, onto which other important cellular constituents, such as proteins and pigments, may interact through either ionic or van der Waals attraction, or both. Muehlethaler's interpretation is of special interest, in view of the experiments using the pigmented bilayer lipid membrane as a model for the thylakoid membrane (Tien, 1968; 1970).

4.1.1. The rationale for studying model systems (Background and rationale)

Before describing pigmented bilayer lipid membranes as a model for the thylakoid membrane, it is of relevance to mention certain past experimental models that have been used for understanding physico-chemical aspects of photosynthesis. These past investigations which preceded the work reviewed here, have greatly stimulated our thoughts and inspired our experiments. In particular, mention should be made of the use of chlorophyll-coated metallic electrodes in aqueous media. This system, which exhibits photopotentials, has been investigated intensively by Evstigneev and Terenin (1951). Eley and Snart (1965) and Nelson (1957) observed spectral photoconductivity in compressed disks or films of chlorophyll and its derivatives and Arnold and Maclay (1959) reported the photoconductivity of thin films of chlorophyll and carotene as well as films of dried chloroplasts. Although these investigations were significant in their own right, their models as far as photosynthesis is concerned are all deficient in at least one respect: they did not contain (as do the thylakoid membranes) lipid, proteins and pigments organized in the form of ultrathin *bilayer* membranes separating two aqueous phases.

It must be also stated that little was known about the thylakoid membrane when the structural-functional relationships are considered at the molecular level (see the flow diagram). Concerning the primary events of

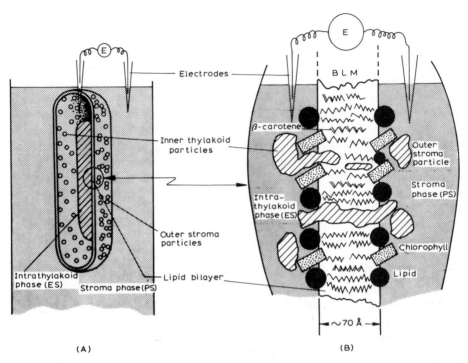

Fig. 4.1. A highly schematical model of a thylakoid illustrating a portion of the lipid bilayer of lammelar membrane structure [patterned after Muehlethaler et al. (1965) with the notations suggested by Branton et al., 1975]. (A) Showing a hypothetical situation for measurements of membrane electrical properties with one microelectrode inserted transversely through the thylakoid membrane and the other microelectrode in the stroma phase. (B) An enlarged view of the hypothetical experiment indicated in (A). Also shown in (B) is a model of thylakoid membrane depicting an arrangement of amphipathic lipids with their hydrophobic chains inside and hydrophilic portions in contact with aqueous phases forming the lipid bilayer. This lipid bilayer is visualized as a two-dimensional liquid crystal matrix, onto which pigments, proteins and the like are interpolated (see Fig. 4.2C).

quantum conversion, for example, a number of hypotheses have been proposed, one of which is that the mechanism of light transduction by the photosynthetic apparatus in chloroplasts might be similar to that taking place in semiconductors (Katz, 1949). If the thylakoid membrane, in which the photosynthetic apparatus is believed to be located, functioned as a semi-conductor photovoltaic device (e.g. compared with the silicon solar cell), the most direct kind of experiment would be to place electrodes transversely across it and monitor its voltage when it is excited by light. This hypothetical experiment is shown in Fig. 4.1. At present, the minuteness and complexity of the thylakoids preclude such a straight-forward experiment from being carried out, unless many technical difficulties can be overcome (Bulychev et al., 1976).

Thus, in the light of what was known about the thylakoid membrane and the suggested mechanisms of its operation, two important questions had to be answered: (i) "Can the photosynthetic pigments and lipids be organized in the form of a lipid bilayer separating *two* aqueous solutions?", and (ii) "In the event that such pigmented bilayer lipid membranes can be formed, will they exhibit any light-induced phenomena such as photovoltaic effect and photoconductivity?" Fortunately, both questions were answered affirmatively by experiments initiated in 1967. One of the purposes of this paper is to show that experimental bilayer lipid membranes containing chloroplast extracts are almost ideal systems with which to investigate energy transduction and photochemical processes as they might be occurring in the photosynthetic membranes in green plants.

4.1.2. Scope of the chapter

This chapter is concerned with light-induced effects in pigmented bilayer lipid membranes of both planar and spherical configuration as a model for the thylakoid membrane. In particular, the focus is on bilayer lipid membranes containing chlorophyll and/or related compounds, although bacteriorhodopsin modified lipid bilayers are also included since the purple membrane of *Halobacterium halobium* from which bacteriorhodopsin is derived behaves like the photosynthetic thylakoid membrane. Dye-sensitized bilayer lipid membranes (BLM) and photoelectric effects in other BLM systems not directly relevant to photosynthesis have been reviewed elsewhere (Tien, 1976) and therefore will not be covered in this paper. Recent references to other aspects of bilayer lipid membranes and related systems of interest but not central to this review are listed below.

Among the studies of experimental bilayer lipid membranes recently reported, other than light-induced phenomena, are the effect of lipid composition (Bradley et al., 1976; Kunitake and Okahata, 1977; Yampolsky et al., 1977), techniques (Casper et al., 1975; Noll, 1976; Jones, 1975; White and Blessum, 1975), thickness (Engelsen and Koning, 1975; Kolb and Bamberg, 1977), diffusion and permeability (Fahey et al., 1977; Strickholm and Clark, 1977; Wolf et al., 1977), electrical properties (Braun, 1976; Carius, 1976; Coster et al., 1975; Foigel, 1976; Hoffman et al., 1976; Karvaly, 1975; Melnik et al., 1977; Neher and Eibl, 1977), interfacial properties (Brochard et al., 1976; Gaines, 1977; Kremer et al., 1977; Nagle, 1976; Makowski, 1976; Requena and Haydon, 1975; Sanfeld and Steinche, 1975; Suezaki, 1978; Yoshida et al., 1977), excitability and selectivity (Monnier, 1977; Rosenstreich and Blumenthal, 1977; Sugiura and Shinbo, 1976; Tummler et al., 1977), interactions (Dobias and Heckmann, 1977; Henkart and Blumenthal, 1975; Ksenzhek et al., 1976; Laclette and Montal, 1977; Schubert et al., 1977; Volkov et al., 1975; Wobschall and McKeon, 1975), miscellaneous (Brody and Brody, 1977; Bukovsky, 1977; Fawcett and

Maccarich, 1976; Inui et al., 1977; Kudzina et al., 1977; Loschilova and Karvaly, 1977; Barsky et al., 1976; Ohki and Ohki, 1976; Aizawa et al., 1977; Tabushi and Funakura, 1976; Vaillier et al., 1975; Zimmermann et al., 1977), liposomes (Gruppe, 1977; Matsomoto et al., 1977), related systems (Csatorday et al., 1975; Hess, 1977; Szabad et al., 1974; LeBlanc et al., 1974; Salamon and Frackowiak, 1975; Takahashi and Kikuchi, 1976; Villar, 1976; Whitten et al., 1977). The interested reader may also wish to consult a number of perceptive reviews on the general aspects of lipid bilayers and their relation to biomembranes (Aveyard and Vincent, 1977; Baumeister and Hahn, 1976; Buscall and Ottewill, 1975; Lagaly, 1976; Lakshminarayanaiah, 1975; Larsson et al., 1976; Lenaz, 1977; McLaughlin and Eisenberg, 1975; Scott, 1977; Selwyn and Dawson, 1977; Shamoo and Goldstein, 1977).

4.2. METHODOLOGY

Two types of artificial bilayer lipid membrane systems have been employed for the study of light induced reactions involving photosynthetic pigments and thylakoid extracts. The first consists of a planar bilayer lipid membrane (BLM) separating two aqueous solutions where photoeffects can be readily measured electrically. The second system, much less used until now, comprises of lipid microvesicles, also popularly known as "liposomes". In this system a bilayer lipid membrane of spherical configuration encloses a volume of aqueous solution. Since the liposomes are very stable and can be easily made in quantity, they are ideally suited for studies of permeability, conventional spectroscopy, chemical reactions, and oxygen evolution. To make our discussion more informative, a description of each system relevant to this review is given. For more complete technical details, two pertinent publications are available (Tien, 1974; Bangham et al., 1974).

4.2.1. Planar bilayer lipid membranes (BLM)

Membrane formation. Planar BLM interposed between two aqueous phases may be formed in the following manner. A droplet of lipid solution is applied to an aperture, observed through a low power microscope, in a hydrophobic barrier (e.g., a Teflon cup) separating two aqueous solutions. Within minutes, there is rapid drainage of excess lipid solution into the border (Plateau-Gibbs border) and brilliant interference colors are seen. In a while, optically "black" spots appear and enlarge themselves to occupy the majority of aperture area. Once formed, the BLMs are generally stable for one hour or more (some lasting for more than 24 hours), depending on experimental conditions. Fig. 4.2A is a schematic representation of a BLM in formation and its postulated structure (Fig. 4.2C).

A variety of lipid solutions have been formulated, from highly purified

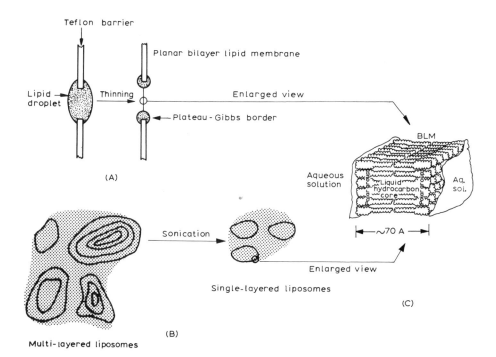

Fig. 4.2. Schematic illustration of experimental processes for forming bilayer lipid membranes. (A) Two stages of thinning of a planar lipid membrane: A lipid droplet is placed on aperture and thins spontaneously to a bilayer lipid membrane (BLM) supported by Plateau-Gibbs border. (B) Formation of multi- and unilayer bilayer lipid membranes (liposomes) by mechanical agitation and sonication. (C) A three-dimensional view of a bilayer lipid membrane illustrating the liquid crystalline state of the lipid bilayer. The lipid chosen is that of phosphatidyl choline (Tien, 1967).

single phospholipids, surface-active agents (detergents) to tissue extracts. Either negative or positive net surface charge can be conferred to the BLM by incorporating an anionic compound or cationic compound in the membrane-forming solution, respectively.

Instrumentation. For the measurement of electrical properties, non-polarizable electrodes (e.g., calomel electrodes) are connected to the solutions via saturated KCl bridges as shown in Fig. 4.3. A circuit for applying various values of voltage and resistance and for monitoring is shown in Fig. 4.4. Commercial electrometers of high input impedance (greater than $1 \times 10^{12} \, \Omega$) are satisfactory. For continuous illumination either tungsten or xenon lamp has been used. In flash experiments, a xenon flash tube of high light intensity together with a good oscilloscope and associated timing circuit are necessary. A chart recorder such as Model VOM 5 (Bausch and Lomb) can be connected in parallel with the oscilloscope to provide a record of the low speed variations in the membrane voltage and an accurate time history

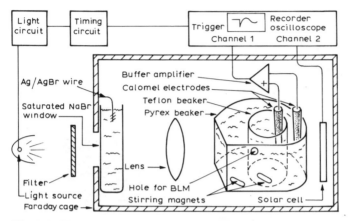

Fig. 4.3. Experimental arrangement for studying light-induced electrical properties in pigmented BLM.

of the BLM under investigation. These and other aspects of the BLM system have been described in detail (Tien, 1974).

Procedure. The basic method of BLM formation can be described by referring to Figs. 4.2A and 4.3. The cell assembly consists of two concentric chambers. The wall of the inner chamber made of Teflon or polyethylene contains a small hole (1—2 mm, diameter) in which the BLM is formed. Prior to BLM formation, both chambers are filled with salt solution (either 0.1 N NaCl or KCl) slightly above the hole. Lipid solution of appropriate composition is introduced in the hole. This can be done easily with the aid of a Hamilton microsyringe. Since the size of the BLM formed is small (usually less than 50 mm^2), the formation characteristics of the membrane are observed using a 20—60 x microscope with reflected light.

Generally, the voltage drop across the BLM is measured, and from a knowledge of the applied voltage, the BLM resistance (R_m) is readily calculated. The capacitance (C_m) of a BLM may be measured by a dc transient (charge leakage) method, using standard resistors. The two most commonly studied light-elicited phenomena in BLM are the photovoltaic effect and photoconduction. Instead of using white light, the photovoltaic effect of a BLM can also be investigated as a function of wavelength with the aid of a monochromator, thereby obtaining the photovoltage (action) spectrum of the BLM (see Fig. 4.10). To investigate fast photoprocesses, such as mechanisms of charge carrier generation and charge carrier lifetime in pigmented BLM, a flash unit is used (Fig. 4.3).

4.2.2. Spherical bilayer lipid membranes (Liposomes)

Liposomes are simply the liquid—crystalline vesicles obtained when amphipathic lipids such as phospholipids are dispersed in excess water. By

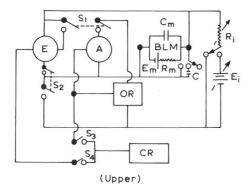

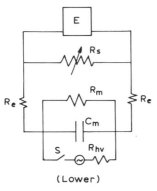

Fig. 4.4. Upper: Circuit diagram for measuring the electrical properties of BLM. R_i, a bank of resistors (10^5 to 10^{10} Ω); E_i, voltage source; C, 0.01 mfd capacitor; E, electrometer; A, picoammeter; S_1, S_2, S_3, S_4 are switches; C_m, BLM capacitance; E_m, BLM voltage; R_m, BLM resistance; OR, overload protection circuit; CR, chart recorder. Lower: Equivalent circuit for a pigmented BLM under illumination. R_s, external shunt resistors (see text for details).

simple mechanical agitation the multilamellar vesicles of various size (microns in diameter) and shape are produced. Upon sonication, these large multi-shelled liposomes tend to break up into a fairly homogeneous population of single-layered vesicles of approximately 250 Å diameter. Fig. 4.2B is a schematic diagram of the procedure for generating lipid microvesicles. Pigments (e.g., chlorophylls) and other photosynthetic materials of interest can be incorporated into the bilayer lipid membrane phase by this method. Techniques of liposome formation and their studies have been authoritatively reviewed (Bangham et al., 1974; Papahadjopoulos, 1973).

For experiments carried out in my laboratory involving acidification, bleaching and O_2 consumption, liposomes containing photosynthetic materials are prepared by drying a chloroform solution of egg lecithin (300 mg) and the thylakoid extract (about 10 mg of chlorophyll per experiment) under nitrogen, in the dark, on the inside of a 50 ml round bottom flask.

The thylakoid extract, consisting of only pigments and lipids, is obtained using a 2 : 1 (v/v) petroleum ether : methanol mixture (for details, see Tien, 1974, p. 546). Thirty ml of the test solution (usually 100 mM KCl, and 10 mM of either tris or potassium acetate at pH 7.5 or 5.0, respectively, to which water soluble electron donors or electron acceptors are added) is vigorously mixed in the dark on a vortex mixer for 15 min. The pigmented liposomes are then sonicated in the dark for 90 sec on the top power setting of a Bronson sonicator (Model W140D). The pigmented liposomes are stored in the dark. Ten-ml samples are removed and added to 140 ml of the test solution described above and mixed thoroughly. This diluted liposome preparation is usually used in the illumination experiments.

For oxygen evolution experiments of broken thylakoids fused with liposomes, the chloroplasts (from 240 g spinach) are broken by blending at top speed in an electric blender in 200 ml of distilled water and ice for 5 min followed by sonication in ice for 10 min. The material is centrifuged at 8300 rpm for 5 min to remove unbroken chloroplasts. The green supernatant is then made to 10 mM tricine, 5 mM potassium ferricyanide and adjusted to pH 7.5. These green broken chloroplast thylakoids are either: (1) added to preformed liposomes described above (with 0.3 g of phospholipid in 30 ml of 10 mM tricine, 5 mM potassium ferricyanide, pH 7.5) or else (2) added to flasks containing 0.3 g of dried phospholipids and are directly incorporated in the process of liposome formation. For the experiments involving addition of preformed liposomes, 50 ml of the broken thylakoids are mixed with 10 ml of liposomes and the 60-ml sample is tested for oxygen. For the experiments incorporating the broken thylakoid directly into liposomes, 30 ml of the chloroplast material and 0.3 g of the phospholipids are used to make the liposomes. To this, 30 ml of 10 mM tricine, 5 mM potassium ferricyanide, pH 7.5 buffer is added and the 60-ml sample is tested for oxygen. The samples, which are purged with nitrogen and sealed, are placed in a constant bath which is surrounded by a bank of lights. The light intensity at the center of the sample chamber is 1.7×10^5 erg cm^{-2} sec^{-1}. The temperature is controlled at $25°C \pm 0.2$. Throughout the experiment the liposomes are constantly agitated with a magnetic stirrer. Oxygen is measured with a Clark-type O_2 electrode (Stillwell and Tien, to be published).

4.3. PROPERTIES OF PIGMENTED MEMBRANES IN THE DARK

4.3.1. Molecular organization of Chl-containing BLM

Since BLM made of pure lipid (e.g., phosphatidyl choline) or oxidized cholesterol in common salt solutions are nonconducting, the physical properties of these simple or unmodified BLMs are strikingly similar to those of a liquid hydrocarbon layer of equivalent thickness with one exception. The interfacial tension of BLM is less than 5 dynes/cm, which is about one order of magnitude lower than that of the hydrocarbon/water interface. This low interfacial tension is due to the presence of polar groups at the interface and is explicably in terms of Gibbs adsorption isotherm (Tien, 1974; Zuezaki, 1978). In view of the postulated liquid-hydrocarbon interior of the BLM, it is not surprising that unmodified BLM have a negligible permeability for ions and most polar molecules. Permeability to water is comparable to that of biological membranes. The permeability to water of Chl-BLM, as determined by an osmotic flow method, is 50 μm/sec, which is in the range of phospholipid BLM but six times larger than that of oxidized cholesterol BLM. The effect of temperature on permeability is linear when the permeability coefficient data are plotted according to the Arrhenius equation. The activation energy has been found to be 4.2 -kcal/mol in the range 17.5— 34°C. The linearity of the plot implies that a change in membrane structure as a function of temperature is perhaps not involved.

Since the chloroplast extract used in the BLM formation contains amphipathic lipids and pigments, one would expect that, as a result of the strong adsorption of these compounds at the solution/bilayer interfaces, a BLM of this type should be more stable. Experimentally, this has been indeed found to be the case. The stability of BLM is related to the bifacial tension of the membrane. Using a maximum bubble pressure technique, the bifiacial tension (free energy) of the Chl-BLM have been measured. Typical values are in the range of 3.8 to 4.5 dynes/cm. This implies that those chlorophyll molecules situated at the biface in the membrane are tightly compressed together. On the basis of this information, it is estimated that the average area occupied per porphyrin group is much less than 75 Å^2 (Tien, 1968).

Similar to other BLMs, the thickness of a Chl-BLM comprises three layers and may be determined from its optical reflectance (Tien, 1967; Engelsen and Koning, 1975). The principal layers and the two polar regions are likely to possess a significant optical anisotrophy. On the basis of optical reflectance measurements, preliminary experiments give a thickness of 70 ± 30 Å, depending on the assumed values of refractive index used in the calculation. Taking the experimentally determined thickness at its face value, the BLM generated from chloroplast pigments are pictured as similar to those of liquid crystals in two dimensions (Fig. 4.2C). In the case of chlorophyll, an amphipathic molecule, we may expect that the hydrophobic portions (phytyl group) of the molecules extend inwards, while the hydrophilic heads (porphyrin) and polar lipid groups are situated at the aqueous solution/membrane interface. Several orientations of the prophyrin head in a "dynamic" state are possible. That is, the plane of the porphyrin head may sometimes lie parallel to the bilayer, sometimes perpendicular, and at other times somewhere in between. An intermediate or slant orientation is favored, based upon the photovoltage polarization data as will be described later. On the basis of these data, it seems probable that the porphyrin plate of a chlorophyll molecule is oriented about 45° to the lipid bilayer (Steinemann et al., 1972; Cherry et al., 1972; Weller and Tien, 1973) as illustrated schematically in Fig. 4.1B. With this picture in mind, the observed thickness of the membrane could be also accounted for by the depth of anchoring of the phytyl chain, and by the amount of other lipids, such as carotene located in the interior of the membrane.

4.3.2. Physical properties of BLM

The electrical properties of BLM can be easily measured by placing suitable electrodes in each chamber (see Fig. 4.3). A BLM can be represented by a parallel capacitance-resistance circuit, as is shown in Fig. 4.4. The intrinsic resistance of an unmodified BLM is extremely high and is generally of the order of $10^8 \, \Omega \cdot cm^2$ or more. The dc conductivity is linear up to a potential difference of about 50 to 100 mV. The dielectric breakdown voltage is 200 ± 50 mV, although a voltage as high as 400 mV has been observed. For a BLM 70 Å thickness, a 70 mV potential difference corresponds to a field strength of 10^5 volts/cm. This high field intensity can be withstood by most BLM for a long period of time without any apparent deleterious effect.

In the dark, the dc resistance (R_m) and capacitance (C_m) of Chl-BLM are, respectively, $10^6 \, \Omega \cdot cm^2$ and $0.5 \, \mu F/cm^2$ (or 10^8 ohms and 5000 pF). The dielectric breakdown voltage of Chl-BLM is generally much less than 200 mV. Current/voltage curves appear to be linear, up to 150 mV. The Chl-BLM exhibit uniform stable resistive and capacitive properties. It will be shown that photovoltage features are determined largely by these characteristic electrical properties. It is interesting to note that unmodified BLM in 0.1 M NaCl, for example, have resistance about 2—3 orders of magnitude higher than Chl-BLM. One explanation that has been offered is that the Chl-BLM are intrinsically more conductive owing to the constituent molecules in the lipid bilayer. It is conceivable that, even in the dark, some charge carriers could be generated thermally and/or by an applied voltage, which may be conducted through the lipid bilayer via the conjugated double bonds of the carotenes and xanthophylls (Tien, 1968).

The dark conductivity of Chl-BLM has been measured as a function of temperature and, as expected, the conductivity increased with increasing temperature. In the temperature range studied (16 to 30°) the membrane conductivity followed the Arrhenius equation yielding an activation energy value of 0.7 eV. It should be noted that almost all rate processes observed experimentally can be fitted into a form of the Arrhenius equation, and in the absence of additional information, it is difficult to give a unique interpretation of the energy of activation. Therefore this energy may be interpreted to mean either the activation energy required for the charge transport, or the creation of charge carriers, or both. For instance, the dark conductivity of BLM in general, and Chl-BLM in particular, could be due to the presence of H^+. Since unmodified BLM are freely permeable to water, but poorly permeable to ions, as evidenced by dc resistance measurements, a simple calculation shows that the amount of water present in the BLM can provide enough charge carriers, in the form of H^+ and OH^- ions, to account for the dark current of the BLM. The experimental activation energy value of 0.7 eV for the chloroplast BLM lends some support for such a protonic conduction mechanism (Ting et al., 1968).

4.4. PROPERTIES OF PIGMENTED PLANAR BLM IN THE LIGHT

Experiments have shown that BLM containing light sensitive pigments such as chlorophylls, retinals, and bacteriorhodopsin exhibit interesting photoelectric effects. When such a BLM is illuminated by light of various wavelengths, a photoelectric (action) spectrum can be obtained which is practically identical to the absorption spectrum. These observations provide strong evidence for the separation of electronic charges (electrons and holes) in the BLM. In this section, we shall describe, under separate headings, the general aspects of photoelectric effects, experiments under continuous illumination, experiments with pulsed light, and spectroscopy of Chl-containing BLM (Berns, 1976; Hong, 1976; Strauss, 1976).

4.4.1. General aspects of photoelectric effects

Broadly speaking, the production of charge carriers in a material under illumination can be considered to be the principal cause underlying all photoelectric effects. These include the photovoltaic effect, the photogalvanic effect, the Dember effect, photoemission, photoconduction, etc. Most of these phenomena have been observed in suitably modified BLM systems.

Listed here are some of the notable features of experimental conditions that experience has shown are important in understanding the different photoelectric responses. The term photoelectric effect (PE) comprises photophysical and photochemical effects. The term photoelectric response is sometimes used to designate how a given photoelectric effect varies from experiment to experiment. For example, the photoresponse of a system will vary with external conditions. Standard symbols are used: mV for millivolts, V for potential difference (also called voltage), depending on context, PV for photovoltage or photopotential (also called photo-emf), J for current, and I for light intensity.

Each BLM system is defined by different lipid, pigment, or protein composition of the BLM or by different composition of the two bathing solutions. For any given BLM system there are a number of different photoelectric effects; some photoelectric effects exist only in BLM systems modified in special ways. An asymmetrical BLM system is one whose two bathing solutions differ in prepared composition (by adding components to one side or the other after formation of the BLM), or one whose bilayers have different sorbed molecules. Three types of asymmetry that affect the photoelectric response are concentration, pH, and electrical potential differences (also called gradients). A photoactive BLM is characterized by the presence or absence of light-sensitive compounds embedded in the lipid bilayer.

In the early studies, pigmented BLM were formed from spinach chloroplast extracts (Chl-BLM) and the magnitude of the observed photo-emf was small, of the order of a few mV. It was soon realized that larger photopotential could be elicited in pigmented BLM under asymmetrical conditions, particularly if the membrane was interposed between two solutions containing different redox couples. For example, the presence of $FeCl_3$ in one side of the bathing solution and ascorbic acid in the other side, a PV of more than 150 mV has been observed across the Chl-BLM (Shieh and Tien, 1974). The PE effect in the presence of electron acceptor Fe^{3+} is independent of the direction of illumination, but the polarity of the PV is determined by the location of Fe^{3+}, being always negative with respect of the acceptor-free side. Furthermore, the PE effects of most pigmented BLM have been found to be dependent on light intensity (Loxsom and Tien, 1972). The relationship between the open-circuit photo-emf (E_{op}) and the light intensity (I) is given by

$$E_{op} = A \log[1 + (I/B)] \tag{1}$$

where A and B are constants for a given BLM at a particular temperature. Under conditions of low light intensity ($B \gg I$), as would be expected, E_{op} has been shown to be directly proportional to I. It should be mentioned that this equation is also obeyed by most semiconductor photovoltaic cells.

Fig. 4.5 shows some typical photo-emf/time curves. In Fig. 4.5B, for instance, the BLM was formed from chloroplast extracts in 0.1 M acetate buffer at pH 5. $FeCl_3$ was added to one side of the bathing solution, whereas the other side contained cystine and cysteine. It can be seen that the photo-response is completely reversible. Significant photo-emfs have been also observed in many other systems (see below). In general, the time constants for the rise and decay of the photo-emf have been found to vary from less than $10 \, \mu sec$ to several seconds, depending on electron acceptors and donors added.

Some experiments report the photoelectric properties of pigmented thin

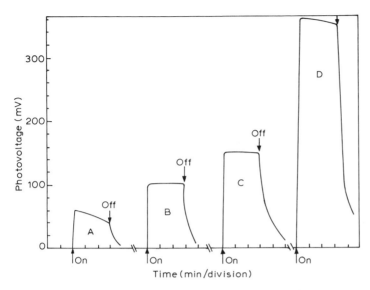

Fig. 4.5. Photoelectric effects in pigmented BLM. The BLMs were formed from chloroplast extract. The bathing solution was 0.1 M sodium acetate at pH 5. (A) The outer bathing solution also contained 0.001 M $FeCl_3$. (B) The same as (A) except the outer solution contained 10^{-3} M $FeCl_3$ and $FeCl_2$ and equal molar (10^{-3}) of cystine and cysteine were added to the inner solution. (C) The Chl-BLM was sensitized by adding water-soluble chlorophyllin to the inner solution. In this case the bathing solution was 0.01 M KCl with equal molar (10^{-3}) $FeCl_3/FeCl_2$ added to the outer solution. (D) Same membrane as in (C) except the membrane was under an external potential of 180 mV (applied through a 10^9 Ω shunt resistor with the chlorophyllin side negative).

lipid membranes (TLM), membranes with the same constituents as BLM, but in different proportions, that do not spontaneously thin to BLM, whose two layers are typically 1000 Å or more apart. There are differences between TLM and BLM with regard to both the observed photoelectric effects and the theories to explain them. TLM may be useful as BLM for the studies of photoelectrochemistry and of photoelectric energy conversion. With respect to photosynthesis, while TLM is a less attractive model system, comparison of its photoelectric effects with those of BLM may guide the formulation of theories to explain BLM. If pigmented TLM could contribute to the observed PE effects, the question naturally arises concerning the Plateau-Gibbs border which surrounds and supports the BLM. This question has been elegantly answered by Hong (1976), who used a focused beam of light to scan across the entire membrane (Plateau-Gibbs border and BLM). Hong has found that the photoresponse is specific to the BLM when a 50-μs argon ion laser (514.5 nm at 50 μJ) was used. However, the Plateau-Gibbs border also contributed significantly to the photoresponse under continuous light.

4.4.2. Experiments under continuous illumination

By continuous illumination we mean the pigmented BLM under investigation is excited in the duration of one second or more at a time until a steady-state photoresponse is achieved. Under this heading we shall describe the effects of physical and chemical parameters such as electric field, temperature, redox couples, quinones, and uncouplers and inhibitors on the Chl-BLM photoresponse. Light-induced water flow across Chl-BLM and asymmetrically pigmented chlorophyll in BLM will also be discussed.

4.4.2.1. Effect of applied field

The electrical field effect is due principally to the ultrathinness of the BLM (<100 Å) and its very high electrical resistivity (many orders of magnitude higher than that of the bathing solution). As already mentioned, a potential difference of even a few millivolts would result in an electric field of more then 100 000 volts/cm across the BLM. Thus, the oppositely charged species, whether generated directly, or indirectly by light in the membrane, should move in the direction of the external field and be collected at the opposite side of the BLM interfaces. Obviously, the field-assisted charge separation will be greater the higher the applied voltage. Fig. 4.6 shows the effect of applied field for Chl-BLM's in the presence of

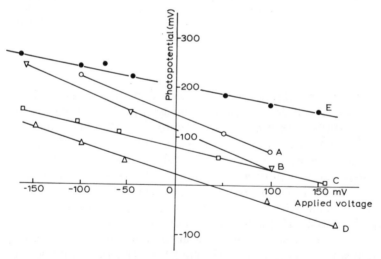

Fig. 4.6. Effect of applied voltage on the photopotentials of pigmented BLM. Chl-BLMs were formed in 0.1 M sodium acetate at pH 5 with the outer solutions containing 0.5 mM FeCl$_3$. The inner solution contained (A) ascorbic acid, (B) riboflavin, (C) catechin, (D) nicotinamideadenine dinucleotide (NAD). In E, oxidized cholesterol BLMs were formed in 10 mM KCl with the outer solution containing 1 mM FeCl$_3$. After the BLM had stabilized for 10 min, 0.1 ml chlorophyllin was added to the inner solution.

various compounds. Also shown is an oxidized cholesterol BLM sensitized by chlorophyllin introduced to one side of the bathing solution. In this case, a photopotential >250 mV was obtained.

4.4.2.2. Effect of temperature

Evidence presented initially by Tien (1968) strongly suggests that the origin of light-induced voltage in these BLMs is electronic rather than thermal. However, as with all new observations, the PE effect in the BLM system was met with some criticisms. The small size of the observed trans-membrane potentials suggested the possibility of heating. Hesketh (1969) performed a similar experiment in which the only significant change was the light source. He introduced two new points worthy of consideration: (1) possible temperature rises and differences caused by heating due to light absorption both in the membrane and in the bathing electrolyte solution, and (2) the suggestion that because the thickness of the BLM is much less than one wavelength, then the probability of an interaction of a given type of molecule with a photon is practically equal at each interface of a BLM, whereas it is likely that photoexcitation is truly negligible on the dark side of the much thicker TLM. There is evidence that consideration (2) is correct, and it is usually assumed in present theories.

Although a temperature gradient generates only a negligible potential difference ($60 \, \mu V/^\circ C$) across the Chl-BLM, the photopotential was found to decrease with increasing temperature and became negligible at about $40^\circ C$. At the same time, the temperature decreased the resistance of the membrane by one order of magnitude when the temperature of the bathing solution increased from 15 to $40^\circ C$. The same behavior was observed when the temperature of only one compartment was varied (Van and Tien, 1970). The decrease of the PE effect may result from an increase in ion permeation which raises the dark conductivity by one order of magnitude. The temperature effect is partially reversible, i.e., when the temperature is lowered, the light effect reappears, but is much smaller in magnitude.

4.4.2.3. Effect of electron acceptors and donors

As would be expected, little photoresponse should be observed if a pigmented BLM were illuminated under symmetrical conditions. The small PE effect detected in the initial studies was apparently owing to: (i) the presence of a small dark voltage across the membrane (e.g., resulting from calomel reference electrodes used), and (ii) the generation of a concentration gradient of charge carriers in the Plateau-Gibbs border (i.e., a Dember effect). In the early investigation of Chl-BLM, apart from the minuteness of the observed photovoltage, the response was sluggish taking many seconds to reach the maximum value. In contrast, if redox compounds (electron acceptors and donors) are present on either side of a pigmented BLM, an open-circuit photo-emf greater than 180 mV has been detected (Tien, 1974).

TABLE 4.1

PHOTO-EMFS OF CHLOROPLAST-BLM IN 0.1 M SODIUM ACETATE OF pH 5 IN THE PRESENCE OF VARIOUS REDOX COMPOUNDS.

Modifier in solution		Photo-emf (mV)
Outer side	Inner side	
$FeCl_3 (10^{-3}$ M)	—	40—60
—	Benzoquinone (3×10^{-4} M)	5—7
$FeCl_3$	Benzoquinone	40—50
—	1,4-Dihydroquinone (HQ)	6—8
$FeCl_3$	HQ (7×10^{-4} M)	80—117
—	Tetrabutylhydroquinone (TQ)	5
$FeCl_3$	TQ	69
—	Chloranil	6
$FeCl_3$	Chloranil	100
—	2-Hydroxy-1,4-naphthoquinone (NQ)	5
$FeCl_3$	NQ	87
$FeCl_3$	2-Sulfonate(K)-NQ (10^{-2} M)	80
$FeCl_3$	4-Sulfonate(Na)-NQ)	65
$FeCl_3$	2-Methyl-NQ	50
$FeCl_3$	2-Amino-NQ	30
$FeCl_3$	Flavin mononucleotide (FMN)	138—167
$FeCl_3$	Riboflavin (8×10^{-5} M)	79—99
$FeCl_3$	NaI (5×10^{-3} M)	74—127
$FeCl_3$	$Na_2S_2O_4$	44—120
$FeCl_3$	Tannic acid (4.5×10^{-7} M)	113—133
$FeCl_3$	Flavin mononucleotide (FMN)	138—167
$K_4Fe(CN)_6$	Methyl viologen (4×10^{-2} M)	73

Fig. 4.5A shows the PE response of a Chl-BLM with 10^{-3} M $FeCl_3$ present in the bathing solution on one side of the membrane. The Fe^{3+}-free side was positive. The light-induced photovoltage was more than doubled on addition of an equal molar cysteine/cystine to the opposite side plus 10^{-3} M $FeCl_2$ added to the Fe^{3+} side (Fig. 4.5B). Note that the waveforms are different. The photovoltage attained a steady value when different redox couples were present in opposite sides of the Chl-BLM. A variety of redox compounds affecting the photoresponse of Chl-BLM have been investigated (Shieh and Tien, 1974) and some of the results are summarized in Table 4.1.

Illani and Berns (1972; Bern, 1976) have also experimented with Chl-BLM and observed sustained photo-current under continuous illumination. Two different redox couples, $FeCl_2$—$FeCl_3$ and Ce^{3+}—Ce^{4+} were used. Their data and others for steady PV can be adequately described by the equivalent circuit shown in the lower diagram of Fig. 4.4 (Miller and Tien, 1974). Electrically, the pigmented BLM is equivalent to a capacitor (C_m) shunted by a resistor (R_m). When voltage is applied to this system, either through an external voltage source or via the action of light, the voltage transients appearing across the effective membrane R_m–C_m combination obey the following relationship: for charging

$$V(t) = V_0(1 - e^{-t/RC_m}) \tag{2}$$

and for discharging

$$V(t) = V_0 e^{-t/RC_m} \tag{3}$$

Where V_0 is the maximum voltage appearing across the BLM, $V = Q/C_m$ and R is the Thevenin equivalent resistance with respect to C_m. For the Chl-BLM this value is given by R_m in parallel with the internal or series resistance of the EMF shunted across the membrane (i.e., $R = R_S R_m/(R_S + R_m)$). In the simple case of a voltage discharge when the charging element has been removed, R_S is nearly infinite and thus $R = R_m$ and the discharge rate constant is simply $1/R_m C_m$, typically $5\ \text{sec}^{-1}$. For a potential applied through a resistance ($R_S = 10^8$ ohm) R becomes 5×10^7 ohm ($R_m = 10^8$ ohm) and the charging rate constant equals $10\ \text{sec}^{-1}$. Illani and Berns (1972) used only steady-state data in their analysis and showed that to a first approximation light intensity affects only $R_{h\nu}$, which increases with I; redox potential affects only PV, which increases with redox potential; and temperature affects only R_m, which decreases with temperature. The dependence of PV on redox potential and its independence of temperature strongly suggest electronic conductance in the bilayer part of the BLM.

4.4.2.4. Effect of quinones and flavins

In addition of redox compounds, a variety of quinones were tested on Chl-BLM. Some of these quinones are the well known artificial electron donors and acceptors (Table I) with the exception of 2-hydroxy-1,4-naphthoquinone, which generated a photo-emf with the polarity opposite to that of $FeCl_3$, the others were not effective when used alone. In the case of chloranil, it was not clear whether the compound was photoreduced.

Flavins in the form of flavin mononucleotide and riboflavin are effective electron donors when used in conjunction with $FeCl_3$. Further, the light response can be very complicated if the compound is sensitive to H^+. The photoresponse of a Chl-BLM in the presence of riboflavin in the pH range 4 to 6 has been investigated. Below pH 5.5, the photoresponse is biphasic and predominately in the positive direction. At pH 6 and above, the photoresponse is in the negative direction, with a very slight initial deflection toward the opposite polarity. A simple explanation is that riboflavin in the forms of semiquinone and hydroquinone can function as either an electron acceptor or donor depending on the pH of the bathing solution. The possibility that light-sensitive riboflavin, when associated with the BLM, can participate in photoreactions should not be ruled out (Galston, 1977).

4.4.2.5. Effect of uncouplers and inhibitors

In the postulated scheme of photosynthesis, chemical compounds such as 2,4-dinitrophenol (DNP) are thought to be able to uncouple electron transfer from adenosine triphosphate (ATP) formation in the electron-transport chain. The effects of the uncouplers, dinitrophenol (DNP), p-trifluoromethoxycarbonylcyanide phenylhydrazone (FCCP), and dicoumarol on the electrical conductivity of Chl-BLMs have been tested. With FCCP concentrations below 10 μM, little or no change in the conductivity could be measured in the dark, but with higher FCCP concentrations, up to 100 μM, the conductivity increased rapidly. Similar curves were obtained with DNP and dicoumarol, although the rise took place at different concentrations. The order of effectiveness at the same concentration is FCCP > dicoumarol > DNP. The effect of FCCP is such that the optimum occurs roughly at the pK value of the uncoupler, which suggest that H^+ is the charge transported across the membrane. The uncoupler, which has appreciable lipid solubility, serves as the carrier of hydrogen ions. The pH dependence of the uncoupler effect does not appear to be related to the composition of the lipid solution used for BLM formation, since lecithin BLMs display similar characteristics (Liberman et al., 1970).

The effects of 3-(p-chlorophenyl)-1,1-dimethylurea (CMU) and 3-(3,4-dichlorophenyl)-1,1-dimethylurea (DCMU) have also been tested on Chl-BLMs (Tien, 1974). For both inhibitors, the photo-emf decreased with increasing concentration (Fig. 4.7), and the dark transmembrane voltage (data not shown) increased linearly with concentration. Furthermore, the membrane resistance decreased by one order of magnitude in the presence of 4 mM DCMU or 5 mM CMU in the outer chamber. Both DCMU and CMU

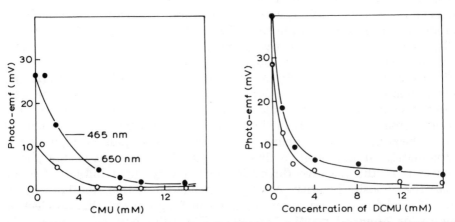

Fig. 4.7. Dependence of photo-emf of Chl-BLM as a function of inhibitor CMU or DCMU concentration. The Chl-BLM were formed in 0.1 M sodium acetate at pH 5 with CMU or DCMU in the inner chamber and 2.5×10^{-4} M $FeCl_3$ in the outer chamber.

are well-known poisons of photosynthesis and in the present BLM system may prevent charge separation in the membrane.

4.4.2.6. Water permeability

Chl-BLMs exhibit significantly higher water permeability than, for example, BLMs formed from oxidized cholesterol; and, when the membrane is illuminated, an increase in water flow is observed. A typical curve of volume change versus time, showing light-induced water flow, can be described as follows. The initial rate was 0.04 μl/min, and when the light was turned off the flow resumed its initial rate. Application of an electrical potential difference across such a BLM causes electro-osmosis, the flow rate increasing with increased applied voltage. The direction of water flow has been shown to depend on the polarity of the field. The net accumulation of water was in the compartment where the potential was negative. The light-induced voltage across the BLM is believed to be the additional driving force responsible for the water flow. Externally applied voltage should therefore accomplish the same purpose, as has been experimentally demonstrated (Tien, 1974).

Light-induced water flow could be decreased or abolished completely by an inhibitor or uncoupler of photophosphorylation. High concentrations of uncouplers such as DPN and FCCP tended to rupture the BLM. When 10 μM of DNP or 0.1 μM of FCCP was added to the aqueous phase, both the photovoltaic effect and light-induced water flow were abolished. The inhibitor DCMU at 0.1 μM, however, was only partially effective.

4.4.2.7. Chlorophyllin-sensitized BLM

Certain BLM, such as those formed from oxidized cholesterol (Robinson and Strickholm, 1978) which are otherwise not photoactive, can be sensitized by water-soluble compounds such as dyes, inorganic salts present in the bathing solution. For example, an oxidized cholesterol BLM has been made photoactive by incorporation of water-soluble chlorophyllin (Huebner and Tien, 1973). This system is of some interest in that the photoresponse has been found to be much greater than that of Chl-BLM (see Fig. 4.5C). Further, the presence of chlorophyllin lowered the BLM resistance by an order of magnitude. Evidently, the compound not only interacted with the BLM but served a dual role, that of acting as light absorbing pigment in the membrane and as an electron donor. Interestingly, the photopotential of chlorophyllin-sensitized BLM is influenced by the presence of β-carotene in the membrane. Curve A of Fig. 4.8 shows the response of a chlorophyllin-sensitized BLM. Note that it takes >15 sec to reach a steady-state photovoltage. In Curve B, obtained under similar conditions except the β-carotene was incorporated into the BLM, the observed photovoltage is more than doubled and the response time is shortened to one-fourth the previous value. A simple explanation is that the presence of

136

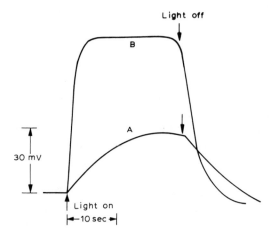

Fig. 4.8. Effect of β-carotene on the time course of the Chl-BLM photoresponse. Cell arrangement: 10 mM KCl, 1 mM $FeCl_3$/oxidized cholesterol BLM/10 mM KCl, chlorophyllin. (A) with no β-carotene in BLM, (B) with β-carotene.

the π-bond system of carotene provides a low-resistance pathway for electron transfer (Tien, 1968; Mangel et al., 1975). Future studies with other carotenoids such as xanthophylls would be of interest.

4.4.3. Experiments with pulsed light

Hitherto we have reviewed only the photoeffects of BLM elicited by continuous illumination (duration of seconds and longer). The possibility of obtaining information on fast photoprocesses operating in the millisecond and microsecond range is therefore obscured owing to prolonged excitation. In this section we shall review the results of experiments of pigmented BLM excited by light pulses of less than 8-μsec duration. The first flash experiments on Chl-BLM were invesigated as a function of: (a) gradient of redox potential using $FeCl_3$, (b) pH gradient, and (c) external electrical potential, V_e (Huebner and Tien, 1972). Any one of these conditions was sufficient to give a PE potential across the membrane. The value of microsecond illumination is that it can resolve the PE response into distinct components, each due to a different, kinetically independent mechanism. These are nearly indistinguishable under stimulation lasting longer than the relevant response times. Three phases of the PV were observed (Fig. 4.9). The first, component A, appeared able to follow I(t), the rise and fall of the flash. The other two, components B and C, were relaxation currents. One had a time constant of approximately 20 msec and the other, about 1 second. Thus three separate charge transport mechanisms were indicated, and in particular, the ability of the first phase of PV to follow the course of the flash strongly

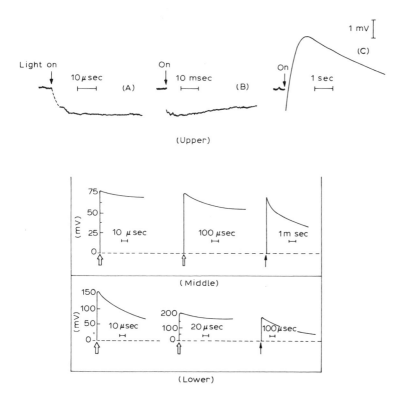

Fig. 4.9. Flash excitation of a Chl-BLM. Upper: Time course of the photoresponse of a Chl-BLM in 0.1 M Na acetate at pH 5.5. The bathing solution on one side contained 1 mM $FeCl_3$. Middle: successine traces of an oxidized cholesterol—lecithin BLM sensitized by chlorophyllin (1 mg/l) in one side of the bathing solutions and the other side contained 1.5 mM $(NH_4)_2Ce(NO_3)_6$. Lower: The flash-induced responses of a Chl-containing BLM. One side of the bathing solution contained 10 mg/l chlorophyllin and 1 mM $FeCl_2$ and the other side 2 mM $(NH_4)_2Ce(NO_3)_6$. The width of the arrows indicate the duration of excitation.

suggested either an electronic current or charge separation (i.e., the formation of a dipole layer as discussed by Ullrich and Kuhn (1972) and by Trissl (1975)). The slowest component was almost certainly ionic, probably H^+, for it was strongly dependent on pH and less so on the other ionic concentrations. None of the PV components, A, B, or C, was due to the decay of the charge on C_m through R_m; which is also slow and blends with component C. It has been identified as a fourth component, D (Miller and Tien, 1974).

Since the capacitance of a BLM may be over two orders of magnitude greater than that of a TLM, yet the amplitudes of the voltage change produced by component A and B are comparable in the two types of membranes, the amount of charge transported by these components must vary nearly proportionally with the membrane capacitance and inversely with the membrane thickness. Thus, only a small fraction of components A and B in BLM can result from the Plateau-Gibbs border. The fact that component C does not occur in TLM, while both TLM and BLM have such borders or comparable areas, eliminates the possibility that component C is a border effect. Thus the border can be neglected. There has been speculation that fluorescence from chlorophyll in the border may be a source of error in quantitative measurements of PE versus intensity, and in some such experiments (Hong, 1976; Mauzerall and Hong, 1975) much trouble is taken to shield the border, or in some cases to focus laser light so that the beam area is less than the black area. Evidence that the border contribution is truly negligible would reduce the effort required for quantitative measurements, but whether these results can be extended to other systems is not certain.

Some conjectures were made and tested with regard to the origin of the various components. Assume that the currents from components A, B, and C, which are all that contribute to the photoelectric effect, add algebraically. Initially, $J_A(t)$ appears dominant, and assuming that it is proportional to light intensity $I(t)$ gives, after integrating the current into the capacitance C_m and allowing for simultaneous leakage through R_m, a PV(A) that will rise as the time integral of $I(t)$ and fall as the discharge of the membrane in the dark.

The assumption of electronic, versus ionic, charge separation is further strengthened by the ability of PV(A) to follow the fast (microseconds) response, and the high mobility, 10^{-2} cm²/volt, deduced from the fact that the transport time for component A charge carriers in TLM approaching 1 micron thick is not greater than that of BLM.

In considering Components B and C, it is convenient to consider that the pulsed illumination is instantaneous and results in the formation of excitons. Their dissociation will result in charge transport, provided suitable membrane asymmetries exist. Assuming the currents J_B and J_C are controlled by a single dissociation process at a rate proportional to the density of excitons, the photocurrent leads to the generation of a voltage that starts from zero and returns to zero as the sum of exponentials:

$$V(t) = V^0(e^{-at} - e^{-t/R_m C_m}) \tag{4}$$

where "a" is the unknown dissociation rate constant, and V^0 is an experimental parameter. This function fitted the data ($V(t)$, R_m and C_m) fairly well, but was apparently not closely enough to eliminate coincidence or the suitability of other possible models. The time constants calculated were $1/a = 50$ msec and $1/a_C = 0.2$ sec. A good fit would require modifications based on more knowledge of the molecular mechanism involved (Huebner and Tien, 1973).

Miller and Tien (1974) further investigated Chl-BLM by pulsed light. As usual, repeated pulses allowed a stable display on an oscilloscope. Time resolution was of the order of milliseconds, so that only components B, C, and D were observed. For this work Component B was termed fast, and Component C, slow. The complete dependence of PV(B) upon the $FeCl_3$ concentration gradient, its millisecond response time, and the agreement of its action spectrum with the absorption curve of chlorophyll concurred with the earlier hypothesis that it is due to the reduction of Fe^{3+} by photoelectrons from excited Chl. The kinetics indicate that C_m becomes charged, which in turn means that the charge separation giving rise to PV(B) must be a transmembrane current, in contrast to the formation of a dipole layer at only one interface (or the formation of separate dipole layers at each interface).

PV(C) was affected by a pH gradient and by V_e. It was also studied by the addition of various agents that quenched it, the most effective being ascorbate. This allowed distinction between PV(C) and PV(D), the $R_m C_m$ decay component. From a comparison of data with calculations based on the equivalent circuit (Fig. 4.4), it was judged that PV(C) is due to an increase in membrane conductivity, with kinetic features containing information concerning Chl metastable states in the Chl-BLM.

Hong (1976) introduced a new measuring circuit for flash studies of PE for the determination of kinetics. He used magnesium mesoporphyrin in pigmenting the BLM, which was also used by Lutz et al. (1974). Electron transfer from excited pigment is extremely fast. The redox reactions at both interfaces are coupled by diffusion of the mobile pigment molecules, and the photocurrent flows predominantly through the transmembrane resistance R_s, apparently related to the energy barrier the pigment molecules must cross. The response to pulsed light is specific to the black region of BLM because the Plateau-Gibbs border is far too thick. In his analysis, Hong introduced a new term, C_p, as a chemical capacitance arising from photogenerated charge pigment molecules in membranes, and suggests that it may be of fairly general occurrance.

The physical and chemical behavior of Hong's model is reminiscent of the hopping model proposed by Trissl and Laeuger (1972). In both cases the current is carried not by electrons but by ions hopping across an energy barrier in the lipid interior. It appears that on the basis of the usual Boltzmann thermal energy distributions, tunneling might be possible, although it has a very low probability for an object as massive as an ion.

Regrading the circuit developed by Hong, which he calls a tunable voltage clamp, it uses feedback amplifiers to maintain the two terminals of the salt bridge/calomel electrodes at constant potential, plus some other refinements to permit optimization of the time constants and gain of the system, and an instrumental time constant as low as 1.5 μsec. Such methods circumvent the large $R_m C_m$ time constant that can easily hide the transient phenomena.

The flash-induced photoelectric effects in these BLMs under asymmetrical conditions (ΔC, ΔpH, and V_e) can be explained in terms of a simple model where the photoactive BLM is pictured as similar to a barrier-layer photo-voltaic cell comprising a current generator in parallel with a resistance and a capacitance (see Fig. 4.4). A detailed discussion on the mechanism of light transduction will be given in section 4.6.

4.4.4. Spectroscopy of Chl-containing BLM

In the preceding sections, we have considered various photoeffects in BLMs. From the results obtained, it is obvious that the presence of photo-active species, either incorporated into the membrane or at the solution-BLM interface, is a prerequisite. To demonstrate conclusively the presence of the photoactive species involved, measurement of action spectra is in principle, the simplest approach. Experimentally, photoelectric action spectra can be easily obtained, whereas standard absorption spectra of the BLM are more involved but positive results have been reported (Brown, 1977).

4.4.4.1. Photoelectric spectroscopy

Experimental techniques for obtaining photovoltage action spectra of BLMs are essentially the same as those developed for research in photo-chemistry and photobiology. Briefly, the set-up consists of a powerful light source, heat filters, and a monochromator. The light emerging from the monochromator is focused on the BLM. The light-induced membrane voltages picked up by the contacting electrodes, after amplification, are recorded as a function of wavelengths (Van and Tien, 1970).

In a system consisting of Fe^{3+}, Fe^{2+}/Chl-BLM/ascorbic acid, dehydro-ascorbic acid, resolution of the photoelectric spectrum was as good as that of an absorption spectrum obtained with the BLM-forming solution. Not only did the two peaks at 430 and 670 nm correspond, but also the shoulders at 450 and 620 nm were clearly seen. Fig. 4.10 shows the spectra obtained in the presence of the above redox couples. Although the wave-lengths for the maxima are the same, the relative height of the peaks varies considerably. After adjustment for the spectrum of the light source, the curves of PV versus wavelength had peaks nearly congruent with those of the absorption curve of the BLM forming solution, with a small shift of one peak. The black part of BLM produced a bigger PV than the Plateau-Gibbs border.

Variations of spectra with light intensity and applied voltage have also been investigated. To date no evidence of the so-called electrochromic effect in Chl-BLMs under a field strength greater than 10^5 V/cm has been observed, although such an effect has been reported in the thylakoids of chloroplasts (Junge and Schmidt, 1971; Schmidt, Reich and Witt, 1972).

Photovoltage spectroscopy has been used for the determination of the orientation of the chlorophyll porphyrin ring in Chl-BLM (Weller and Tien, 1973). Bascially the photovoltage spectra using polarized light are mea-

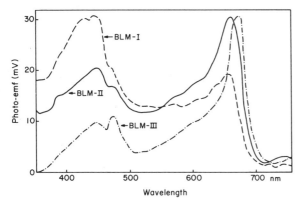

Fig. 4.10. Photovoltage spectra of Chl-BLMs. BLM-I, after 0.5 h; BLM-II, after 2 h; BLM-III, after 4 h.

sured. From polarization data, the directions of the transition moments of the different peaks are determined, from which the angle of the porphyrin ring with respect to the plane of the BLM is deduced. To deduce the angle from the experimental data, the usual assumption is made in that the blue and red transition moments are mutually perpendicular to each other and are located in the plane of the porphyrin ring. In BLMs formed from chloroplast extracts, the chlorophyll porphyrin ring is tilted at about $45 \pm 5°$ from the plane of BLM as calculated from the polarization data. The results, obtained by photovoltage spectroscopy, together with those deduced from absorption polarization data are given in Table 4.2.

Since an action spectrum for the in vivo photoelectric potential is not known, at least from direct measurements, then no comparison can be made with that of the Chl-BLM system. It is possible, however, to compare the absorption spectra with the absorption spectra of chloroplast pigments in living cells. For example, the red absorption at 675 nm has been reported by Rabinowitch and Govindjee (1969). They have suggested that the broad nature of the red band in vivo is due to the presence of more than one type of chlorophyll a, Chl a-670, Chl a-680, and Chl b (650-nm). In the case of Chl-BLM the red absorption bands of chloroplast pigments in Chl-BLM (Fig. 4.10) are wider than the corresponding bands in solution. This probably indicates the following: (i) membrane structure is a crucial factor for maintaining more than one form of Chl a, (ii) in a Chl-BLM newly formed from fresh lipid solution, two forms of Chl a may exist, (iii) the finding that the red absorption maximum is strong at 670 nm in Chl-BLM prepared from fresh solution but is only a small shoulder in that from aged solution suggests that the Chl a having an absorption band at the longer-wavelength side in the absorption bands of the two Chl a forms might be unstable.

TABLE 4.2.

ORIENTATION OF THE CHLOROPHYLL-PORPHYRIN RING IN VARIOUS BLM
FROM ABSORPTION AND PHOTOVOLTAGE SPECTROSCOPY

BLM	Angle	Technique	Reference
Chl a + dioleollylecithin	$46 \pm 3°$	Absorption	Steinemann et al. (1972)
Chl a + phosphatidylserine	$44 \pm 3°$	Absorption	Steinemann et al. (1972)
Chl a + phosphatidylethanolamine	$49 \pm 5°$	Absorption	Steinemann et al. (1972)
Chl b + dioleoyllecithin	$42 \pm 4°$	Absorption	Steinemann et al. (1972)
Chl a + egg lecithin	$48°$	Absorption	Cherry et al. (1971, 1972)
Chl b + egg lecithin	$51°$	Absorption	Cherry et al. (1971, 1972)
Chloroplast extract (Chl a + Chl b)	$45 \pm 5°$	Photovoltage	Weller and Tien (1973)

Fig. 4.10 shows the photoelectric action spectra, calibrated for constant incident energy, taken from a single BLM at 0.5, 2, and 4 h after the addition of 1 mM $FeCl_3$ to the bathing solution on one side. They are of three distinctive types, designated as BLM-I, BLM-II, and BLM-III. BLM-I is observed with a Chl-BLM formed from freshly prepared lipid solution (aged <1 h). This type of the spectrum has a red/blue peak ratio greater than unity. BLM-II is a transition between BLM-I and BLM-III. BLM-II was obtained from Chl-BLMs formed with either fresh or aged lipid solution. This type of spectrum has a red/blue peak ratio less than unity. The red peak position shifted with time from 670 to 660 to 655 nm. It leveled off after 70 min and increased slightly after 2 h. This time dependence of spectral change was also seen with Chl-BLMs formed from aged lipid solution (e.g., 61 days at 4° in the dark). Chl-BLMs formed from aged lipid solution gave only BLM-III. An increase with age in the resistance of Chl-BLMs from 0.1 MΩ cm^2 (BLM-I) to 0.5 (BLM-II) to 0.75 (BLM-III) was observed (Tien, 1974).

It has been postulated that the changeable substance causing the increase of Chl-BLM resistance is Chl a_{670}, a pigment proposed to have two different forms, designated A and B. In freshly prepared membrane-forming solution, form A is dominant. Form A is thought to undergo a transformation (or bleaching) to form B in the dark and also catalyzed by light:

$$A_{670} \underset{\text{dark}}{\overset{\text{dark or light}}{\rightleftharpoons}} B \text{ (probably Chl } a_{660} \text{ or deactivated Chl } a_{670}) \tag{5}$$

This may be one of the reasons why the action spectrum of a BLM formed with fresh solution, when scanned from long wavelength to short wavelength, shows a gradual shift in the maximum of the red peak from 670 to 660 nm. If the spectrum is scanned from short wavelength to long wavelength, however, the pigment is quickly transformed (or bleached) by the short-wavelength light from form A to form B, before the scan reaches the red band, and therefore no apparent blue shift is observed. Since the transformation (or chemical reaction) could proceed also in the bulk phases, in the aged membrane-forming solution, form A of the pigment would not be detectable: this is why no BLM-I type of spectrum was observed with membranes formed from aged solution. The hypothesis accommodates the finding that the absorption spectra of Chl-BLM formed with fresh solution have a maximum absorption band at 670 nm in the red region, whereas those of Chl-BLM formed with aged solution have a maximum at 660 nm.

4.4.4.2. Absorption spectroscopy

Many investigators have wondered what would be the final concentration of a photoactive compound in the BLM, if it were added to the BLM-forming solution. Application of standard spectroscopic techniques should be an obvious approach. However, for a single BLM, it can be shown that the amount of photoactive species present is about the limit detectable by conventional absorption spectroscopy. To increase the absorption, a "brute force" approach has been used by forming a stack of BLMs in the path of the light beam. Cherry et al. (1971) have constructed a cell with a number of Teflon plates containing circular holes for BLM formation. The whole cell fits into the sample space of a conventional spectrophotometer.

The procedure used by Cherry et al. for measuring absorption spectra is as follows. The series of apertures is first carefully aligned in the spectrophotometer and the cell is filled with the KCl solution. A baseline is then run over the spectral range to be examined. BLMs are formed across the apertures. The absorption spectrum is then measured, the BLMs broken and the baseline rechecked. The results obtained by Cherry et al. demonstrate the feasibility of using absorption spectroscopy to estimate the composition of the BLMs. In their later experiments, Cherry et al. (1972) have also obtained polarized spectra of chlorophyll-containing BLM. The data indicate that there is some degree of orientation of the porphyrin rings of chlorophyll in the BLM and that the transition moments of red and blue bands must lie in different directions (see Table 4.2).

While absorption spectroscopy of a single BLM is not feasible using the conventional equipment, Steinemann et al. (1971) have overcome the problem of the very low light absorption of a single BLM by using an improved double-beam spectrophotometer. They showed chlorophyll a to be 3×10^{13} per cubic centimeter with the porphyrin rings 20 Å apart. Steinemann et al. (1972) have also developed a method for measuring the dichroism of BLM containing Chl *a* and Chl *b*. They have found that Chl possess a preferential orientation in the BLM. For three different kinds of phospholipid BLMs, the angles between the porphyrin plane and the membrane were found to be in the range of 44 to 49°. The results of Steinemann et al. are summarized in Table 4.2.

Another question of interest is the importance of the interaction, perhaps cooperatively, of chlorophyll molecules in the form of aggregates, having crystalline properties to some degree. Csorba et al. (1975), citing the observation by Thomas et al. (1970) and Inoue et al. (1972) on the observation of an aggregated Chl form absorbing at about 740 nm (where crystalline Chl has a characteristic absorption band), and the observation by Tien (1968) of an absorption band at about 745 nm in the photoelectric action spectrum of Chl-BLM, performed an experiment on the existence of crystalline chlorophyll in BLM. Although the BLM forming conditions seemed to meet the requirements for the formation of crystalline-like Chl, they sought more direct information by a detailed measurement of absorption spectra. They found that a minimum concentration of Chl *a* in lecithin solution gave a step response in the absorption, which they concluded signified the formation of crystalline aggregates in the BLM. On the contrary, Mangel (1976a) found, using phosphatidyl choline—Chl *a*—carotene BLM, that the photocurrent is proportional only to the second power of Chl concentration. This is inconsistent with the cooperation of large aggregates of chlorophyll. Although such aggregates may be necessary for some functions of the thylakoid membrane, they appear not to be necessary for some types of PE in certain chlorophyll BLM systems.

4.4.4.3. Fluorescence spectroscopy

Fluorescence of BLM-containing chlorophylls and carotenoid pigments was reported shortly after the discovery of light-induced effects (Van and Tien, 1970). The fluorescence techniques are useful in probing into the mechanism of photon capture and the initial photochemical processes. For example, by incorporating fluorescent dyes such as ANS (8-anilino-1-naphthalenesulfonic acid) into otherwise nonfluorescent membrane, information may be obtained on lipid—protein interactions, chemical composition, conformation, and physical state of the membrane (Smekal et al., 1970; Pohl and Teissie, 1975; Vaillier et al., 1975).

In a short communication, Alamuti and Laeuger (1970) estimated the chlorophyll concentration in the BLM from the fluorescence intensity. They have found that the fluorescence intensity of the BLM increases with the molar ratio of chlorophyll to lecithin in the BLM-forming solution and reaches a maximum at about 0.3. If the concentration assumed to be present at the molar ratio 0.1 and below, the number of chlorophyll molecules per square centimeter of the BLM is 6.5 to 40×10^{11}. Since the interior of the BLM is belived to be liquid-hydrocarbon-like (Tien, 1967), the most likely arrangement would be that the hydrophilic porphyrin plate is at the solution-BLM interface and the hydrophobic phytol chain anchored in the membrane phase (Ting et al., 1968; Steinemann et al., 1972; Cherry et al., 1972). With this in mind together with the aforementioned number of chlorophylls per square centimeter, the average distance between the porphyrin plates at the surface of the BLM is estimated to be about 70 Å.

It should be mentioned that the concentration of chlorophyll a in the BLM reaches a constant if its concentration in the forming solution exceeded a certain level (Cherry et al., 1971; Steinemann et al., 1971). Trosper (1972) was unable to form a BLM from chlorophyll a alone. The membrane would not drain to the black state without the addition of glycerol mono-oleate. Using fluorescence measurements, only a small percentage of chlorophyll a was found in the BLM. The conductance of the chlorophyll a BLM did not differ significantly from that of a BLM made from glycerol mono-oleate alone. No reproducible photoconductance change was observed with the chlorophyll a BLM. The conductivity was unaltered upon the addition of cytochrome but the BLM became unstable.

The fluorescence of BLMs has been utilized to investigate energy transfer (Strauss and Tien, 1973). Generally, the efficiency of energy transfer between donor and acceptor molecules can be found most conveniently by measuring the sensitized fluorescence yield of the acceptor resulting from irradiation of the donor. This direct approach is difficult to measure, and because direct excitation of the acceptor (Chl a) could not be avoided, owing to its spectral overlap with the donors (carotenoids and Chl b). Instead, the 730-nm fluorescence intensity is followed as a function of the excitation wavelength. This fluorescence is mainly due to Chl a, with a smaller contribution from Chl b, and none from carotenoids. The relative efficiency of energy transfer from the accessory pigments to Chl a could then be found by comparing the fluorescence excitation spectrum of the system being studied with the absorption spectrum of the mixture of pigments present. The peaks at 664—666 nm which are due to Chl a only were normalized relative to each other before making the comparison. In case of 100% energy transfer efficiency, the fluorescence excitation spectrum should coincide with the absorption spectrum; a smaller transfer efficiency should cause a decreased excitation spectrum in those regions where the donors absorb light. However, in the absence of energy transfer, as, for example, in a very dilute pigment solution, the fluorescence excitation spectrum should be identical with the absorption spectrum of the energy acceptor only. Transfer efficiencies determined in this way are unaffected by the magnitude of the fluorescence yield of the acceptor since each spectrum serves as its own internal standard. Therefore, a valid comparison of energy transfer efficiencies in BLM and in solutions could be made despite the fact that these systems are likely to differ considerably in the fluorescence yield of Chl a, due to concentration quenching.

The average intermolecular distance R between energy donors and acceptors in a particular chlorophyll containing BLM-forming solution studied by Strauss and Tien (1973) is 61 Å, if the theory of Forster is assumed to be valid (Forster, 1956). This value of R in the BLM is not known. However, the state of the hydrocarbon chains in a BLM may be a two-dimensional liquid so that R should be of the same order of magnitude as in the membrane-forming solution, or even greater, provided that the pigment molecules are randomly distributed over the area of the membrane. As already cited above, Alamuti and Laeuger (1970) have obtained R values of similar magnitude from comparative fluorescence intensities of BLM and bulk solutions of Chl a in lecithin. The only energy transfer mechanism possible at such intermolecular distances is dipole—dipole inductive resonance, as suggested by Forster (Strauss, 1976).

4.4.5. Bacteriorhodopsin-containing BLM

Closely related to the thylakoid membrane is the purple membrane of halophilic bacteria (*Halobacterium halobium*), which appears to function either as a photoconverter or a photodetector depending on environmental conditions (for a more extensive coverage of this topic see chapter 6 of this book and also Oesterhelt, 1976). This remarkable membrane system has been shown to contain only one protein (Kushwaha, Kates and Martin, 1975), a rhodopsin-like pigment, which is called bacteriorhodopsin (BR). As discussed in chapter 6, this pigment is arranged in the membrane (lipid bilayer) in a highly ordered hexagonal lattice in patches with vectorial properties (Henderson and Unwin, 1975). Experimental observations to date may be summarized as follows. The intact cells thrive in concentrated NaCl solution and in the dark, and under anaerobic conditions in the light, acidify the medium. The acidification of the medium by anaerobic cells in the light is abolished by uncouplers only, whereas with respiring cells both uncouplers and electron transfer chain inhibitors are effective. As long as they are exposed to light, both anaerobic and starved cells containing purple membrane, generate and maintain a pH difference across the cell membrane. Furthermore, it has been reported that starved cells consumed reduced amounts of O_2, requiring even less O_2 when illuminated (Oesterhelt and Stoeckenius, 1973). Moreover, growing cells from late-log phase culture use less O_2 when illuminated than when in the dark. If the culture is vigorously aerated with air the purple membrane fails to form in significant amounts, whereas the purple patches form in the membrane if the culture is aerated with O_2-poor air. If aeration of a growing culture is stopped entirely, the production of purple membranes increases some 50 times. Another interesting finding is that the cells synthesize ATP under anaerobic conditions in the light but fail to do so in the dark. Before describing the generally accepted explanation and an alternative interpretation to account for these interesting observations, related experiments using model membranes which contain bacteriorhodopsin should be given.

Attempts to solubilize bacteriorhodopsin (BR) with the use of common surfactants has generally resulted either in denaturation or in the formation of micelles. BR exists in two forms, dark-adapted pigment with λ_{max} at 558 nm and light-adapted pigment with λ_{max} at 568 nm. Photochemistry of BR has been investigated by Tokunaga et al. (1976), Kung et al. (1975), and by Dencher et al. (1976). So far only purple membranes (or their derivatives) have been incorporated into experimental bilayer lipid membranes. BR-containing liposomes showed light-induced alkalinization of the medium, which was reversed in the dark and inhibited by uncouplers such as 2,4-dinitrophenol (Racker and Hinkle, 1974). Just the opposite occurred in intact cells. This has been explained by the presumption that purple membrane pathces were preferentially oriented inside-out in the liposomes. Eisenbach et al. (1977) have found that light-induced proton release and uptake by BR-containing liposomes show biphasic kinetics. Furthermore they have suggested that the biphasic proton movement results from a combination of a faster process of association (or dissociation) and a slower process which reflects transport across the lipid bilayer. In either intact cells or BR-containing liposomes, it is difficult if not impossible to measure membrane potentials. Therefore, incorporation of BR into planar membranes have been carried out for potential measurements by several investigators (Shieh and Packer, 1976; Dancshazy and Karvaly, 1976). Dancshazy and Karvaly have made the first successful attempt to incorporate the protein—pigment complex of *H. halobium* into bilayer structure and to observe some photoelectric processes of the cell membrane of *H. halobium*. The significant difference between chlorophyll-BLM and a reconstituted Halobacterium membrane is that while the former required the incorporation of relatively simple hydrophobic pigments into a bilayer, the latter aimed at the embedding of protein—pigment complexes into, or their binding with, a lipid bilayer structure, in an operationally active state. The procedure for incorporating of BR has been worked out by Dancshazy and Karvaly (1976).

Karvaly and Dancshazy (1977) have observed both photovoltaic effect and photoconductivity in BR-containing BLM. The chamber on the pigment side was electrically negative in the light. Additionally, they have obtained photovoltaic action spectra of the pigmented BLM, which revealed some differences from the absorption spectrum of BR in aqueous solution. An interesting feature of the photovoltaic response of BR-BLMs is that at sufficiently high light intensities (but far below the photoelectric saturation) the photo-electromotive force is less than the sum of the photopotentials under 500 and 554 nm excitations of considerably lower quantum flux. (The monochromatoc exciting light beams were cut out from the white-light by using metal interference filters). The photovoltaic action spectra of BLMs with protein—pigment complexes only roughly follow the absorption spectra of corresponding membrane fragments in aqueous solution and reveal marked systematic deviations in several wavelength regions. In the case of BR-BLM, the photovoltaic action spectrum matches the absorption spectrum of BR in the long-wave region fairly well; deviations appear at 486 nm, 512 nm and 547 nm. These differences coincide with the absorption bands of carotenoids indicating a possible role that the carotenoids may play in shielding or quenching the photoelectric process. Packer et al. (1977) have studied a modified BLM fused with BR-containing liposomes as well as the usual BR-BLM. In all cases, they have found that the presence of BR generates large and stable photopotential. Using a related system, Blok et al. (1977) have incorporated BR into lipid-impregnated Millipore filters and reported that photovoltages up to 215 mV were obtained.

Returning now to the mode of operation in intact cells of *H. halobium*, the generally accepted explanation which is given in more detail in chapter 6 is that bacteriorhodopsin functions as a proton pump translocating H^+ ions from the interior (cytoplasm) to the outer medium and thus generates an electrochemical gradient of protons (Oesterhelt and Stoeckenius, 1973; Oesterhelt, 1976; Renthal and Lanyi, 1976; Danon and Caplan, 1976). The energy stored in this gradient across the plasma membrane may then be used for ATP synthesis according to the chemiosmotic hypothesis of energy coupling (Mitchell, 1968). Although such an explanation is plausible and persuasive, it is by no means compelling. One of the most troublesome questions that has been raised (Tien, 1977b) is that the direction of H^+ movement. If it is granted that the primary function of bacteriorhodopsin of the purple membrane is to pump protons, it would seem to make more sense to move protons from the bathing medium to the cell interior rather than in the opposite direction as proposed. This can be easily seen from an energetic viewpoint, since it is definitely more efficient to acidify the cytoplasm than the bathing medium, for the interior volume of the cell is negligibly small as compared with the external volume of the bathing solution. If so, this interpretation is in apparent contradiction with the observed facts. However, a closer examination of experimental data reveals that upon illumination of an anaerobic cell suspension, a fast *rise* in pH of about 0.1 pH unit is first observed. This rise (alkalinization) is followed by a larger, slower drop in pH (acidification) which has been always reported as evidence for proton pumping (Oesterhelt and Stoeckenius, 1973; Bogomolni and Stoeckenius, 1974; Danon and Caplan, 1976). These findings are reminiscent of the observation on chloroplast thylakoids (Jagendorf and Hind, 1963; Dilley, 1972), whereupon switching on the light an alkalinization of the medium initially resulted. Furthermore, ATP synthesis by outward proton movement via the thylakoid membrane was demonstrated by the acid—base experiment (Jagendorf, 1977), which has been often cited as evidence in support of Mitchell's hypothesis. In the case of the purple membrane, as with the thylakoid membrane, the pigment is the key component in the energy transducer which appears to use a similar mechanism. That is, light absorbed by the pigment embedded in the lipid bilayer leading eventually to a proton gradient across the membrane with the inside of the cell containing more protons. In addition, in the thylakoids it is generally assumed (Junge et al., 1977; Diner and Joloit, 1977) that water oxidation results in the release of four protons per O_2 evolved also taking place at the interior side of the membrane (see eq. 13). Is there any evidence for oxygen evolution in *H. halobium*? It has been reported (Oesterhelt and Stoeckenius, 1973) that exposing the cells to light greatly reduced respiration as measured by oxygen consumption. This, however, may be interpreted to mean that oxygen production occurs as a result of water photolysis similar to that occurring in the chloroplast thylakoids (see section 4.6 on Mechanisms of light transduction). The protons, generated by pumping and/or by water oxidation, flow outwardly through the phosphorylating enzyme, in accordance with Mitchell's hypothesis, to acidify the medium, thereby giving rise to the observed pH change and ATP production (see Fig. 4.18).

The proposed scheme applies equally well to *H. halobium* in the dark. It has been reported (Oesterhelt and Stoeckenius, 1973) that the cells in the dark synthesize ATP when oxygen is available. If the same phosphorylating enzyme is used, as there is no reason to doubt otherwise, the oxygen must be used instead of light to pump protons *inwardly* and thus generate the necessary electrochemical potential gradient. In essence, it seems highly probable that *H. halobium* behaves similar to a mitochondrion in the dark (oxidative phosphorylation) and like a chloroplast in the light (photophosphorylation). Since the cells of *H. halobium* do not possess outer membranes (or envelopes), the electrochemical potential gradient of protons, created by energy transduction (light or substrate oxidation), must be outwardly directed, in order to be consistent with Mitchell's chemiosmotic theory.

In sum, it appears that the purple membrane employs mechanisms not unlike those of the thylakoid membrane and the cristae membrane. The interpretation given here differs from the generally accepted explanation in three respects: (i) if bacteriorhodopsin acts as a light-driven proton pump, it moves protons inwardly to generate an electrochemical gradient for phosphorylation; it is far more efficient to acidify the inside of the cell than that of the ocean comprising of the bathing medium, (ii) in the light, the respiration rate as measured by O_2 consumption is not inhibited but rather O_2 is evolved, and (iii) substrate oxidation in the dark should initially make the cell interior more acidic, if the same phosphoryling enzyme is used for ATP synthesis. If (ii) is true, one should be able to detect oxygen production when the cells are illuminated. Such experiments are currently underway in our laboratory.

The successful incorporation of BR into BLM and the photoelectric-spectroscopic investigations on them confirm that BR is really operating as a high-efficiency photoelectric transducer. The method used may give impetus to incorporate charged proteins into lipid bilayers and may make even broader the application of the BLM technique in studying the function of proteins in membrane processes as well (Tien and Mountz, 1975). The data on the wavelength dependence of photovoltaic responses clearly demonstrate the unique power of photoelectric spectroscopy in studying photobiological phenomena at the molecular level (Van and Tien, 1970). Although these studies are just beginning, the results available to data are very promising for the future of model investigations on visual processes, photophosphorylation, photoelectric conversion, etc.

4.5. PROPERTIES OF PIGMENTED LIPOSOMES IN THE LIGHT

Concurrent with the observation of photoelectric effect in Chl-containing BLM (Tien, 1968), Chapman and Fast (1968) reported photochemical studies of Chl-containing lipid microvesicles or liposomes. The lipid microvesicles are much more stable as compared with planar BLM; they can be made in large quantity and hence are amenable to study by electrochemical techniques. For examples, attempts to detect light-induced pH changes and oxygen evolution have been carried out with these pigmented liposomes (Tien, 1971, 1972). In this section we shall summarize the experiments of our laboratory and review relevant papers on Chl-containing liposomes (Tomkiewicz and Corker, 1975; Oettmeier et al., 1976; Mangel, 1976b). References to earlier work may be found in a recently published paper (Stillwell and Tien, 1977).

4.5.1. Chl-containing liposomes

It is well known that the absorption maximum of Chl *a* in acetone is at 663 nm compared with 678 nm in vivo (Vernon and Seely, 1966). This difference in absorption maxima has been attributed to solvent effects, state of aggregation of the tetrapyrole moieties or specific interactions with lipids and proteins (Boardman, 1968). Initially, we were interested in finding out the reason(s) for the differences in chlorophyll absorption spectra and photochemical activities in vivo and in organic solvents (Beall and Tien, unpublished study, 1969—1970). It was reasoned that the liposomes made from plant lipids and chlorophyll could be used in absorption studies to give an absorption spectrum closely similar to that in vivo, if the postulated structure of the thylakoid membrane were correct. The reasoning was that the lipid bilayer of the liposomes would provide a favorable site so that the phytol chain of the chlorophyll would anchor while the hydrophilic porphyrin ring would orient at the membrane/solution interface (Ting et al., 1968). With these thoughts in mind, the following experiments were performed. The Chl-containing liposomes were prepared by sonication of a lipid—chlorophyll mixture for 45 min in ice-cold 0.1 M phosphate buffer at pH 7.5. The sonicated mixture was passed through a Sephadex G-50 column previously equilibrium with the same buffer. The liposomes collected were those contained in the first two ml of the fastest green band to pass down the column. The absorption spectrum was taken of the Chl-liposomes and the peak was at 673 nm (Beall and Tien, unpublished study, 1969—1970). The results strongly suggested that the Chl in liposomes was in an environment more similar to that in vivo than when it was simply dissolved in organic solvents. More recently, in an elegant study Oettmeier et al. (1976) have reported that liposomes containing Chl a exhibit an absorption maximum at 670 nm indicating the presence of monomeric Chl. This is further supported by the ESR spectrum of Chl^+ free radical ($\Delta H = 9.1$ G and a g-value of 2.0022). Earlier, Tomkiewicz and Corker (1975) demonstrated that the photochemistry of Chl in liposomes is different from the photochemistry in homogenous solution and reported a g-value of 2.0026 ± 0.0003 with a line width of 9.0 ± 0.5 G. On the basis of their results Tomkiewicz and Corker have concluded that the lipid/water interface is unique in that it imparts some stability to the photochemical charge separation.

Besides confirming a 10-nm red shift of the Soret band with respect to organic solvents in Chl-liposomes, Mangel (1976b) has reported the appearance of absorbance peaks beyond 700 nm, implying the presence of Chl aggregates (Tien, 1968; Csorba et al., 1975). Moreover, Mangel has shown that liposomes containing Chl and β-carotene are capable of photo-induced electron transfer in the presence of a redox gradient, consistent with the PE effects in Chl-BLM as already discussed.

Much earlier, Sahu and Tien (unpublished results, 1971—1972; Tien, 1972) attempted to test the idea of water splitting by light by incorporating photosynthetic pigments into liposomes, very much as has been done for the planar BLM. Preliminary experiment using a Clark type electrode showed some O_2 evolution from Chl-containing liposomes. However, later experiments with better temperature control indicated otherwise. The apparent oxygen evolution as sensed by the electrode was due to heating effects on the electrode itself. Our failure notwithstanding, Toyoshima et al. (1977), using a similar system, reported light-assisted oxidation of water in chlorophyll-containing liposomes. The rate of oxygen production, as calculated from the initial slope, was 4.2×10^{-4} per mol at 10^5 lux. Under similar conditions as those reported by Toyoshima et al. but employing rigid temperature controls (within $\pm 0.2°C$), however, we observed oxygen *consumption* instead, as the pigments in the liposomes became irreversibly photooxidized (Stillwell and Tien, 1977). What follows is a description of the behavior of Chl-containing liposomes in the light in the presence of oxygen and a variety of compounds.

Chl-containing liposomes were prepared and studied by the procedure

152

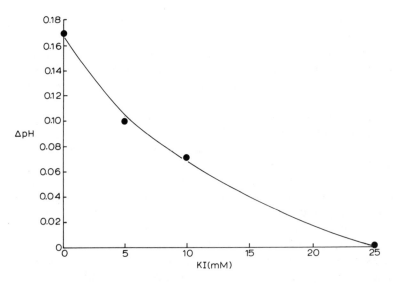

Fig. 4.11. Inhibition of acidification by KI when chlorophyll-liposomes were irradiated with white light. Liposomes were made in 2 mM tris, 2 mM potassium carbonate, 0.01 mM $MnCl_2$, 150 mM KCl, pH 7.5 buffer. Two ml of the stock microvesicles were added to 58 ml of buffer containing 2 mM tris, 2 mM potassium carbonate, 0.01 mM $MnCl_2$, pH 7.5 to which was added enough KCl and KI to bring the final salt concentration to (A) 150 mM KCl, 0 mM KI, (B) 145 mM KCl, 5 mM KI, (C) 140 mM KCl, 10 mM KI and finally (D) 125 mM KCl, 25 mM KI. Additional KI did not change the ΔpH (Stillwell and Tien, 1977).

described in the Methodology section. Upon illumination with white light, under aerobic conditions, a slight acidification of the Chl-containing microvesicle suspension was noted. This ΔpH was measured at pH 7.5, 6.25 and 5.0 in both unbuffered solutions and solutions slightly buffered with tris, potassium phosphate and potassium acetate, respectively. Potassium iodide was found to inhibit this acidification shown in Fig. 4.11. A very large acidification (about 4 times that of the unmodified chlorophyll-lipid microvesicles) was noted at pH 7.5 in the presence of ascorbic acid. This large ΔpH was not inhibited at all by potassium iodide and appears to be a different reaction than the one producing a small ΔpH in the absence of ascorbic acid. No acidification could be measured when the Chl-containing microvesicles were kept in the dark for several hours or were sealed under nitrogen and illuminated. Also, the egg lecithin microvesicles without pigments resulted in no acidification upon illumination. From these experiments it was concluded that photooxidation of the chlorophyll-lipid microvesicles resulted in production of protons. The reaction depended on the presence of air, light and chlorophyll (Stillwell and Tien, 1977).

Upon illumination with white light the Chl-containing microvesicles rapidly bleached from green to yellow. During the bleaching, the chlorophyll

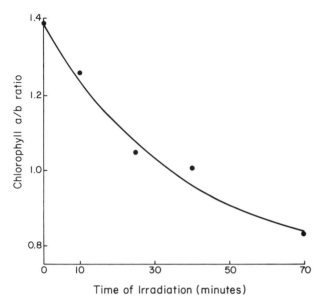

Fig. 4.12. Chlorophyll *a/b* ratio when chlorophyll-lipid microvesicles (liposomes) were illuminated with white light. Microvesicles were made in 100 mM KCl, 10 mM tris, pH 7.5 buffer and contained initially 40 mg of egg lecithin and 1.3 mg of chlorophyll in the 60 ml sample.

a to chlorophyll *b* ratio continuously decreased as shown in Fig. 4.12. The photobleaching could not be reversed by the addition of ascorbate. These microvesicles at pH 7.5 were more resistant to oxidation than were identical chlorophyll-lipid microvesicles at pH 5.0. At pH 7.5 they remained very much the same in terms of oxygen consumption, even after aging 23.5 hours in the dark, while at pH 5.0 a noticeable change occurred in 3.4 hours.

With pigmented liposomes, the so-called Krasnovsky and Mehler reactions using various electron donors and oxygen as the electron acceptor have been demonstrated. A number of electron donors used by Krasnovsky (1969) were tested on Chl-containing liposomes. The rate of oxygen reduction in the presence of these donors resulted in the following series: control (no added electron donor) = KI = ferrocyanide < hydroquinone < thiourea < cysteine < NADH < Fe^{+2} < ascorbic acid < phenylhydrazine.

The initial objective of the experiments was to demonstrate photo-assisted oxidation of water in Chl-containing liposomes (see Eq. 13). However, from the results presented above, it is obvious that several complications have arisen in the undertaking. The Chl-containing liposomes readily *consume* oxygen. In addition, the pigments are not resistant to bleaching degrading rapidly in the light. It is possible that oxygen is produced in this system but can not be detected due to the rapid bleaching and Mehler-type reaction

(Mehler, 1951). It appears that before oxygen evolution can be detected in the pigmented microvesicles either some control on the rate of pigment photobleaching and oxygen uptake must be accomplished, or a different approach is necessary such as by incorporating oxygen evolving complexes into liposomes.

4.5.2. Incorporation of oxygen-evolving complexes

As described above, Chl-containing liposomes under the condition tested did not evolve oxygen but rather consumed oxygen at a rapid rate, in contrast to the results of similar experiments reported by Toyoshima et al. (1977). Since Toyoshima et al. did not employ any temperature controls, it seems possible that their oxygen was in reality an artifactual heating effect on the oxygen electrode. We now believe that oxygen evolution from simple chlorophyll-phospholipid vesicles may be difficult unless high light intensity and/or suitable redox compounds, yet to be discovered, are used or the oxygen evolving complexes (OEC) isolated from chloroplasts are incorporated into the lipid bilayer separating two aqueous phases (e.g., liposomes). In pursuit of the latter approach, a new series of experiments have been undertaken (Stillwell and Tien, to be published), which are designed to answer the following questions:
(1) Is a sealed membrane, topologically separating two aqueous solutions, a requirement for O_2 evolution? and
(2) Does the reconstituted system of OEC-containing liposomes respond to compounds known to alter O_2 production in vivo?
The first question stems from the belief, coupled with some evidence to support it, that the thylakoid membrane, in the form of a sealed vesicle, is an indispensable structure for O_2 evolution and photophosphorylation (Gregory, 1977; Jagendorf, 1977). Indeed, it has been found that osmotically shocked or broken thylakoids did not readily generate O_2 (Stillwell and Tien, 1978). If, however, these broken thylakoids were sonicated, some of the O_2 evolving ability was revived (Fig. 4.13). The large increase in O_2 evolving capacity upon sonication is attributed to the formation of resealed vesicles from broken thylakoids, even though the reconstituted vesicles were not stable and their ability to evolve O_2 diminished within about two hours. In contrast to the above findings, broken thylakoids when fused with phosphatidyl choline—liposomes generated not only much larger quantity of O_2 but also sustained the ability of oxygen evolution for 7 hours or more.

Experimentally, two methods of incorporation of oxygen evolving complexes (OEC) from broken thylakoids were used. In the first method, the OEC were added to preformed liposomes. Fifty ml of the broken thylakoids (containing OEC) were mixed with 10-ml liposomes and the mixture was tested for O_2. The second method consisted of adding the OEC to a flask

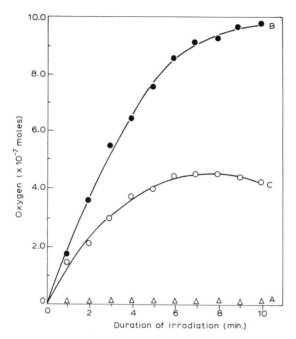

Fig. 4.13. Evolution of oxygen during each step in the preparation of broken chloroplast thylakoids; (A) After blending in water, (B) After sonication, and (C) After centrifugation. Illumination was with white light at 1.7×10^5 ergs cm^{-2} sec^{-1}.

containing 0.3 g dried phospholipids and the complexes were directly incorporated into the lipid bilayer in the process of liposome formation. The temperature of the reaction chamber was controlled at $25°C \pm 0.2$ by a constant temperature circulating bath. The pH was determined with an accuracy of 0.002 pH units by a Beckman pH meter (Expandomatic Model). An Oxygen Monitor (Model 53, Yellow Springs Instrument Co.) was used for O_2 measurements.

Fig. 4.14 shows the ability of O_2 evolution by the broken thylakoids fused with liposomes as a function of time. The "hump" in the curve is interpreted to mean the incorporation of OEC into liposomes. As the fusion progressed, the ability of the system to evolve O_2 increased initially. After that, the "normal" decrease of O_2 evolution, presumably due to the degradation of the OEC, became predominant. The problem of not measuring a true time zero was overcome by breaking the thylakoids directly in the reaction chamber. The thylakoids were ruptured and resealed in high ionic strength buffer and were then injected into a buffer of low ionic strength containing benzoquinone. As the high ionic strength, sealed vesicles entered the low ionic strength buffer they swelled and burst. This was measured by a

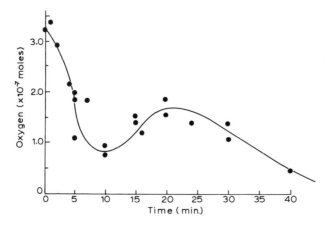

Fig. 4.14. Oxygen evolution when thylakoid vesicles made in high ionic strength buffer were injected into a low ionic strength, oxygen free buffer containing 0.2 mM benzo-quinone, 10 mM potassium phosphate, 5 mM KCl, pH 7.5 at 25°C. The breaking and resealing of the vesicles in the reaction chamber was followed by oxygen generation.

decrease in oxygen evolution for the first 10 min. However, the broken thylakoids would spontaneously reseal which resulted in an increase in oxygen evolution for about 20 min. After that time, oxygen evolution decreased as the thylakoids aged.

The oxygen evolving complexes are known to be sensitive to a varity of compounds. Accordingly, we have tried chloride, bicarbonate and manganese on the reconstituted system (see chapter 8). Fig. 4.15 shows effect of 0 to 5 mM quantities of these compounds on oxygen evolution in the fused liposome—thylakoids. Chloride increased oxygen evolution throughout the range and even was an effective stimulant at 25 mM. Mn^{2+} stimulated most at about 2 mM. Above this concentration oxygen evolution was greatly inhibited as the phospholipids of the liposomes congealed. Bicarbonate had very little effect on oxygen evolution (curve C, Fig. 4.15).

In Table 3 compounds known to inhibit in vivo oxygen generation were

TABLE 4.3.

THE EFFECT OF COMPOUNDS WHICH INHIBIT OXYGEN EVOLUTION ON THE FUSED LIPOSOME—THYLAKOID.

Compound	Type	Concentration for 50% inhibition of O_2 (mM)
DCMU	Inhibitor of PS II	0.0001
Hydroxylamine	Poison (removes Mn^{2+}?)	0.64
2,4-Dinitrophenol	Uncoupler	0.98
KSCN	Chaotropic agent	19.8

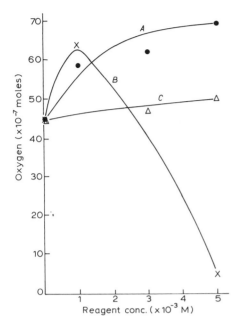

Fig. 4.15. The effect of compounds which stimulate oxygen evolution on the fused liposome-thylakoids: (A) KCl, (B) Mn^{2+}, (C) HCO$_3^-$.

tested on the liposome—thylakoid system. DCMU, the powerful inhibitor of Photosystem II (Joliot and Kok, 1975), caused a 50% reduction in oxygen at 10^{-4} mM. Hydroxylamine and 2,4-dinitrophenol inhibited oxygen in the millimolar range. KSCN, a nonspecific chaotropic agent, inhibited oxygen at about 20 mM. From the results of Table 4.3 and Fig. 4.15, it has been suggested that liposome—thylakoid complexes behave similar to intact thylakoids in relation to the effect of stimulators and inhibitors of oxygen evolution (Stillwell and Tien, 1978).

In concluding this section, it is worth noting that, due to the enhanced stability of the oxygen evolving complexes when incorporated into liposomes, the method described may be used for isolating and characterizing this very elusive and unstable entity (oxygen evolving complexes) associated with Photosystem II.

4.6. MECHANISMS OF LIGHT TRANSDUCTION

In the first report describing the BLM photoelectric effect, three possible mechanisms were proposed. Categorically, these mechanisms can be summarized as follows. (1) Redox electrode mechanism—electronic charges are generated as a result of light excitation which cause one side of the BLM biface reducing and the other side oxidizing. (2) Charge injection mechanism—electronic charges are injected into the BLM as a result of light excitation as in the case of anthracene crystals. (3) Permeability mechanism—absorption of light by the BLM causes a permeability change to protons and other ions; formation of photoproducts some of which are mobile charge carriers (e.g., protons). Insofar as Chl-containing BLMs are concerned, the most plausible mechanism is based on the redox electrode model in which the oxidation and reduction of chlorophylls are involved (Tien, 1968). The model rests on the following assumptions and factors.

(1) For the appearance of a BLM photoeffect, the chlorophyll and accessory pigments reponsible must be present either in or adjacent to the membrane/solution interface.

(2) Upon absorption of light, the Chl is in an excited state which, for the observed photoeffects, must generate charge carriers by direct electron emission, by exciton dissociation, or both.

(3) The dissociation of excitons is influenced by the external field, electron donors and acceptors; a greater number of electrons will be captured on the side containing more powerful electron acceptors.

(4) The charge carriers thus generated may travel a considerable distance with a velocity determined by the wavelength of the light absorbed.

(5) The concentration of added modifiers (electron acceptors and donors) is assumed to be much greater than that of the photoactive species in the membrane.

(6) A photocurrent is produced when either positive holes (or ion-radical) move from one side of the bilayer to the other side, or a change occurs in the rate of accumulation of charge on the membrane capacitance (capacitance current).

(7) Generation of the BLM photoeffect requires some asymmetrical conditions, such as the presence of added modifiers, a difference in light intensity, or a gradient of hydrostatic pressure, temperature, or electrical potential.

Before explaining the redox electrode hypothesis, one may recall the usual picture of a BLM separating two electrolyte solutions consisting of a liquid hydrocarbon core (~50 Å) sandwiched between two polar regions which contain the sites for ion binding. These sites, if charged, determine the interfacial potentials, and constitute one of the two major factors governing the charge transport, the other factor being the hydrophobic layer. When electrons and holes are generated in the BLM, they will polarize the surroundings and attract charges of opposite sign across the lipid bilayer. These charges will, in turn, attract the electrons (or holes) toward the oppositely charged interface, resulting in the reduction of the interfacial potential. Similarly, it can be shown that the interfacial barrier for charge transport can also be reduced by an externally applied voltage across the BLM.

When a Chl-BLM separating two aqueous solutions containing an electron

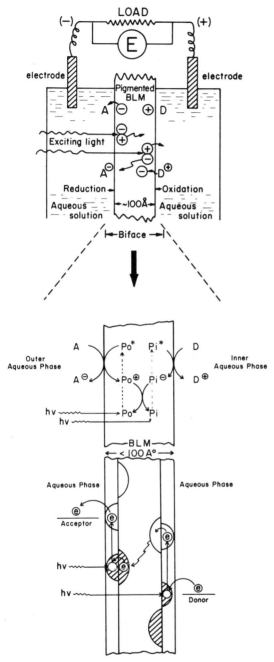

Fig. 4.16. Photoelectric bilayer lipid membrane. The BLM system is depicted as similar to that of a p—n junction solar cell. Upper: Schematic illustration of the basic electronic processes in a photoactive BLM separating two aqueous solutions (Tien and Verma, 1970). Lower: Mechanisms of light transduction and charge transfer in pigmented BLM. P_0 and P_i denote pigment molecules adjacent to the outer and inner solution, respectively. P^*, excited pigment molecule after absorption of a photon ($h\nu$); P^+, oxidized pigment molecule; A, electron acceptor; A^-, reduced electron acceptor; D, electron donor; D^+, oxidized electron donor; $\bigcirc$, electron; ⊜, hole (see text for details).

acceptor on one side and an electron donor on the other side is illuminated, a photovoltage and/or a photocurrent can be generated. In the redox-electrode model diagrammed in Fig. 4.16, two solution/BLM interfaces act as electrodes taking part in coupled redox reactions initiated after the absorption of a quantum by a chlorophyll (Tien and Verma, 1970). Mechanistically, the scheme can be described by the following equations:

$$Chl + h\nu \rightarrow Chl^*, \tag{6}$$

$$Chl^* + A \rightarrow Chl^+ + A^-, \tag{7}$$

$$Chl^+ + D \rightarrow Chl + D^+, \tag{8}$$

where Chl^* is in the excited singlet or triplet state, Chl^+ is the oxidized ion radical of the pigment, A and D are the electron acceptor and donor, and A^- and D^+ are their reduced and oxidized equivalents. An analogous case starting with the donor may be written as follows:

$$Chl + h\nu \rightarrow Chl^*, \tag{9}$$

$$Chl^* + D \rightarrow Chl^- + D^+, \tag{10}$$

$$Chl^- + A \rightarrow Chl + A^-. \tag{11}$$

Although eqs. (6) to (8) and eqs. (9) to (11) have the same net result, namely,

$$A + D \xrightarrow[Chl]{h\nu} A^- + D^+, \tag{12}$$

writing them out stresses that the chlorophyll can function both as electron donor and acceptor depending on the relative strength of redox compounds present in the aqueous solutions. If D, shown in Fig. 4.16 and in eqs. (8) and (10), is water, then

$$4 Chl^+ + 2 H_2O \xrightarrow{4 h\nu} 4 Chl + 4 H^+ + O_2 \tag{13}$$

If an ion radical or a positive hole are produced, as indicated in the scheme, water could conceivably be oxidized to H^+ and oxygen (Tien, 1968). The protons thus generated would influence the observed photoelectric effects; for instance, they should migrate towards the negative side and cause a depolarization of light-induced photopotential. The result shown in Fig. 4.5A where no powerful donor other than water was present may be interpreted by the mechanism just given in which the decrease of observed photovoltage as a function of time could be due to the translocation of photogenerated protons across the membrane.

It seems certain that the absorption of light by the Chl-BLM leading eventually to the observed photoelectric effects results initially in the formation of excitons. The excitons once created by optical absorption are assumed to be able to move throughout the whole membrane which is considered as an ultrathin layer of liquid crystals (Tien, 1967). If so, the energetics of light-induced PE process may be described in terms of the band theory of solids. An excellent example would be the semiconductor/electrolyte photovoltaic cell, upon which the redox electrode model is ultimately based. To a casual reader the connection between the pigmented membrane system and semiconductor photovoltaic cell might seem far-fetched. Therefore, a brief digression on semiconduction and biology is in order.

Originally, the idea that energy in the chloroplast grana might be transfer by conduction bands was suggested by Szent-Györgyi (1941). At present, the existence of mobile charge carriers in the form of electrons and holes in films of dried chloroplasts and chromatophores seem established (Bogomolni and Klein, 1975). During the intervening years, the suggestion that photosynthesis is an electronic process has been taken up and extended by several investigators (for reviews and references see Tien, 1972, 1974). Among the early proponents are two proposals put forth by E. Katz (1949). According to Katz, two types of process might occur in chlorophyll when a photon is absorbed. In the first type of process, the formation of an exciton is suggested. An excited chlorophyll can transfer its excitation energy to its neighbour and subsequently return to the ground state. In this scheme, the state of excitation may wonder through the two-dimensional crystal by repeated transfer. The exciton is not localized at any instant at a given chlorophyll molecule but is spread out over the entire structure. In the manner described, energy is transferred by migration of an exciton (electron-hole pair) without separation of charges. The second process assumes that the two-dimensional chlorophyll arrays are photoconductive. The absorption of a photon promotes one electron into the conduction band and leaves one "hole" in the valence band. The light-generated electrons and holes are free to move around. They are also considered to be spread out over the whole "unit" although their lifetimes and mobilities may be different. In the second scheme, it is assumed that the electrons can be transferred to an electron acceptor which is thereby effectively reduced, and the holes can be transferred to an electron donor which is thereby effectively oxidized. The obvious advantage of the second process is that it depicts how the light absorbed by the chlorophyll can give rise to a reductant (electron) and an oxidant (hole) both of which are mobile and will move in their own way thereby carrying out reduction and oxidation at different sites. Fig. 4.16 (lower portion) shows such a scheme in which the energy levels of the electron acceptor and donor are assumed to lie, respectively, below the excited state and above the ground state of the photoactive species in the BLM.

Upon the absorption of a quantum by the photoactive pigment, an excited electron whose energy level is above the acceptor level is produced, thereby permitting a thermodynamically favored electron transfer. Meanwhile, since the level of the electron donor is higher than that of the vacant site, a transfer of an electron is also possible from the donor as illustrated in Fig. 4.16. The end result is, of course, a reduced oxidizing agent and an oxidized reductant as has been summarized in eqs. (11) and (8), respectively.

The next task is to provide some answers to the three most important questions: (a) the mechanism of charge generation in the membrane upon exposure to light, (b) the mechanism of charge separation, and (c) the mechanism of charge transport. If the final process is energy storage, an additional mechanism must be provided for affecting redox reactions in the bulk solutions bathing the membrane.

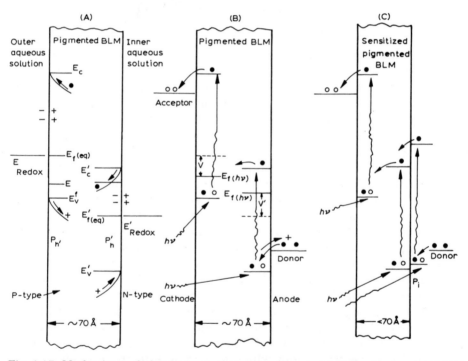

Fig. 4.17. Mechanisms of charge generation, separation and transfer in pigmented bilayer lipid membranes (BLM). The BLM is considered to be liquid—crystalline in structure. A pigmented BLM/redox solution interface is likened to that of a semiconductor/electrolyte interface. Notations in the figure are: E_c, conduction band; E_F, Fermi level; $E_{F(eq)}$, Fermi level at equilibrium; E_{redox}, redox potential energy level; E_v, valence band; P_h, pigments in the hydrophobic phase at the outer interface. The corresponding symbols for the inner phase are primed ($'$). (A) A pigmented BLM separating two redox solutions in the dark, (B) Same as in (A) but under illumination, (C) A sensitized pigmented BLM in the light, P_i stands for a water soluble pigment (or dye) adsorbed at the interface.

The origin of the photopotential observed in a pigmented BLM rests on the following assumptions.

(1) The BLM is considered liquid crystalline; it is capable of electronic conduction.

(2) The pigmented BLM is treated as a bipolar semiconductor electrodes; one side is oxidizing (n-type), the other reducing (p-type).

(3) The density of electron states in a BLM can be described with the band model of solids.

(4) The energy terms in redox solutions are related to the same reference state, namely the free electron in vacuum.

The left-hand side of Fig. 4.17A shows the energy levels of a "p-type" BLM/solution interface containing an electron acceptor whose redox potential is denoted by E_{redox}. At equilibrium in the dark, the electrochemical potential ($\bar{\mu}$) between the two phases are equal, i.e., E_{redox} is equal to the Fermi energy in the membrane. Both energy terms are related to the energy level of an electron in vacuo at infinity (assumption #4) and is given by

$$E_{redox}(\text{volts}) = E_F + 4.5 \qquad (14)$$

where 4.5 (in eV) refers to the energy level for the normal standard hydrogen electrode. Analogously, the energy levels of an "n-type" BLM/solution interface, shown in the right-hand side of Fig. 4.17A, can be similarly defined. In this case E'_{redox} must lie below E'_F of the membrane to form a space charge layer. Thus, in the language of solid-state physics, a Schottky-type barrier or p/n junction is assumed to exist at each interface (Myamlin and Pleskow, 1967). Absorption of light with energy greater than E_v-E_c by the pigments in the membrane creates excitons which are subsequently separated because of the potential difference across the interface. The separated charges move in the opposite direction. Specifically, the p-type membrane/solution interface becomes a photocathode with electrons moving towards the interface to the bulk solution leading to a reduction reaction and holes towards the interior of the membrane, as shown in Fig. 4.17B. In order for the reduction to occur, electrons can only be transferred into the aqueous solution if an acceptor energy level exists in the solution which is below or equal to E_c at the interface. As a result of the charge movement which disturbs the equilibrium at the interface, $E_{F(eq)}$ is no longer equal to E_{redox} but moves downward to a new position at $E_{F(h\nu)}$. As shown in the left-hand side of Fig. 4.17B, the observable photovoltage is indicated by V with its maximum value equal to $E_{F(h\nu)}-E_{redox}$. For the interface at the right-hand side, which is an n-type and acts as a photoanode, the same description holds apart from substituting holes for electrons, changing the direction of movements, having an oxidation reaction instead, etc. Since there are two interfaces, the overall maximum photovoltage observable would be the sum of V + V' as shown in Fig. 4.17B.

To carry the above scheme one step further, pigments with different spectral characteristics should be associated with each interface (or environment) for more efficient light utilization (also see argument presented in chapter 1). This can be easily done with Chl-BLM, by adding water-soluble chlorophyllin to one side of the bathing solution. The result is shown in Fig. 4.5C. Clearly the sensitivity of the photoelectric BLM has been extended. Finally to account for the marked increase in efficiency of the electron transfer, the decrease in distance that the charge carriers have to move is suggested. This is shown in Fig. 4.17C in the form of reduced membrane thickness. As a result of sensitization, both the height and the width of the energy barriers are lowered. This explanation can be supported by the further enhancement of the photovoltage when the sensitized pigmented BLM was under anodic bias as shown in Fig. 4.5D.

The question as to the mechanism by which charge carriers move through the lipid bilayer, is partially answered by the nature of postulated carriers (electrons and holes). The most plausible mechanism is the quantum-mechanical tunneling assuming the height of the insulating lipid bilayer to be 1 eV (see chapter 3 and also Gutmann, 1968; Hopfield, 1974). For practical considerations at room temperature such a high energy barrier reduces the probability of the electron jumping over to a negligible value. It is noteworthy that carotenoids (e.g., β-carotene and xanthophylls) with a delocalized π-electron system might act as electronic conductors across the lipid bilayer, which are abundantly present in the photosynthetic membranes and in the extracted chloroplast pigments used in the pigmented BLM (Tien, 1968).

4.7. RELATION TO PHOTOSYNTHESIS

Biophysical chemical studies have suggested that the thylakoid membrane of the chloroplast consists of an ultrathin layer of lipids (presumably a lipid bilayer) with sorbed proteins and pigments organized in a lamellar structure about 100 Å thick. Although the precise functions of the thylakoid membrane are still obscure, it is believed that the membrane is the locus of the primary photophysical and photochemical processes. For examples, light-induced potential differences have been observed in green cells by Nagai and Tazawa (1962), Nishizaki (1963), and Barber (1968). Luttge and Pallaghy (1969) detected light-induced electrical transients in green cells related to photosynthetic electron transport. More recently, using ultrafine tipped microelectrodes, Bulychev et al. (1976) have found two types of photo-induced potential changes in isolated chloroplasts. These recent findings are the most exciting developments in membrane plant physiology, if the electrodes were indeed successfully placed transversely across the thylakoid membrane (see Fig. 4.1). However, physico-chemical characterization of the thylakoid membrane at the molecular level is still very difficult owing to the complexity and minuteness of the thylakoid. In addition, from a biophysical point of view, the use of isolated thylakoids to understand the membrane aspects of photosynthetic processes at the molecular level is not adequate. For example, to test Mitchell's hypothesis, the measurement of the translocation of ions, such a protons across a thylakoid, is obviously complex, if not impossible. The interactions between various fluxes (electron and hole transport, ion and water movements, etc.), and their conjugate forces (electrical potential and osmotic pressure, etc.) across the thylakoid membrane, are far too intricate to be amenable to a detailed analysis. Therefore, in my opinion an understanding of the thylakoid membrane would be best approached by studies of simpler, well defined model systems such as Chl-containing planar BLM and liposomes as described in the preceding sections.

Recalling now the three outstanding problems of photosynthesis (quantum conversion, photophosphorylation, and oxygen evolution) listed in the Introduction, Mitchell's suggestion (1966; 1968) appears to be the most predictive and the best working hypothesis that can be tested experimentally. Fig. 4.18 depicts the primary processes of photosynthesis that might take place across a pigmented bilayer lipid membrane. As suggested by Mitchell (1966; 1968), the light-initiated charge separation is vectorial in the membrane and thus generates an electrochemical gradient of protons ($\Delta\mu_H$). In terms of membrane electrochemistry, the free energy difference per unit area is given by

$$\Delta G = -\tfrac{1}{2}\sigma\Delta\Psi \qquad (15)$$

where σ is the charge density of the Gouy layer and $\Delta\Psi$ $(=\Psi_i-\Psi_0)$ is the potential difference, assuming that the membrane behaves as a parallel plate condenser. Further, each membrane/solution interface can be considered as a Gouy-Chapman double layer (Tien, 1974) and its surface potential is given by (Davies and Rideal, 1961)

$$\Psi_G = \left(\frac{2kT}{e}\right)\sinh^{-1}\left(\frac{134}{A\sqrt{C_i}}\right) \qquad (16)$$

where k is the Boltzmann constant, T is temperature, e is electronic charge, A is the area occupied per ionogenic group in the membrane, and C_i is the bulk aqueous electrolyte concentration. Using Ψ_i and Ψ_0 to denote Ψ_G at the inner and outer interface, respectively, $\Delta\Psi$ is the potential difference measured with electrodes in the bulk solutions. Therefore, photochemical reactions may be studied by monitoring $\Delta\Psi$, provided of course that a change in charge density elicits a change in Ψ. This change in $\Delta\Psi$ may be transient or steady. Indeed, the bulk of the photoelectric phenomena

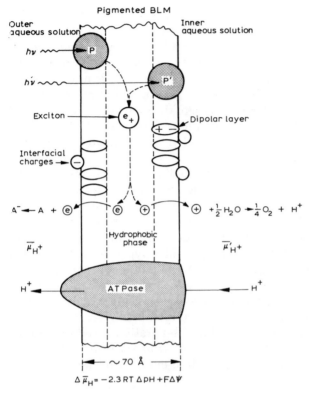

Fig. 4.18. Schematic representation of a modified pigmented BLM. The diagram depicting the three basic processes initiated by light as suggested by the chemiosmotic hypothesis of Mitchell (see text for explanation).

described in this paper and those reported by Bulychev et al. (1976) can be understood in terms of the Gouy-Chapman theory of the electric double layer, as suggested by eq. (16).

The generation of electrochemical free energy in photosynthesis is predicated on the integrity of the thylakoid membrane. The density of fixed negative charges must be greater in the inner side of the thylakoid. These seem to be essential requisites in applying Mitchell's hypothesis. Energy transducing entities endowed with vectorial properties are embedded in the membrane (see Fig. 4.1). It is believed that there are roughly two kinds of entities: one kind performs the task of electron and/or proton transport and the other carries out ATP synthesis, and they are coupled in some obscure manner. Energy provided by light generates a proton gradient by translocating H^+ ions from the outside to the inside, and by oxidizing water with the accompanying evolution of O_2. The proton motive force thus created is discharged by channeling H^+ ions through a phosphorylating enzyme to form ATP, as shown in Fig. 4.18.

The chemiosmotic hypothesis, while it is very elegant and cogent, does not concern itself with the primary event, that is the mechanism of transduction of light into electrical and/or chemical energy. Also not considered in the chemiosmotic hypothesis is the mechanism of water oxidation and O_2 evolution. With the experimental study on pigmented bilayer lipid membranes now in progress, it is timely to address the problem of water oxidation (see also chapter 8).

To split water a minimum energy of 1.23 eV is necessary, which could be supplied by a quantum of red light. However, in the scheme shown in Fig. 4.17A, B, various energy losses have to be considered, which are inherent in the proposed process of quantum conversion. Some of these are the internal resistance of the lipid bilayer, and the overpotential associated with each interface. We have specifically assumed that the outer and inner membrane interface function, respectively, as a photocathode and photoanode. These electrode overpotentials are unavoidable and subject to the usual considerations of electrochemistry. A more urgent question that so far has not been raised is "How can one ever hope to have BLM or for that matter thylakoid membranes about 70 Å thickness of sufficient dielectric strength to withstand 1230 mV, a minimum potential needed to electrolyze water, assuming such a large photovoltage could be generated?"

Faced with this dilemma, it has been suggested (Tien, 1977a) that, for the photooxidation of water in thylakoid membranes and in pigmented BLM, a light-induced potential difference much less than 1.23 V should be sufficient, if the membrane behaves as a semiconductor electrode as described in the last section. The reasoning is as follows. First, let us take a closer look at the pigmented membrane/redox solution interfaces (Fig. 4.17B). Upon illumination with photons of energy equal to or greater than the band gaps ($E_c - E_v$ and $E'_c - E'_v$), excitons are generated. These excitons (say, 1.8 eV) may decay by a variety of processes (Northrop and Simpson, 1958; Choi and Rice, 1963; Kearns, 1963; Lyon and Morris, 1963), one of which is exciton-exciton fusion generating higher energetic species than are available in the exciting photon. (For example, when anthracene is excited by high intensity 1.8 eV red light, it emits 3.4 eV blue light (Avakian et al., 1963).) It seems likely that such a process of upgrading light energy can take place in pigmented lipid membranes. Referring to Fig. 4.18, excitons thus formed under the influence of the electric field can be separated into electrons and holes. If so, electrons move to the left of the pigmented membrane where they are captured by electron acceptors to produce a reduction. On the other hand, holes drift towards the interior of the lipid bilayer. In a similar manner, an oxidation reaction takes place at the inner (or right-side) membrane/redox solution interface where electron donors are located. In this case, electrons move to the left; holes move to the right of the pigmented membrane and are injected into the bathing solution. The interesting result is a sum of the two photoevents operating in concert (see Fig. 4.17B). Moreover, the photoinduced chemical reactions are carried out by one electron-hole pair for each two photons absorbed (eq. 13). At present, the most unequivocal experimental evidence in support of the proposed scheme is the observation by Fujishima and Honda (1972), who have reported O_2 evolution by light at a semiconductor TiO_2 electrode at potentials less than 500 mV. They have also used eq. (13) to describe this most exciting finding.

As described in this chapter, the use of artificial bilayer lipid membranes (BLM and liposomes) containing thylakoid extracts are almost ideal systems for investigating photoelectrical and photochemical processes. Although many of the results obtained are qualitative and preliminary, the pigmented bilayer lipid membranes have opened a new vista to the study of the intricate membranous aspects of photosynthesis, particularly the primary events of solar energy conversion (Calvin, 1976; Hall, 1976; Gregory, 1977). Whether the pigmented BLM system can contribute further towards our eventual understanding of the mechanisms of photosynthesis remains to be seen. However, one thing seems certain, that model systems will be increasingly used in future research. Therefore, I would like to conclude this paper by quoting a remark made by Sager (1958) — "The test of mastery of the essential principles of chloroplast biology would be the ability to make a chloroplast." — Even though to date no artificial chloroplast has been assembled, much understanding and progress have been achieved in clearly identifying the major problems of the membrane aspects of photosynthesis. With the observation of photoelectric effects and light-induced redox reactions in lipid bilayers containing photosynthetic constituents, the study of artificial membrane systems has now provided us with certain insights hitherto unattainable.

4.8. ACKNOWLEDGEMENTS

Much of this chapter is based on experiments already published or to be published. I thank all my associates, past and present, who have contributed to the work reviewed here. Original investigations were supported by grant GM-14971 from the National Institutes of Health.

REFERENCES

Aizawa, M., Tomono, S. and Suzaki, S. (1977) J. Membr. Sci., 2, 289—301.
Alamuti, N. and Laeuger, P. (1970) Biochim. Biophys. Acta, 211, 362—364.
Anderson, J.M. (1975) Biochim. Biophys. Acta, 416, 191—235.
Arnold, W. and Maclay, H.K. (1959) Brookhaven Symp. Biol., 11, 1—9.
Arntzen, C.T., Armond, P.A., Briantais, J.M., Burke, J.J. and Novitzky, W.P. (1977) Brookhaven Symp. Biol., 28, 316—337.

Avakian, P., Abramson, E., Kepler, R.G. and Caris, J.C. (1963) J. Chem. Phys., 39, 1127.
Aveyard, R. and Vincent, B. (1977) Prog. Surf. Sci., 8, 59—102.
Avron, M. (1975) in Proc. 3rd Int. Congr. Photosynth., 3 volumes. Elsevier, Amsterdam.
Bangham, A.D., Hill, M.W. and Miller, N.G.A. (1974) in Methods in Membrane Biology (E.D. Korn, ed.), Vol. 1, Ch. 1. Plenum Press, New York.
Barber, J. (1968) Biochim. Biophys. Acta, 150, 618—625.
Barber, J. (ed.) (1977) Primary Processes of Photosynthesis, Vol. 2, 508 pp. Elsevier North-Holland, Inc., New York.
Barsky, E.L., Dancshazy, Z., Dracher, L.A., Ilina, M.D., Jasaitis, A.A., Kondrashin, A.A., Samuilov, V.D. and Skulacher, V.P. (1976) Biol. Chem., 215, 7066—7071.
Baumeister, W. and Hahn, M. (1976) Prog. Surf., 11, 227—340.
Benson, A.A. (1966) J. Am. Oil Chemists' Soc., 43, 265—270.
Berns, D.S. (1976) Photochem. Photobiol., 24, 117—140.
Blok, M.C., Hellingwerf, K.J. and van Dam, K. (1977) FEBS Lett., 76, 45—50.
Boardman, N.K. (1968) Adv. Enzymol., 30, 1—79.
Bogomolni, R.A. and Klein, M.P. (1975) Nature, 258, 88—89.
Bogomolni, R.A. and Stoeckenius, W. (1974) J. Supramol. Struct., 2, 775—780.
Bradley, R.J., Howell, J.H., Romine, W.O., Carl, G.F. and Kemp, G.E. (1976) Biochem. Biophys. Res. Commun., 68, 577—584.
Branton, D., Bullivant, S., Gilula, N.B., Karnovsky, M., Moor, H., Muehlenthaler, K., Northcote, D.H., Packer, L., Satir, B., Speth, V., Staehlin, L.A., Steere, R.L. and Weinstein, R.S. (1975) Science, 190, 54—56.
Braun, H.P. (1976) Biochim. Biophys. Acta, 443, 609—612.
Brochard, F., Degennes, P.G. and Pfeuty, P. (1976) J. Phys. (Paris), 37, 1099—1104.
Brody, S.S. and Brody, M. (1977) Photochem. Photobiol., 26, 57—59.
Brown, J.S. (1977) Photochem. Photobiol., 26, 319—326.
Bukovsky, J. (1977) J. Biol. Chem., 252, 8884—8889.
Bulychev, A.A., Andrianov, V.K., Kurella, G.A. and Litvin, F.F. (1976) Biochim. Biophys. Acta, 420, 336—351.
Buscall, R. and Ottewill, R.H. (1975) Sep. Colloid Sci., 2, 191—245.
Calvin, M. (1976) Photochem. Photobiol., 23, 425—444.
Carius, W. (1976) J. Colloid Interface Sci., 57, 301—307.
Caspers, J.M., Dellers, M. and Jaffe, J. (1975) J. Phys. Sci. Instr., 8, 569—570.
Chapman, D. and Fast, P.G. (1968) Science, 160, 188—189.
Cherry, R.J., Hsu, K. and Chapman, D. (1971) Biochem. Biophys. Res. Commun., 43, 351—358.
Cherry, R.J., Hsu, K. and Chapman, D. (1972) Biochim. Biophys. Acta, 288, 12—21.
Choi, S. and Rice, S.A. (1963) J. Chem. Phys., 38, 366.
Coster, H.G.L. and Zimmermann, U. (1975) J. Membr. Biol., 22, 73—90.
Csatorday, K., Lehoczki, E. and Szalay, L. (1975) Biochim. Biophys. Acta, 376, 268—273.
Csorba, I., Szabad, J., Erdei, L. and Fajszi, Cs. (1975) Photochem. Photobiol., 21, 377—378.
Dancshazy, Z. and Karvaly, B. (1976) FEBS Lett., 72, 136—139.
Danon, A. and Caplan, S.R. (1976) Biochim. Biophys. Acta, 423, 133—140.
Davies, J.T. and Rideal, E.K. (1961) Interfacial Phenomena, Ch. 2. Academic Press, New York.
Dencher, N.A., Rafferty, Ch.N. and Sperling, W. (1976) Kernforschungsanlage Juelich GmbH, pp. 42—57.
Dilley, R.A. (1972) Methods Enzymol., 24, 68—70.
Diner, B.A. and Joliot, P. (1977) in Encyclopedia of Plant Phys8iology (A. Trebst, M. Avron, eds.), Vol. 5, pp. 188—205. Springer Verlag, Berlin.

Dobias, B. and Heckmann, K. (1977) Bioelectr. Ber., 4, 231—241.

Eisenbach, M., Cooper, S., Garty, H., Johnstone, R.M., Rottenberg, H. and Caplan, S.R. (1977) Biochim. Biophys. Acta, 465, 599.

Eley, D.D. and Snart, S.D. (1965) Biochim. Biophys. Acta, 102, 379—387.

den Engelsen, D. and de Koning, G. (1975) Photochem. Photobiol., 21, 77—80.

Evstigneev, V.B. and Terenin, A.N. (1951) Dokl. Akad. Nauk SSSR, 81, 223—235.

Fahey, P.F., Koppel, D.E., Barak, L.S., Wolf, D.E., Elson, E.L. and Webb, W.W. (1977) Science, 195, 305—306.

Fawcett, W.R. and McCarrich, T.A. (1976) J. Electrochem. Soc., 123, 1325—1331.

Fesenko, E.E. and Lyubarskity, A.L. (1977) Nature, 268, 562—564.

Foigel, A.G. (1976) Biofizika, 21, 463—468.

Forster, T. (1956) Discuss. Faraday Soc., 27, 7—17.

Forti, G., Avron, M. and Melandri, B.A. (eds.) (1972) Photosynthesis, Two Centuries After its Discovery by Joseph Priestley, 3 volumes. Junk, The Hague.

Fujishima, A. and Honda, K. (1972) Nature, 238, 37—38.

Gaines, G.L. (1977) J. Colloid Interface Sci., 59, 438—446.

Galston, A.W. (1977) Photochem. Photobiol., 25, 503—504.

Gregory, R.P.F. (19770 Biochemistry of Photosynthesis. Wiley, New York.

Gruppe, R., Zschatzski, S., Preusser, E. and Goring, H. (1977) Stud. Biophys., 66, 31—46.

Gutmann, F. (1968) Nature, 219, 1359—1360.

Hall, D.O. (1976) FEBS Lett., 64, 6—16.

Henderson, R. and Unwin, P.N.T. (1975) Nature, 275, 28—32.

Henkart, P. and Blumenthal, R. (1975) Proc. Natl. Acad. Sci. U.S.A., 72, 2789—2793.

Hesketh, T.R. (1969) Nature, 224, 1026—1028.

Hess, M. (1977) Naturwissenschaften, 645, 94.

Hoffman, R.A., Long, D.D., Arndt, R.A. and Roper, L.D. (1976) Biochim. Biophys. Acta, 455, 780—795.

Hong, F.T. (1976) Photochem. Photobiol., 24, 155—190.

Hopfield, J.J. (1974) Proc. Natl. Acad. Sci. U.S.A., 71, 3640—3641.

Howell, S.H. and Moudrianakis, E.M. (1967) J. Mol. Biol., 27, 323—333.

Huebner, J.S. and Tien, H.T. (1972) Biochim. Biophys. Acta, 256, 300—306.

Huebner, J.S. and Tien, H.T. (1973) J. Bioenerg., 4, 469—478; J. Membr. Biol., 11, 57—74.

Illani, A. and Berns, D.S. (1972) J. Membr. Biol., 8, 333—356.

Inoue, Y., Ogawa, T. and Shibata, K. (1972) Plant Cell Physiol., 13, 385—389.

Inui, K.I., Tabara, K., Hori, R., Kaneda, A., Muranish, S. and Sezaki, H. (1977) J. Pharm. Pharmacol., 29, 22—26.

Jagendorf, A.T. (1977) in Encyclopedia of Plant Physiology (A. Trebst, M. Avron, eds.), pp. 307—337. Springer Verlag, Berlin.

Jagendorf, A.T. and Hind, G. (1963) N.A.S.—N.R.C., Publ. 1145, 599—610.

Joliot, P. and Kok, B. (1975) in Bioenergetics of Photosynthesis (Govindjee, ed.), pp. 387—412. Academic Press, New York.

Jones, M.N. (1975) Biological Interfaces, pp. 191—234. Elsevier, New York.

Junge, W., Renger, G. and Auslander, W. (1977) FEBS Lett., 79, 155—159.

Junge, W. and Schmidt, R. (1971) J. Membr. Biol., 4, 179—192.

Karvaly, B. (1975) Bioelectrochem. Bioenerg., 2, 124—141.

Karvaly, B. and Dancshazy, Z. (1977) FEBS Lett., 76, 36—41.

Katz, E. (1949) Photosynthesis in Plants, pp. 287—290. Iowa State College Press, Ames.

Kearns, D.R. (1963) J. Chem. Phys., 39, 2697.

Kolb, H.A. and Bamberg, E. (1977) Biochim. Biophys. Acta, 464, 127—141.

Krasnovsky, A.A. (1969) in Progress in Photosynthesis Research, (H. Metzner, ed.), Vol. 2, 709—727.

Kremer, J.M.H., Agterof, W.G.M. and Wiersema, P.H. (1977) J. Colloid Interface Sci., 62, 396—405.

Kreutz, W. (1970) Adv. Bot. Res., 3, 54—64.

Ksenzhek, O.S., Koganov,M.M. and Gevod, V.S. (1976) Dokl. Akad. Nauk SSSR, 227, 220—223.

Kudzina, L.Y., Lukyanenin, A.I., Rotaru, V.K. and Evtodien, Y.V. (1977) Biofizika, 22, 363—364.

Kung, M.C., Devault, D., Hess, B. and Oesterhelt, D. (1975) Biophys. J., 15, 907—917.

Kunitake, T. and Okahata, Y. (1977) Chem. Lett., 1977, 1337—1340.

Kushwaha, S.C., Kates, M. and Martin, W.G. (1975) Can. J. Biochem., 53, 284—299.

Laclette, J.P. and Montal, M. (1977) Biophys. J., 19, 199—202.

Lagaly, G. (1976) Angew. Chem., 15, 575—586.

Lakshminarayanaiah, N. (1975) Electrochemistry (Chem. Soc., London), 5, 132—219.

Larsson, K. and Lundstrom, I. (1976) Adv. Chem. Ser., 152, 43—70.

LeBlanc, R.M., Galinier, G., Tessier, A. and Lemieux, L. (1974) Can. J. Chem., 52, 3723—3727.

Lenaz, G. (1977) in Membrane Proteins and their Interactions with Lipids (R.A. Capaldi, ed.), pp. 47—149. M. Dekker, Inc. New York.

Liberman, Y.A., Topaly, V.P. and Tsofina, L.M. (1970) Biophys. Rev., 15, 67—75.

Loshchilova, E. and Karvaly, B. (1977) Chem. Phys. Lipids, 19, 159—168.

Loxsom, F.M. and Tien, H.T. (1972) Chem. Phys. Lipids, 8, 221—229.

Lutz, H.U., Trissl, H.-W. and Benz, R. (1974) Biochim. Biophys. Acta, 345, 257—262.

Luttge, U. and Pallaghy, C.K. (1969) Z. Pflanzenphysiol., 61, 58—67.

Lyon, L.E. and Morris, G.C. (1963) J. Chem. Phys., 38, 366.

Makowski, L. (1976) J. Theor. Biol., 61, 27—45.

Mangel, M. (1976a) Biochim. Biophys. Acta, 419, 404—410.

Mangel, M. (1976b) Biochim. Biophys. Acta, 430, 459—466.

Mangel, M., Berns, D.S. and Ilani, A. (1975) J. Membr. Biol., 20, 171—180.

Matsumoto, S., Kohda, M. and Murata, S.I. (1977) J. Colloid Interface Sci., 62, 149—157.

Mauzerall, D. and Hong, F.T. (1975) in Porphyrins and Metalloporphyrins (K.M. Smith, ed.), 2nd edn., pp. 701—725. Elsevier, Amsterdam.

McLaughlin, S. and Eisenberg, M. (1975) Annu. Rev. Biophys. Bioeng., 4, 335—366.

Mehler, A.H. (1951) Arch. Biochem. Biophys., 33, 65—77.

Melnik, E., Latorre, R., Hall, J.E. and Tosteson, D.C. (1977) J. Gen. Physiol., 69, 243—257.

Menke, W. (1967) Brookhaven Symp. Biol., 19, 328—340.

Metzner, H. (ed.) (1969) Progress in Photosynthesis Research, 3 volumes. H. Laup Jr., Tuebingen, Germany.

Metzner, H. (ed.) (1978) Photosynthetic Oxygen Evolution (in press).

Miller, T.E. and Tien, H.T. (1974) Bioenergetics, 6, 1—20.

Miller, K.R. (1976) J. Ultrastruct. Res., 54, 159—167.

Mitchell, P. (1966) Biol. Rev., 41, 445—502; Chemiosmotic Coupling and Energy Transduction. Glynn Research, Bodmin, U.K. (1968).

Monnier, A.M. (1977) J. Membr. Sci., 2, 49—65.

Muehlenthaler, K., Moor, H. and Szarkowskim, J.W. (1965) Planta, 67, 305—315.

Myamlin, V.A. and Pleskov, Yu.V. (1967) Electrochemistry of Semiconductors. Plenum Press, New York.

Nagai, R. and Tazawa, M. (1962) Plant Cell Physiol., 3, 323—333.

Nagle, J.F. (1976) J. Membr. Biol., 27, 233—250.

Neher, E. and Eibl, H. (1977) Biochim. Biophys. Acta, 464, 37—44.

Nelson, R.C. (1957) J. Chem. Phys., 27, 864—872.

Nishizaki, Y. (1963) Plant Cell Physiol., 4, 353—361.

Noll, G.G. (1976) Z. Natuforsch., 31c, 40—44.

Northrop, D.C. and Simpson, O. (1958) Proc. R. Soc. London, Ser. A, 244, 377—389.

Oesterhelt, D. (1976) Angew. Chem., Int. ed., 15, 17—38.

Oesterhelt, D. and Stoeckenius, W. (1973) Proc. Natl. Acad. Sci. U.S.A., 70, 2853—2859.

Oettmeier, W., Norris, J.R. and Katz, J.J. (1976) Z. Naturforsch., 31c, 163—168.

Ohki, S. and Ohki, C.B. (1976) J. Theor. Biol., 62, 389—407.

Papahadojopoulos, D. (1973) in Biological Horizons in Surface Science (L.M. Prince, D.F. Sears, eds.), pp. 159—225. Academic Press, New York.

Packer, L., Shieh, P.K., Lanyi, J.K. and Criddle, R.S. (1977) in Bioenergetics of Membranes (L. Packer, ed.), pp. 149—159. Elsevier North-Holland, Amsterdam.

Park, R.B. and Pon, N.G. (1961) J. Mol. Biol., 3, 1—10.

Pohl, W.G. and Teissie, J. (1975) Z. Naturforsch., 30c, 147—151.

Rabinowitch, P. and Govindjee (1969) Photosynthesis. Wiley, New York.

Racker, E. and Hinkle, P.C. (1974) J. Membr. Biol., 17, 181—189.

Renthal, R. and Lanyi, J.K. (1976) Biochemistry, 15, 2136—2145.

Requena, J. and Haydon, D.A. (1975) J. Colloid Interface Sci., 51, 315—327.

Robinson, R.L. and Strickholm, A. (1978) Biochem. Biophys. Acta, in press.

Rosenstrauss, D.L. and Blumenthal, R. (1977) J. Immunol., 118, 129—136.

Sager, R. (1959) Brookhaven Symp. Biol., 11, 101—117.

Salamon, Z. and Frackowiak, D. (1975) Photosynthesis, 9, 337—339.

Sanfeld, A. and Steinchen, A. (1975) Biophys. chem., 3, 99—106.

Schmidt, S., Reich, R. and Witt, H.T. (1972) in Photosynthesis, Two Centuries After its Discovery by Priestley (Proc. 2nd Int. Congr. Photosynthesis Res., Stresa, 1971), (G. Forte et al., eds.), Vol. 2, pp. 1088—1095. Junk, The Hague.

Schubert, D., Bleuel, H., Domning, B. and Wiedner, G. (1977) FEBS Lett., 74, 47—49.

Scott, A.C. (1977) Neurophysics, Ch. 3. Wiley-Interscience, New York.

Selwyn, M.J. and Dawson, A.P. (1977) Biochem. Soc. Trans., 5, 1621—1629.

Shamoo, A.E. and Goldstein, D.A. (1977) Biochim. Biophys. Acta, 472, 13—53.

Shieh, P.K. and Packer, L. (1976) Biochem. Biophys. Res. Commun., 71, 603—609.

Shieh, P. and Tien, H.T. (1974) J. Bioenerg., 6, 45—55.

Smekal, E., Ting, H.P., Augenstein, L.G. and Tien, H.T. (1970) Science, 168, 1108—1109.

Staehelin, L.A., Armond, P.A. and Miller, K.R. (1977) Brookhaven Symp. Biol., 28, 278—315.

Steinemann, A., Alamuti, N., Brodmann, W., Marschal, O. and Laeuger, P. (1971) J. Membr. Biol., 4, 284—294.

Steinemann, A., Stark, G. and Laeuger, P. (1972) J. Membr. Biol., 9, 177—194.

Stillwell, W. and Tien, H.T. (1977) Biochim. Biophys. Acta, 461, 239—252.

Stillwell, W. and Tien, H.T. (1978) to be published).

Strauss, G. (1976) Photochem. Photobiol., 24, 141—153.

Strauss, G. and Tien, H.T. (1973) Photochem. Photobiol., 17, 425—431.

Strickholm, A. and Clark, H.R. (1977) Biophys. J., 19, 29—48.

Suezaki, Y. (1978) J. Theor. Biol. Biochem. Biophys. Res. Commun., in press.

Sugiura, M. and Shinbo, T. (1976) J. Agric. Chem. J., 50, 91—98.

Szabad, J., Lehoczki, E., Szalay, L. and Csatorday, K. (1974) Biophys. Struct. Mech., 1, 65—74.

Szent-Gyorgyi, A. (1941) Nature, 148, 157—158.

Tabushi, I. and Funakura, M. (1976) J. Am. Chem. Soc., 98, 4684—4685.

Takahashi, F. and Kikuchi, R. (1976) Biochim. Biophys. Acta, 430, 490—500.

Thomas, J., Phondke, G.P., Tatake, V.G. and Gopal-Ayengar, A.R. (1970) Photochem. Photobiol., 11, 85—92.

Tien, H.T. (1967) J. Theor. Biol., 16, 97—110.

Tien, H.T. (1968) Nature, 219, 272—274; J. Phys. Chem., 72, 4512—4519.

Tien, H.T. (1970) Int. Conf. Photosynthetic Unit, May 18—21, Gatlinburg, Tennessee, Abstr. C-11.

Tien, H.T. (1971) in The Chemistry of Bio-surfaces (M.L. Hair, ed.), pp. 233—348. M. Dekker, New York.

Tien, H.T. (1972) in Biennial Review of Surface Chemistry and Colloids, MTP International Review of Science (M. Kevker, ed.), p. 56. Butterworths, London; University Park Press, Baltimore.

Tien, H.T. (1974) Bilayer Lipid Membranes (BLM): Theory and Practice. M. Dekker, Inc., New York, 672 pp.

Tien, H.T. (1976) Photochem. Photobiol., 24, 95—116.

Tien, H.T. (1977a) in Proc. Int. Symp. Photosynthetic Oxygen Evolution (H. Metzner, ed.). August 29—September 4, 1977, Tuebingen.

Tien, H.T. (1977b) Proc. 4th International Conference on Bioelectrochemistry, 2—8 October 1977, MBL, Woods Hole, Mass.; Bioelectrochem. Bioenerg. (1978), 5, in press.

Tien, H.T. and Mountz, J. (1976) in The Enzymes of Biological Membranes (A. Martonosi, ed.), Vol. 1, pp. 113—170. Plenum Press, New Jersey.

Tien, H.T. and Verma, S.P. (1970) Nature, 227, 1232—1234.

Ting, H.P., Huemoeller, W.A., Lalitha, S., Diana, A.L and Tien, H.T. (1968) Biochim. Biophys. Acta, 163, 439—450.

Tokunaga, F., Iwasa, T. and Yoshizawa, T. (1976) FEBS Lett., 72, 33.

Tomiewicz, M. and Corker, G. (1975) Photochem. Photobiol., 22, 249—256.

Toyoshima, Y., Morina, M., Motoki, H. and Sudigara, M. (1977) Nature, 265, 187—189.

Trissl, H.-W. (1975) Z. Naturforsch., 30c, 124—126.

Trissl, H.-W. and Laeuger, P. (1972) Biochim. Biophys. Acta, 282, 40—54.

Trosper, T. (1972) J. Membr. Biol., 8, 133—142.

Tummler, B., Maass, G., Weber, E., Wehner, W. and Vogtle, F. (1977) J. Am. Chem. Soc., 99, 4683—4690.

Ullrich, H.-M. and Kuhn, H. (1972) Biochim. Biophys. Acta, 266, 584—596.

Vaillier, J., Vaillier, D., Trentesa, C. and Robert, J. (1975) Cytobios, 13, 23—30.

Van, N.T. and Tien, H.T. (1970) J. Phys. Chem., 74, 3559—3568.

Vernon, L.P. and Seely, G.R. (1966) The Chlorophyll. Academic Press, New York.

Villar, J.-G. (1976) Bioenerg. Biomembr., 8, 199—208.

Volkov, A.G., Lozhkin, B.T. and Boguslavsky, L.I. (1975) Dokl. Akad. Nauk SSSR, 220, 1207—1210.

Weller, H. and Tien, H.T. (1973) Biochim. Biophys. Acta, 325, 433—440.

White, S.H. and Blessum, D.N. (1975) Rev. Sci. Instrum., 46, 1462—1466.

Whitten, D.G., Hopf, F.R., Quina, F.H., Sprintschnik, G. and Sprintschnik, H.W. (1977) Pure Appl. Chem., 49, 379—388.

Wobschall, D. and McKeon, C. (1975) Biochim. Biophys. Acta, 413, 317—321.

Wolf, D.E., Schlessinger, J., Elson, E.L., Webb, W.W., Blumenthal, E. and Henkart, P. (1977) Biochemistry, 16, 3476—3483.

Yampolsky, G.P., Rangelov, N.I., Bobrova, L.E., Platikanov, D.N. and Ismailov, V.N. (1977) Biofizika, 22, 939—941.

Yoshida, T., Okuyama, M. and Itoh, T. (1977) B. Chem. S.J., 50, 1399—1402.

Zimmermann, U., Beckers, F. and Coster, H.G.L. (1977) Biochim. Biophys. Acta, 464, 399—416.

Photosynthesis in relation to model systems, edited by J. Barber
© Elsevier/North-Holland Biomedical Press 1979

Chapter 5

Transduction of Light into Electric Energy in Bacterial Photosynthesis: a study by means of orthodox electrometer techniques

V.P. SKULACHEV

Department of Bioenergetics, A.N. Belozersky Laboratory of Molecular Biology and Bioorganic Chemistry, Moscow State University, Moscow, U.S.S.R.

CONTENTS

Abbreviations
CCCP, 2,4,6-trichlorocarbonyl cyanide phenyl hydrazone; NQ, 2,4-naphthoquinone; PMS, phenazine methosulfate; PP_i, inorganic pyrophosphate; TMPD, N,N,N',N'-tetra-methyl-p-phenylenediamine; CoQ_6, ubiquinone-30.

5.1. INTRODUCTION

Transformation of light energy into a difference of transmembrane electro-chemical potentials of H^+ ions ($\Delta\bar{\mu}_{H^+}$) occupies the central position in the conversion and stabilization of energy by photosynthetic organisms (Mitchell, 1966, 1968, Skulachev, 1977a, b). It is an electric potential difference across a membrane ($\Delta\psi$) that proves to be the primary energy form produced by the $\Delta\bar{\mu}_{H^+}$ generator (Mitchell, 1968, 1977). In this chapter my objective is to review the experimental results obtained in my laboratory involving direct measurements of $\Delta\psi$ formation using photoactive systems obtained from the photosynthetic bacterium *Rhodospirillum rubrum*.

5.2. PROTEOLIPOSOMES ASSOCIATED WITH A PLANAR PHOSPHOLIPID MEMBRANE

A method for voltmetric measurement of the electrogenic activity of a natural bacteriochlorophyll—protein complex was first elaborated for a system which had previously been used in a study of bacteriorhodopsin (Drachev et al., 1974a, b; Barsky et al., 1975). Purified reaction center complexes from *R. rubrum* chromatophores were mixed with cholate solutions of soya-bean phospholipids (azolectin) supplemented with ubiquinone 30

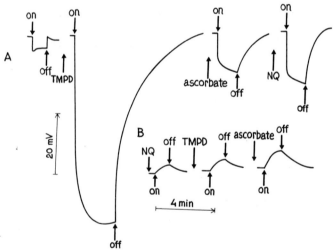

Fig. 5.1. Light-induced electric responses of proteoliposomes containing isolated reaction center complexes (A) or isolated antenna complexes (B) incorporated into a planar azolectin membrane. Incubation mixture: 50 mM Tris · HCl buffer (pH 7.5), 20 mM $MgSO_4$ in A or 30 mM $CaCl_2$ in B, proteoliposomes reconstituted with CoQ_6. $A_{870nm} = 0.05$ in A and $A_{880nm} = 3.4$ in B. Additions: 0.5 mM TMPD, 5 mM sodium ascorbate, 0.1 mM 1,4-naphthoquinone (NQ) (From Drachev et al., 1976a).

(CoQ$_6$) in some experiments. The mixture was dialyzed for 18 h at 3°C to remove cholate. The protein—phospholipid vesicles (proteoliposomes) formed during dialysis were attached by Ca^{2+} ions to one of the surfaces of a thick (about 1000 Å) planar azolectin membrane separating two electrolyte solutions of identical composition. Illumination of such system was found to produce some electric potential difference across the planar membrane (minus on the proteoliposome-free side of the membrane) (Drachev et al., 1975; Barsky et al., 1976). The photoeffect greatly increased (up to 60 mV) after addition of TMPD (all additions were made to both compartments separated by the planar membrane, to avoid asymmetry other than that induced by proteoliposome attachment). Addition of ascorbate significantly lowered, and 1,4-naphthoquinone slightly increased, the light-produced voltage (Fig. 5.1A).

When bacteriochlorophyll antenna complexes, instead of bacteriochlorophyll reaction center complexes, were used as the protein component

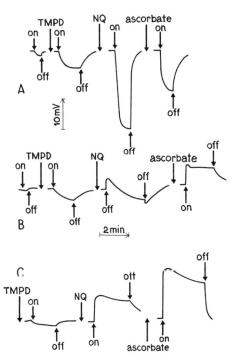

Fig. 5.2. Light-induced electric responses of proteoliposomes containing isolated reaction center complexes and isolated antenna complexes incorporated into a planar azolectin membrane. (A) A$_{870nm}$ = 0.05, A$_{870nm}$: A$_{880nm}$ = 1 : 13. (B) A$_{870nm}$ = 0.04, A$_{870nm}$: A$_{880nm}$ = 1 : 65. (C) A$_{870nm}$ = 0.01, A$_{870nm}$: A$_{880nm}$ = 1 : 325. Incubation mixture: 50 mM Tris · HCl buffer (pH 7.5), 30 mM CaCl$_2$ and proteoliposomes. Additions: 0.5 mM TMPD, 0.1 mM NQ and 5 mM sodium ascorbate (From Drachev et al., 1976a).

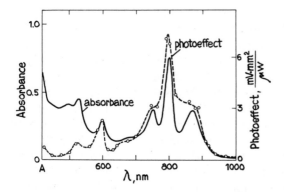

Fig. 5.3. Action spectrum of the photo-electric effect of reaction center proteoliposomes attached to planar phospholipid membrane. Incubation mixture: 0.2 M sucrose, 0.05 M Tris—HCl (pH 7.5), 5 mM $MgSO_4$, 0.03 M $CaCl_2$, 5×10^{-4} M TMPD, and in one of compartments, proteoliposomes (1.2×10^{-7} M bacteriochlorophyll P-870) (from Barsky et al., 1976).

of the reconstitute proteoliposomes, the maximal photovoltage was more than 10-fold lower, the direction of the electric vector was the opposite (plus on the proteoliposome-free side of the membrane) and the addition of ascorbate enhanced the effect (Fig. 5.1B).

Further experiments showed that the photoelectric activity of the antenna proteoliposomes is most probably due to contamination by reaction center complexes. The opposite photoeffect shown in Fig. 1B, was found to increase on addition of a very small amount of reaction center complexes to the mixture for antenna proteoliposome reconstitution. So, at a reaction center-to-antenna ratio equal to 1 : 325, the maximal photovoltage was about 15 mV (Fig. 5.2C) compared with 5 mV for the antenna proteoliposomes reconstituted without added reaction center complexes (Fig. 1B).

With a reaction center-to-antenna ratio equal to 1 : 13, the photoeffect had the characteristics similar to that without antenna, i.e. (1) the proteoliposome-free side of the membrane was electrically more negative, and (2) ascorbate increased the signal size (Fig. 5.2A). An intermediate reaction center-to-antenna ratio (1 : 65) gave a small photo-effect having a polarity dependent on the presence or absence of ascorbate (Fig. 5.2B).

The action spectrum of the photo-effect mediated by reaction center proteoliposomes is given in Fig. 5.3.

5.3. CHROMATOPHORE-PLANAR MEMBRANE SYSTEM

It was found that not only reconstituted vesicles (proteoliposomes) but also a natural membrane system (*R. rubrum* chromatophores) can be studied

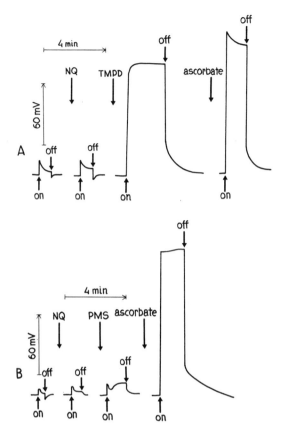

Fig. 5.4. Photoelectric generation by *R. rubrum* chromatophores associated with the planar azolectin membrane. Incubation mixture: 50 mM Tris · HCl buffer (pH 7.5), 10 mM MgSO$_4$ and chromatophores (A$_{880nm}$ = 1.9; here and below, the length of the optical pathway is 1 cm). Additions: 0.1 mM NQ, 0.5 mM TMPD, 5 mM sodium ascorbate and 0.01 M PMS (From Drachev et al., 1976a).

if attached to the planar azolectin membrane (Drachev et al., 1976a; Barsky et al., 1976). Photoresponses of chromatophores qualitatively resemble those of proteoliposomes at low reaction center-to-antenna ratios. However, much higher voltages were obtained with the former. Usually they were between 100 and 200 mV reaching sometimes 220 mV. In the light, the effect required a quinone (CoQ$_6$, vitamin K$_3$ or 1,4-naphthoquinone) together with a redox mediator (TMPD or PMS) and was stimulated by ascorbate, especially in the samples with PMS (Fig. 5.4). None of these additions were found to be necessary for the membrane potential generation by chromatophores coupled with hydrolysis of inorganic pyrophosphate (Fig. 5.5) or ATP (not shown).

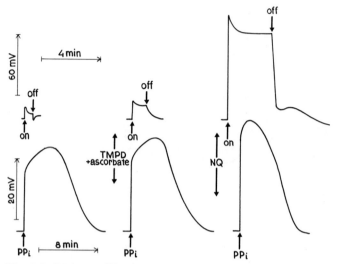

Fig. 5.5. Light- or PP_i-induced electric responses in the system "chromatophores—planar membrane". Incubation mixture: 50 mM Tris · HCl buffer (pH 7.5), 10 mM $MgSO_4$ and chromatophores (A_{880nm} = 1.4). Additions: 0.5 mM TMPD, 5 mM sodium ascorbate, 0.1 mM NQ, 0.1 M PP_i (From Drachev et al., 1976a).

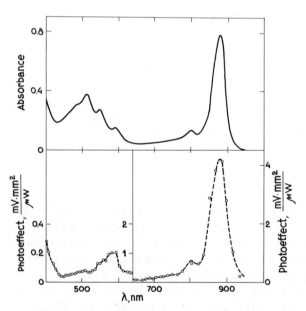

Fig. 5.6. The absorption spectrum of *R. rubrum* chromatophores (solid line) and the action spectrum of the photoelectric response in the system "chromatophores—planar membrane" (broken line). Incubation mixture: 50 mM Tris · HCl buffer (pH 7.5), 10 mM $MgSO_4$, chromatophores (A_{880nm} = 1.7), 0.5 mM TMPD, 0.1 mM NQ, 5 mM sodium ascorbate.

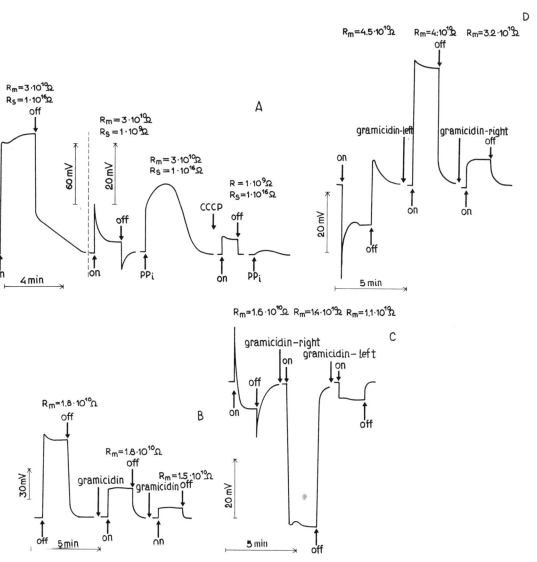

Fig. 5.7. (A) Shunting effects of an external electric resistance and uncoupler CCCP on the light- and PP_i-induced electric responses in the system "chromatophores—planar membrane". Incubation mixture: 50 mM Tris—HCl buffer (pH 7.5), 10 mM $MgSO_4$, chromatophores (A_{888nm} = 1.6), 0.02 mM PMS, 5 mM sodium ascorbate, 0.1 mM NQ. Additions: 0.1 mM PP_i, 1×10^{-6} M CCCP. R_m, electric resistance of the planar azolectin membrane; R_s, shunting electric resistance. (B) Effect of gramicidin on the light-induced responses in the system "chromatophores—planar membrane". Incubation mixture: 50 mM Tris—HCl buffer (pH 7.5), 45 mM KCl, 7.5 mM $MgSO_4$, 0.5 mM TMPD, 3 mM sodium ascorbate, 0.1 mM NQ, chromatophores (A_{880nm} = 1.5). Addition: 1.5×10^{-7} M gramicidin to the chromatophore-containing compartment. (C and D) Effect of gramicidin on the light-induced responses of planar membrane made of a mixture of chromatophores, azolectin and decane. Incubation mixture: 50 mM Tris · HCl buffer (pH 7.5), 0.5 mM TMPD, 3 mM sodium ascorbate, 0.1 mM NQ. Addition: 1.5×10^{-7} M gramicidin. R_s value in B, C and D is 1×10^{16} ohm. Positive charging of the left compartment is shown as electric potential increase.

The action spectrum of the photoeffect in this system is demonstrated by Fig. 5.6.

In Fig. 5.7 the effects of shunting the chromatophore electric generators by an external resistance, by a protonophore uncoupler and by gramicidin A are shown.

Shunting the planar membrane with an external resistance was found to decrease the photo-effect and change its form (differentiation). The protonophore uncoupler, CCCP ($1 \cdot 10^{-6}$ M), also lowered the light-induced response but without differentiation. This was accompanied by a strong decrease in the planar membrane resistance. Unlike CCCP, gramicidin A had no measurable effect on the electric resistance of the thick planar membrane but inhibited the photoresponse, apparently due to an increase in the ion permeability of the thin chromatophore membrane. Gramicidin proved effective only when added to the compartment with the chromatophores (Drachev et al., 1976a).

Of some interest are the experiments with thick planar membrane made of a mixture of chromatophores, azolectin and decane. The electric responses in such a system are a combination of the effects of chromatophores on the right and left hand sides of the planar membrane. The direction and the form of the photoeffect vary from membrane to membrane being determined, most probably, by the ratio of the amounts of chromatophores exposed to the water phases on opposite sides of the membrane. As one can see in Fig. 5.7C and 5.7D, the responses to switching on and off the light are rather small and biphasic due to operation of two sets of photobatteries producing electric potential differences of opposite polarity. Gramicidin added to the right (Fig. 5.7C) or the left (Fig. 5.7D) compartment increases the asymmetry of the system and thus causes an increase in the photo-effects by short circuiting the chromatophores that are exposed to the gramiciding-containing solution. As would be expected the photo-electric responses were inhibited by gramicidin additions to both compartments of the experimental cell (Drachev et al., 1976a).

5.4. CHROMATOPHORES ASSOCIATED WITH THE SURFACE OF A PHOSPHO-LIPID-IMPREGNATED TEFLON FILTER

One of the practical difficulties arising when a protein is incorporated into a planar phospholipid membrane, is that the electric and mechanical stability of the membrane decreases after incorporation. Fortunately, the proteoliposome and chromatophore-generated electric potentials can be measured not only in black (thin) but also in coloured (thick) membranes. The destabilizing effect of the protein-containing material in the thick membrane is not so pronounced as in the thin one. Nevertheless, even in the experiments with the thick membrane, one cannot, e.g., replace the incuba-

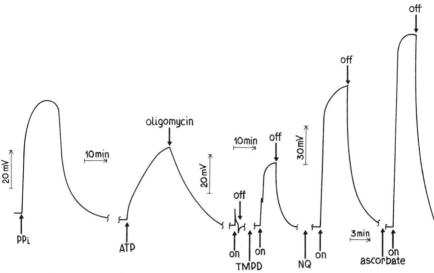

Fig. 5.8. Formation of electric potential difference by three types of generators in the chromatophore—Teflon filter system. Incubation mixture: 0.05 M Tris—HCl (pH 7.5), 2 mM EDTA, 0.02 M MgCl$_2$, and on one side chromatophores (A$_{880nm}$ = 2.5). Additions: 6 × 10^{-4} M PP$_i$, 1 mM ATP, oligomycin (40 γ/ml), 5 × 10^{-4} M TMPD, 2 × 10^{-4} M NQ, 5 mM sodium ascorbate (From Drachev et al., 1978).

tion mixture used in the incorporation procedure by a fresh solution without disrupting the membrane. Besides, the membrane usually breaks when a transmembrane electric potential difference reaches the level of about 400 mV.

To stabilize this system, we substituted a phospholipid-impregnated Teflon filter for planar membrane (Drachev et al., 1976a, 1978). Teflon filters of 0.4 mm thickness and with 4.5-nm pores occupying 70% of the filter area were impregnated with decane solution of azolectin (70 mg/ml). The filter covered a 1-cm aperture in a partition of a two-compartment Teflon cell. Addition of chromatophores or proteoliposomes to one of the compartments resulted in formation of a photoactive system if Ca^{2+} or Mg^{2+} ions were present to stimulate association of the vesicles with filter. In the case of chromatophores, an electric potential difference across the filter was found to be generated not only by light but also by inorganic pyrophosphate or ATP pulses (Fig. 5.8).

5.5. CHROMATOPHORES ASSOCIATED WITH THE INTERPHASE BETWEEN WATER AND DECANE SOLUTION OF AZOLECTIN

The photoeffect of chromatophores associated with a Teflon filter was found to persist after the excess of non-bound chromatophores was removed

184

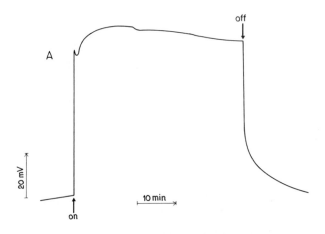

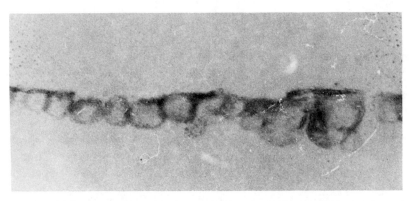

Fig. 5.9. Chromatophores in biphasic system "decane solution of azolectin/water". (A) Light-induced electric response. Water phase: 0.05 M Tris—HCl (pH 7.5), 5×10^{-4} M TMPD, 4×10^{-4} M NQ, 5×10^{-3} M sodium ascorbate; 5×10^{-2} M MgSO$_4$, and chromatophores (A$_{880nm}$ = 2.0). Lipid phase: decane solution of azolectin (150 mg/ml). (B) Cross-section of the interphase-associated chromatophores whose electric response is shown in Fig. 9A; ×125 000. (From Drachev et al., 1978).

by replacing the solutions in the experimental cell. The light-induced voltage in such a system was constant during several days of storage of the chromatophore-treated filter in the dark. However, in the light, rather rapid decrease of the voltage took place. The decrease of the photo-effect was slowed down by EDTA presumably by preventing photo-oxidation of the lipids. Nevertheless, even in the presence of EDTA some lowering in the light-induced potential occurred when the system was illuminated for 20—40 min. The photo-effect returned after storage of the filter in the dark for 1 h. The reason for this photo-induced inhibition is not clear but could be due to the formation of light-dependent pH gradients in the pores of the filter in which chromatophores are incorporated.

To stabilize the photo-effect under long-term light exposures, we used a biphasic system, such as decane solution of azolectin/water (Drachev et al., 1978). In this case the surface to which chromatophores are attached should be much smoother.

A decane solution of azolectin (100 mg/ml) was applied onto the surface of 0.05 M Tris—HCl solution (pH 7.5) to form 1 cm lipid phase over the water phase. Then one Ag/AgCl electrode was immersed into the water phase and the other into the lipid phase. The latter electrode was connected with the lipid phase via an agar bridge containing Tris—HCl. The water phase was supplemented with $R.$ $rubrum$ chromatophores and 0.05 M $MgCl_2$. A period of 2 h was required to associate the chromatophores with the interphase phospholipid monolayer before a photo-effect was measured. The magnitude of the potential difference between the two electrodes decreased by less than 10% within 40 min if the water phase was stirred (Fig. 5.9A). An electron microscope study of this system revealed a chain of vesicles localized in the interphase region (Fig. 9B).

5.6. CHROMATOPHORES ASSOCIATED WITH AN AZOLECTIN-IMPREGNATED COLLODION FILM: A FLASH STUDY

All the above described models proved inadequate for studying the early electrogenic events in bacterial photosynthesis, so a new technique had to be developed. An azolectin-impregnated collodion film was found to measure electric pulses as fast as $0.3\,\mu s$. Apparently, advantages of this film over coloured planar membrane or Teflon filter are due to a lesser thickness of the phospholipid layers in its pores and, hence, to a higher electric capacity. The collodion film was formed on the water surface by adding 1% amyl-acetate solution of nitrocellulose. The film was transferred to air with a ring, dried, impregnated with a decane solution of azolectin (100 mg/ml), and fixed in the Teflon cell used in those experiments in which Teflon filters had been used for the supporting matrix.

Chromatophores attachment to one of the film surfaces was induced by adding 40 mM $MgCl_2$. The electric potential difference across the film was monitored with Ag/AgCl electrodes connected to an operational amplifier (Analog Devices 48K) and a 58-11 storage oscillograph.

Results of a typical experiment performed by Drs. L.A. Drachev and A.Yu. Semenov in this laboratory are demonstrated in Fig. 5.10. It is seen that a 15-ns laser flash generates a 20-mV potential difference. The kinetics of the main portion of the photoresponse was shorter than $0.3\,\mu s$ and could not be measured with the techniques used. This means that the observed electrogenesis is due to the primary events of charge separation in the photosynthetic reaction center rather than to electron transfer from the primary

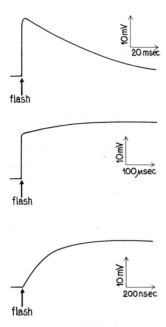

Fig. 5.10. Photoelectric responses induced by a 15-ns laser flash in the chromatophore—collodion film system. Incubation mixture: 0.05 M Tris—HCl (pH 7.3), 0.04 M $MgSO_4$, 2×10^{-4} M PMS, 4×10^{-4} M vitamin K_3, 4 mM sodium ascorbate and in one of the compartments, chromatophores ($A_{880nm} = 1.6$).

electron acceptor to CoQ or from cytochrome c to oxidized bacteriochlorophyll (the latter two processes are much slower than $0.3\,\mu s$; see Dutton et al., 1975, 1977; Parson and Monger, 1977; Jones, 1977 and chapter 3 of this book). Such observations indicate that the light-induced primary charge separation is directed across, rather than along, the chromatophore membrane.

5.7. CONCLUSIONS AND PERSPECTIVES

The above data demonstrate that electric potential formation by bacterial photosynthetic $\Delta\bar{\mu}_{H^+}$ generators can be directly measured with orthodox electrometer techniques. As shown by the detailed analyses carried out by my group, vesicles associated with a planar membrane or filter surface, retain their intravesicular water space, so that formation of an electric potential between two measuring electrodes is a consequence of charge separation across the vesicle membrane (Drachev et al., 1974a, b; 1976a, b, c; Barsky et al., 1975, 1976). This assumption explains the effect of gramicidin, described above (for discussion, see Drachev et al., 1976a) and agrees with the electron microscope data (see Fig. 5.9B).

The main advantage of the described techniques over the previously used indirect methods of $\Delta\psi$ measurement, is first of all a possibility to detect directly, i.e. by a voltmeter, an electric potential difference. In realizing the complexity of biological systems a direct method can be very gratifying. It is fortunate that both native (chromatophores) and reconstituted (proteoliposomes) systems can be investigated by the methods described above.

In addition to establishing the fact that a certain system is electrogenic, the approach outlined above allows the following essential parameters to be elucidated:

(1) Direction of the electric field generated.

(2) Time constants of the process in toto and of its partial reactions.

(3) Localization of electrogenic stages in a multistep reaction sequence.

As to the measured $\Delta\psi$ values, they can be considered as the lower limit of the electromotive force of a generator.

A biphasic lipid/water system may be convenient not only for long-term light exposures but also for investigating the effects of water-insoluble lipid-soluble compounds on the $\Delta\bar{\mu}_{H^+}$ generation mechanisms. Such compounds can be added directly to the lipid phase without any contact with water.

The filter technique represents a way to immobilize chromatophores or proteoliposomes very gently without their electrogenic activity being lost. The resulting system is rather stable against mechanical and electric loads. This feature may be used for constructing a photobattery converting solar energy into electricity by means of biological catalysts. Moreover, systems such as chromatophores possessing both photosynthetic $\Delta\bar{\mu}_{H^+}$ generators and reversible H^+-ATPase when attached to a membrane, can be used to develop a technology of ATP production at the expense of solar energy (Skulachev, 1976).

Without any doubt, large-scale ATP synthesis will be an important problem in any food production using purified and immobilized enzymes, ribosomes, etc. It is noteworthy that H^+-ATPase of chromatophores can synthesize GTP as well as ATP in the presence of nucleoside diphosphatekinase.

5.8. ACKNOWLEDGEMENTS

The author wish to thank Dr. L.A. Drachev and Dr. A.Yu. Semenov for useful discussion, Ms. N. Goreyshina and Ms. T. Konstantinova for help in preparation of the manuscript, and Ms. T.I. Kheifets for correcting the English.

REFERENCES

Barsky, E.L., Dancshazy, Z., Drachev, L.A., Il'ina, M.D., Jasaitis, A.A., Kondrashin, A.A., Samuilov, V.D. and Skulachev, V.P. (1976) J. Biol. Chem., 251, 7066—7071.
Barsky, E.L., Drachev, L.A., Kaulen, A.D., Kondrashin, A.A., Liberman, E.A., Ostroumov, S.A., Samuilov, V.D., Semenov, A.Yu., Skulachev, V.P. and Jasaitis, A.A. (1975) Bioorg. Khim. USSR, 1, 113—126.
Drachev, L.A., Jasaitis, A.A., Mikelsaar, H., Nemecek, I.B., Semenoc, A.Yu., Semenova, A.Yu. and Skulachev, V.P. (1976a) Biochim. Biophys. Acta, 440, 637—660.
Drachev, L.A., Frolov, V.N., Kaulen, A.D., Liberman, E.A., Ostroumov, S.A., Plakunova, V.G., Semenov, A.Yu. and Skulachev, V.P. (1976b) J. Biol. Chem., 251, 7059—7065.
Drachev, L.A, Jasaitis, A.A., Mikelsaar, H., Nemecek, I.B., Semenov, A.Yu., Semenov, E.G., Severina, I.I. and Skulachev, V.P. (1976c) J. Biol. Chem., 251, 7077—7082.
Drachev, L.A., Jasaitis, A.A., Kaulen, A.D., Kondrashin, A.A., Liberman, E.A., Nemecek, I.B., Ostroumov, S.A., Semenov, A.Yu. and Skulachev, V.P. (1974a) Nature, 249, 321—324.
Drachev, L.A., Kondrashin, A.A., Liberman, E.A., Nemecek, I.B., Ostroumov, S.A., Semenov, A.Yu., Skulachev, V.P. and Jasaitis, A.A. (1974b) Dokl. Akad. Nauk SSSR, 218, 481—484.
Drachev, L.A., Kaulen, A.D., Samuilov, V.D., Severina, I.I., Semenov, A.Yu., Skulachev, V.P. and Chekulaeva, L.N. (1978) Biofisika (USSSR).
Drachev, L.A., Kondrashin, A.A., Samuilov, V.D. and Skulachev, V.P. (1975) FEBS Lett., 50, 219—222.
Dutton, P.L., Petty, K.M., Bonner, H.J. and Morse, S.D. (1975) Biochim. Biophys. Acta, 387, 536—556.
Dutton, P.L., Prince, R.C., Tiode, D.M. and Petty, K.M. (1977) Brookhaven Symp. Biol., N 28, 213—237
Jones, O.T.G. (1977) Symp. Soc. Gen. Microbiol., 27, 151—183.
Mitchell, P. (1966) Chemiosmotic Coupling in Oxidative and Photosynthetic Phosphorylation. Glynn Research, Bodmin.
Mitchell, P. (1968) Chemiosmotic Coupling and Energy Transduction. Glynn Research, Bodmin.
Mitchell, P. (1977) FEBS Lett., 78, 1—20.
Parson, W.W. and Monger, T.G. (1977) Brookhaven Symp. Biol., 28, 195—212.
Skulachev, V.P. (1976) FEBS Lett., 64, 23—25.
Skulachev, V.P. (1977a) Usp. Sovrem. Biol., 84, 163—173.
Skulachev, V.P. (1977b) FEBS Lett., 74, 1—9.

Photosynthesis in relation to model systems, edited by J. Barber
© Elsevier/North-Holland Biomedical Press

Chapter 6

The Properties of Bacteriorhodopsin and its Incorporation into Artificial Systems

TH. SCHRECKENBACH

Institut für Biochemie, Röntgenring 11, 8700 Würzberg, F.R.G.

CONTENTS

6.1. INTRODUCTION

It is well accepted that membranes play a central role in the energy conversion processes in the living cell a fact which is clearly emphasised in the various chapters of this book. Of paramount interest are the highly specialized membranes which are involved in the two most fundamental biological processes, respiration and photosynthesis. In both these processes electron transport is mediated by membrane bound enzymes and carriers, the architecture of which is extremely complex. Experimental studies designed to reveal details of their structure and function have involved a wide variety of physical, chemical and biochemical techniques (for reviews see Martinosi, 1976 and Chapter 2 of this book).

Within the last several years the purple membrane from the extreme halophile *Halobacterium halobium* and its retinal containing protein, bacteriorhodopsin, have attracted the attention of many groups working in the field of bioenergetics. The highly specialized purple membrane serves as a relatively simple model for the study of energy transduction by biological membranes (Oesterhelt and Stoeckenius, 1971; Blaurock and Stoeckenius, 1971; Oesterhelt and Stoeckenius, 1973). Unlike photosynthesis and respiration the function of bacteriorhodopsin does not rely on transport of electrons but on the vectorial transport of protons which leads to the establishment of an electrochemical proton gradient across the cell membrane. This gradient serves as an energy pool which can be utilized by various energy consuming processes within the cell. The energy for proton translocation is provided by light which is absorbed by the chromophore of bacteriorhodopsin.

Fig. 6.1 illustrates the main aspects of the purple membrane which are under investigation at present. The problems listed are not restricted to the function of bacteriorhodopsin in the halobacterial cell alone but also extend to fields of more general interest e.g. the biosynthesis, assembly and structure of biological membranes, and to the structure and function of other membrane-bound transport proteins. Of further interest are the chemistry and photochemistry of retinal-protein complexes, the classic examples being the photoreceptor proteins of vertebrate and invertebrate organisms. The final goal of these studies is to understand the molecular mechanism of light energy conversion and its coupling to other energy transducing membrane processes.

Some recent articles on bacteriorhodopsin reviewed in detail the experimental progress being made towards an understanding of how this protein is involved in light induced energy conversion (Henderson, 1977; Oesterhelt et al., 1977, Stoeckenius, 1976; Oesterhelt, 1976; Lanyi, 1974). Therefore the first part of this article will give only a concise and summarizing description of the structural and functional properties of the purple membrane and bacteriorhodopsin, including new developments in the field. The latter half

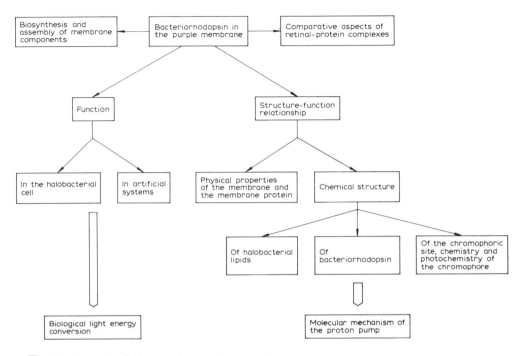

Fig. 6.1. Aspects of the purple membrane system.

of the article will give a more detailed description of the use of bacterio-rhodopsin in artificial systems (see also Schreckenbach, 1976; Montal, 1976).

6.2. PROPERTIES AND BIOLOGICAL FUNCTIONS OF THE PURPLE MEMBRANE AND BACTERIORHODOPSIN

6.2.1. Specific aspects of the halobacterial cell

Extreme halophilic bacteria are aerobic organisms which require in excess of 15% salt for optimal growth (for review on halophilism and physiology of halobacteriaceae see Larsen, 1967 and Dundas, 1976). In the laboratory halobacteria are cultured in a medium containing 4 M NaCl at a temperature of around 40°C. The growth of other microorganisms in this medium is severely restricted. The halobacterial cell however has managed to remain structurally and functionally intact and obtains enough metabolic energy for growth in spite of the fact that respiration is inhibited due to the low solubility of oxygen in saline.

The unusual properties of the natural habitat of halobacteria have necessitated the evolution of some unusual metabolic and structural features of the cell. Halobacteria contain a respiratory chain, and additionally a second and independent bioenergetic system which drives energy consuming processes at the expense of light.

Halobacterial cells are stable at high ionic strength, in distilled water they readily lyze. This process releases the cell contents and the various components of the cell envelope such as the glycoproteins of the cell wall and the cell membrane fragments (Oesterhelt and Stoeckenius, 1974; Blaurock et al., 1976). The membrane fragments from the common bacterioruberine-free mutant strain R_1M_1, when subjected to sucrose density gradient centrifugation, yields four intensely coloured bands, red, yellow, brown and purple. The brown membrane has been identified as the biosynthetic precursor of the purple membrane (Sumper et al., 1976; Sumper and Herrmann, 1976a; Sumper and Herrmann, 1976b). The various membrane fragments are completely devoid of fatty acids. Instead the alkyl side chain of the lipid moieties is constructed solely from the C_{20} isoprenoid alcohol dihydrophytol. The intense colours of all the membrane fragments are due to carotenoid pigments. Retinal is found in the purple and brown membrane, β-carotene is found in the yellow membrane and lycopene in the red membrane. The latter membrane fragment is often referred to as RM 340. The membrane fragments isolated from the wild type organism also contain an abundance of the isoprenoid pigment bacterioruberine. In summary, it is evident that isoprenoid biosynthesis must be of fundamental importance for both structure and pigmentation of the halobacterial cell membrane.

6.2.2. Molecular parameters of the purple membrane and of bacteriorhodopsin

A preparation of a purple membrane fraction from *Halobacterium halobium* was first described by Stoeckenius and Rowen (1967) and McClare (1967). The closely related extreme halophile *Halobacterium cutirubrum*, which incidentally has been used to establish the structure of halobacterial lipids, yields a similar if not identical purple membrane fraction as that isolated from *Halobacterium halobium* (Kushwaha et al., 1975; Kushwaha et al., 1976).

The purple membrane can be purified by a well established routine method (Oesterhelt and Stoeckenius, 1974). Under the electron microscope purple membrane fragments from *Halobacterium halobium* appear as oval or round sheets with a diameter of about 0,5 μm. Their high buoyant density (1.18 g/cm^3) is explained by the fact that the protein amounts to 75% of the membrane's dry weight, the remaining 25% being lipid material. The bacteriorhodopsin molecules each span the width of the membrane and are organized in trimeric units (Fig. 6.2). These trimers are distributed in the plane of the membrane forming a two-dimensional hexagonal array

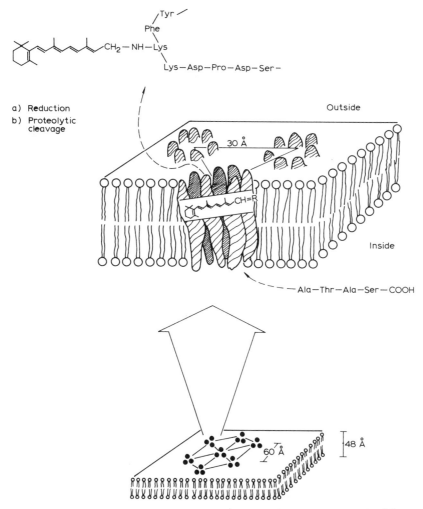

Fig. 6.2. Three-dimensional structure of the purple membrane and of bacteriorhodopsin according to Henderson (1975), Blaurock (1975), Henderson and Unwin (1975). Sequence of the retinyl peptide according to Bridgen and Walker (1976), of the C-terminal peptide according to Wildenauer and Khorana (1977). For more detailed information see section 2.2.

(Blaurock and Stoeckenius, 1971; Henderson, 1975; Blaurock, 1975). This arrangement does not allow translational or rotational motion of the protein (Razi Naqui et al., 1973; Sherman et al., 1976; Cherry et al., 1977).

Formation of the crystalline array of the purple membrane is dependant on the presence of retinal in the bacteriorhodopsin chromophore and on an energy requiring modification step, the nature of which is unknown (Sumper et al., 1976).

The three-dimensional structure of bacteriorhodopsin has been partially elucidated by electron microscopy. Each molecule (molecular weight 26 000) consists of seven α-helices which are more or less oriented perpendicular to the membrane surface (Henderson and Unwin, 1975). The retinal moiety is located at a 23° angle relative to the plane of the membrane (King, 1977; Bogomolni et al., 1977; Mao et al., 1977; Ebrey et al., 1977). The three retinal molecules within a bacteriorhodopsin trimer show strong exciton coupling and are therefore likely to be closely neighboured (Heyn et al., 1975; Becher and Ebrey, 1976).

Electron microscopic studies presented by Neugebauer and Zingsheim (1977) indicated that the membrane has an asymmetric structure. The two membrane faces were clearly distinguished by use of metal treatment and ferritin labelling. The faces differ in surface charge density; by treatment with Ag^+ ions one face showed crystalline order. Further studies have clarified the orientation of the two faces in vivo (Zingsheim, H.P., Henderson, R. and Neugebauer, D.-Ch., 1978).

The primary structure of bacteriorhodopsin has partially been determined (Fig. 6.2). A retinyl peptide was isolated and sequenced by Bridgen and Walker (1976). The C-terminal peptide was sequenced by Wildenauer and Khorana (1977) who additionally claimed that this peptide was located on the cytoplasmatic face of the membrane. At present several groups are attempting to determine the amino acid sequence of the whole protein. This project has turned out to be a very difficult task due to the extreme lipophilic properties of both the protein and the resulting peptides. Preliminary results of sequence studies have been presented by Keefer and Bradshaw (1977), but more recently Ovchinnikov et al. (1977) have made considerable progress in the elucidation of the primary structure of bacteriorhodopsin.

6.2.3. The bacteriorhodopsin chromophore; retinal-protein interaction

The physiological function of bacteriorhodopsin as a proton pump depends totally on the presence of the native chromophore which is known as the purple complex. The λ_{max} value of the purple complex is approximately 565 nm depending on the isomeric configuration of the retinal moiety, since both 13-*cis* and all-*trans* retinal have been isolated from the native protein (Oesterhelt et al., 1973; Pettei et al., 1977). Fig. 6.3 summarizes the chemical and photochemical reactions of the chromophore, the most prominent of which is the photochemical cycle. The existence of several spectral intermediates in this reaction has been demonstrated by use of flash photolysis and low temperature spectroscopy (Lozier and Niederberger, 1977). The exact reaction sequence of the intermediates of the photochemical cycle still remains unknown. Release and uptake of protons from and to the membrane is kinetically coupled with the formation and

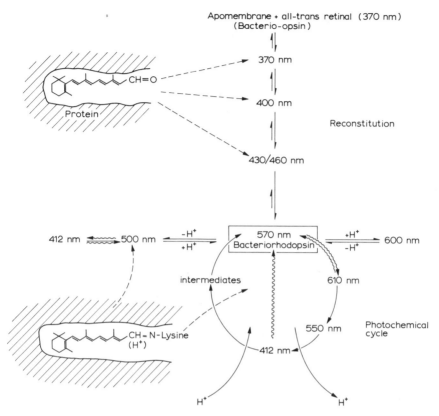

Fig. 6.3. Chemistry and photochemistry of the bacteriorhodopsin chromophore. The schemes on the left illustrate non covalent and covalent attachment of retinal, respectively. For further information and references see section 2.3.

decay of the 412-nm intermediate respectively (Oesterhelt and Hess, 1973). Under physiological conditions and with saturating light levels the chromophore undergoes about 100 photochemical cycles per second; each bacteriorhodopsin molecule therefore can transport up to 100 protons s^{-1} through the membrane. Reformation of the purple complex from the 412-nm intermediate is accelerated by 412-nm light (Oesterhelt and Hess, 1973; Oesterhelt et al., 1975). The regeneration kinetics of this process indicate that the 412-nm chromophore is composed of two or more species (Hess and Kuschmitz, 1977).

At extreme pH values two different chromophore species are derived from the purple complex. A species absorbing at 600 nm can be obtained at acidic pH values and is possibly an intermediate of the photochemical cycle under certain conditions (Fischer, 1978). At alkaline pH a 500-nm chromophore is formed which shows a photochemical reactivity which is similar to that of the purple complex (Schreckenbach et al., 1977).

The molecular events taking place during the photochemical cycle and the molecular mechanism of the proton pump are completely unknown. A key understanding of these processes would certainly be obtained once the structure of the chromophoric site is known. This task however is a difficult one because the purple complex does not react with the typical reagents used for the chemical analysis of retinal-protein complexes (e.g. sodium borohydride, hydroxylamine). Some of the chromophore species derived from the purple complex however have shed some light on the properties of the retinal binding site of bacteriorhodopsin. A most important reaction for the study of retinal-protein interaction is the reconstitution of the purple complex from apomembrane and retinal. Apomembrane is obtained from the purple membrane by reaction with hydroxylamine in the presence of light (Oesterhelt and Schuhmann, (1974). The membrane then contains the chromophore-free protein bacterioopsin in a non-crystalline array. The reconstitution of bacteriorhodopsin from bacterio-opsin and retinal occurs via several intermediates (Fig. 6.3) and restores both crystallinity to the membrane and all functions of the native chromophore. The reconstitution reaction allows the formation of analog chromophores by use of retinal analogs and retinal isomers. This reaction has proved to be a powerful tool for the study of structure—function relationships in bacteriorhodopsin (Schreckenbach et al., 1977; Oesterhelt and Christoffel, 1976; Marcus et al., 1977; Tokunaga et al., 1977; Schreckenbach, Walckhoff and Oesterhelt, in press; see also below).

The binding site of retinal in the pruple complex has not yet been determined by chemical means. Resonance raman studies point towards the presence of a protonated retinal-lysine schiff base (Lewis et al., 1974). Reduction of several chromophores derived from the purple complex and chemical analysis of the reduced chromophore species has revealed that retinal is bound non-covalently to the protein in the 370-nm, in the 400-nm and in the 430/460-nm chromophore (Schreckenbach et al., 1977; Schreckenbach, Walckhoff, and Oesterhelt, in press). In the 500-nm chromophore as well as in a non-defined intermediate of the photochemical cycle the retinal moiety is bound covalently to the ϵ-amino group of a lysine residue (Schreckenbach et al., 1977; Schreckenbach and Oesterhelt, 1977).

The known properties of the binding site are summarized in Fig. 6.4. The results presented have been obtained by the evaluation of chemical and spectroscopic analyses of various reduced and non-reduced chromophore species. Most results have been obtained from work focussed on the conformation and reactivity of retinal isomers and retinal analogs in the chromophoric site. Little information is available on specific amino acids being involved in the reactions observed. Only two types of amino acid have been shown to interact with the retinal molecule. A lysine residue can bind

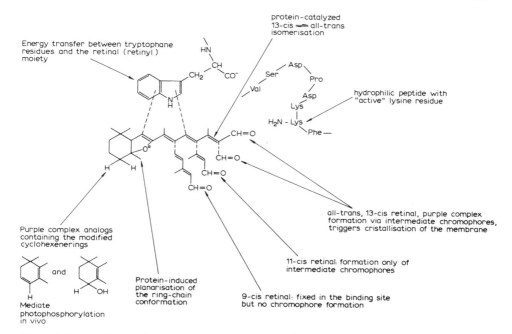

Fig. 6.4. Properties of the chromophoric site. For further information and references see section 2.3.

the retinal in a schiff base linkage (see above and Bridgen and Walker, 1976) and aromatic residues presumably tryptophanes interact noncovalently with the retinal moiety (Schreckenbach et al., 1978).

The function of retinal analogs can be studied in the isolated membrane or in intact cells. Bacteria cultured in the presence of nicotine do not synthesize retinal but nevertheless contain reduced amounts of bacterio-opsin. Upon addition of retinal or an adequate retinal analog to these cells the corresponding purple complex is formed. Using this approach Oesterhelt and Christoffel (1976) and Christoffel (1976) have shown that both 3-4 dehydro retinal and 4 hydroxy retinal when bound in the chromophoric site mediate photophosphorylation in vivo (see Fig. 6.4) whereas 5,6-epoxy retinal does not. By the combination of selective chemical modification of the protein and further use of retinal analogs in the reconstitution reaction some aspects of the molecular mechanism of the proton pump will hopefully be revealed in the near future.

6.2.4. The function of bacteriorhodopsin in the intact cell

The physiological function of bacteriorhodopsin is that of a light-driven proton pump. Illumination of the halobacterial cell results in an electrogenic extrusion of protons and thereby the establishment of an electrochemical proton gradient across the cell membrane (see also Chapter 4). The chemiosmotic theory of Mitchell (Mitchell, 1968) describes this gradient by two components, the pH gradient (ΔpH) and the membrane potential ($\Delta\psi$). Both constitute the proton-motive force (pmf) which according to Mitchell's theory can serve as the driving force for energy consuming processes such as ATP synthesis, ion transport and bacterial movement (for a short review see Skulachev, 1977 (see also Chapters 4 and 5, and also Volume I of this series)). Purple membrane synthesis in the bacterial cell is induced by lack of oxygen (Oesterhelt and Stoeckenius, 1973; Sumper et al., 1976). Light energy conversion by bacteriorhodopsin can thus be regarded as an emergency energy source whenever the availability of oxygen is restricted. A quantitative correlation between respiration and light energy conversion was experimentally determined by Oesterhelt and Krippahl (1973) who clearly demonstrated light inhibition of respiration.

Photophosphorylation in intact cells has been studied by Danon and Stoeckenius (1974), Oesterhelt (1975), Bogomolni (1977) and Hartmann and Oesterhelt (1977). Quantitative aspects of this process were reported recently by Hartmann et al. (1977). Photophosphorylation in halobacteria is not sensitive to inhibitors of electron transport. However, it is sensitive to uncouplers and drugs which lead to dissipation of the light induced membrane potential. Photophosphorylation mediated by bacteriorhodopsin depends therefore only on proton movement and represents a new mechanism for the conversion of electromagnetic energy into the energy of a chemical bond.

The existence of both light induced membrane potential and pH gradient in intact cells was demonstrated by use of membrane penetrating cations and weak acids (Michel and Oesterhelt, 1976; Bakker et al., 1976; Belyakova et al., 1975). A maximal pmf of 284 mV was calculated for illuminated cells pretreated with the ATPase inhibitor dicyclohexyl-carbodiimide (Michel and Oesterhelt, 1976).

Under physiological conditions the halobacterial cell is exposed to high ion gradients (Na^+ outside: 4 M, inside: 1.3 M; K^+ outside: 27 mM, inside 3 M). To maintain these gradients the cell must continually extrude Na^+ ions and take up K^+ ions. Light driven ATP independent uptake of K^+ ions was described by Wagner and Oesterhelt (1976) and Wagner, Hartmann, and Oesterhelt, (1978). Light driven extrusion of sodium ions was shown to be closely coupled with various other membrane-mediated processes and is described in more detail in section 6.3.2.1.

Fig. 6.5 illustrates the processes which according to present knowledge contribute to either the creation or the consumption of the electrochemical proton gradient, $\Delta\mu H^+$, in halobacteria. Processes which lead to an increase in $\Delta\mu H^+$ are respiration and light energy conversion by bacteriorhodopsin. The consumption of $\Delta\mu H^+$ is coupled to ATP synthesis, uptake of amino

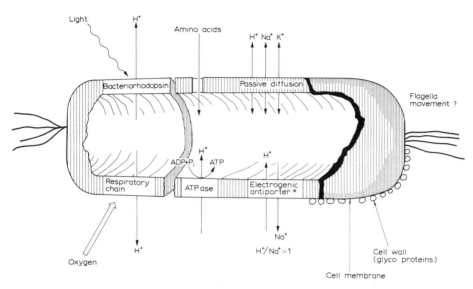

Fig. 6.5. The halobacterial cell membrane; ion fluxes and transport processes related to the electrochemical proton gradient ($\Delta\mu H^+$). Also see section 2.4.

acids (MacDonald and Lanyi, 1977), extrusion of Na^+ ions by a postulated H^+/Na^+ antiporter (Lanyi and MacDonald, 1977) and to the passive diffusion of ions. It could also be assumed that the movement of flagella might be driven by $\Delta\mu H^+$. The chromophore of bacteriorhodopsin certainly is involved in the phototactic behaviour of *Halobacterium halobium* (Hildebrand and Dencher, 1975).

It is evident that studies using intact cells do not allow the exact analysis of individual processes related to the formation of $\Delta\mu H^+$ nor the mode by which they are coupled to each other. The kinetics of pH changes of cell suspensions when illuminated varies significantly depending upon experimental conditions and this has given rise to a multiplicity of interpretations (for comparison see Hartmann and Oesterhelt, 1977, Bogomolni, 1977 and chapter 4 of this book).

6.3. THE FUNCTION OF BACTERIORHODOPSIN IN ARTIFICIAL SYSTEMS

6.3.1. Choice of the system; stability of bacteriorhodopsin

As pointed out in chapter 4 by Tien the reconstitution of artificial membranes is a powerful tool for investigating the mechanism of membrane mediated energy transduction. A wide variety of membrane proteins have been incorporated into lipid vesicles or planar lipid films e.g. Na^+/K^+-ATPase, H^+-ATPase, Ca^{2+}-ATPase and components of the respiratory chain and rhodopsin. Thus the fundamental biological processes of active transport, electron transfer and phototransduction are accessible to an easier interpretation in reconstituted systems (for a review see Montal, 1976).

The studies of bacteriorhodopsin contained within artificial membranes can be conveniently divided into two groups depending on whether lipid vesicles or planar films are used. Lipid vesicles or liposomes are closed structures with a lipid bilayer wall and are often used as simple models to simulate the function of bacteriorhodopsin in the intact bacterium. This can be done with bacteriorhodopsin as the only functional component or by the combination of bacteriorhodopsin with membrane proteins from other sources, such as ATPases. Lipid vesicles containing protein molecules within their bilayer are referred to as proteoliposomes.

Vesicles are very small structures and are therefore only accessible from the outside; the introduction of an electrode would be impossible. Results can only be obtained by indirect methods, mainly by use of membrane permeable compounds as probes for both membrane potential and pH gradient. As described in detail in chapter 4 a most successful technique has been to incorporate purple membrane into several types of film which are used to separate two phases or compartments (e.g. water/water or water/air). The advantage of the technique is that each phase is accessible to measurement of pH and membrane potential and additionally that the contents of each compartment can easily be changed. Furthermore there is in principle no restriction to an infinite area of the artificial membrane.

Most of the artificial lipid membranes described here are relatively small and labile. Consequently some of the work reported has aimed toward the construction of mechanically stable larger membranes which might be useful to biotechnology. It is possible that in the forseeable future a membrane, containing bacteriorhodopsin, might be used to construct a small light driven battery (Skulachev, 1976). These efforts are encouraged by the most unusual stability of the purple membrane and of bacteriorhodopsin compared with other biological membrane systems. The purple membrane is easily isolated from a bacterial culture by simple centrifugation techniques within two days. From a 10-l culture approximately 500 mg of purple membrane is obtained; as a monolayer this amount of membrane would cover about 120 m². Purple membranes can be dehydrated to a certain extent without loss of photochemical activity (Korenstein and Hess, 1977). When completely dried down on a solid support the purple colour remains unchanged for months. The membrane functions equally well in distilled water or saline and is stable in the pH range 1 to 10. It is very resistant against protease attack and under appropriate conditions is stable in the presence of polymerizing agents and certain organic solvents. One has to keep in mind the lability of a normal protein in order to appreciate the exceptional properties of bacteriorhodopsin.

An effective net proton transport through an artificial membrane is only achieved when the majority of the bacteriorhodopsin molecules are oriented in the same direction. Orientation of purple membrane sheets perpendicular to a magnetic field has recently been reported by Neugebauer et al. (1977). However, orientation of the membrane with respect to its two faces was controlled only in a few instances.

6.3.2. Lipid vesicles containing bacteriorhodopsin

6.3.2.1. Cell envelope vesicles

Cell envelope vesicles are intended to represent simplified cells. They lack all cyloplasmatic components but still contain many if not all membrane-functions present in the intact bacterium. A method for their preparation was devised by MacDonald and Lanyi (1975) who found that sonication of a cell suspension led to the formation of vesicles which, additionally, could be loaded with different media by osmotic shock treatment. It was inferred that the purple membrane in such vesicles was oriented in the same direction as found in the cell since illumination of a vesicle suspension resulted in acidification of the medium. Under steady state conditions, with saturating light levels, up to 20 protons per bacteriorhodopsin molecule were extruded from the vesicle (Lanyi and MacDonald, 1977). The existence of a light-induced membrane potential in which the interior of the vesicle became negative was demonstrated using a fluorescent cyanine dye (Renthal and Lanyi, 1976). The maximal pmf demonstrated in three vesicles was near 200 mV which correlated well with the values calculated by Michel and Oesterhelt (1976) for intact cells (see section 6.2.4).

Careful analysis of the kinetics of proton extrusion and its dependence on internal Na^+ concentration has led to the postulate that an electrogenic H^+/Na^+ antiporter ($H^+/Na^+ > 1$) is responsible for the transport of Na^+ into the medium (Lanyi and MacDonald, 1976; Lanyi and MacDonald, 1977, also see section 6.2.4). It was further shown that the uptake of nearly all the common amino acids except cysteine into the vesicles was closely coupled to both membrane potential and sodium gradient. It was claimed that the uptake of glutamate was driven solely by the Na^+ gradient wheras influx of the other amino acids was dependent upon a combination of both $\Delta\psi$ and ΔpH (MacDonald et al., 1977).

Four alternative models were suggested by MacDonald and Lanyi (1977) in order to describe the rather complex interaction of bacteriorhodopsin with the H^+/Na^+ antiporter and the various postulated amino acid carriers.

6.3.2.2. Combination of purple membrane and heterologous lipids

Incorporation of bacteriorhodopsin into vesicles was first reported by

Racker and coworkers. Two methods of preparation were devised: (a) a cholate dialysis method (Racker and Stoeckenius, 1974) and (b) a sonication method which did not involve the use of detergents (Racker, 1973). The vesicles prepared by either technique showed light-induced influx of H⁺ ions (maximal value 19 ng H^+/nmol bacteriorhodopsin under steady state conditions). This result indicated that the purple membrane was preferentially oriented in the reverse direction as that found in the bacterial cell. The generation of both membrane potential and pH gradient was shown by Kayushin and Skulachev (1974) who used vesicles prepared by method (a) above. In addition to bacteriorhodopsin, bovine heart mitochondrial ATPase could be incorporated into both types of vesicles which then showed light induced synthesis of ATP (Racker and Stoeckenius, 1974; Racker, 1973). Photophosphorylation in lipid vesicles containing bacteriorhodopsin was also demonstrated when yeast APTase (Ryrie and Blackmoore, 1976), ATPase from spinach chloroplasts (Winget et al., 1977) and the highly purified and crystalline ATPase from the thermophilic bacterium PS 3 was used (Yoshida et al., 1975; Kagawa et al., 1977).

Another membrane component that has been studied in conjunction with bacteriorhodopsin is cytochrome oxidase (Drachev et al., 1976b). Vesicles containing both components showed increased oxygen consumption when illuminated in the presence of ascorbate and cytochrome c (Hellingwerf et al., 1976). This result correlates with the finding of light inhibition of respiration in intact cells since in the vesicles described bacteriorhodopsin caused an inward flux of protons.

In a systematic study Hwang and Stoeckenius (1977) have described the morphology and proton pumping activity of various types of vesicles containing bacteriorhodopsin. These workers compared the properties of proteoliposomes in which both detergent and lipid composition were varied. The properties of the preparations were determined by electron microscopy, X-ray diffraction, pH-measurements, and absorption spectroscopy. The authors presented a table listing the mode of action of the detergent, effect of the lipid added, degree of bacteriorhodopsin orientation, vesicle morphology and pH-response for five different preparations. The pH responses were maximal in proteoliposomes obtained by a procedure using the detergents desoxycholate/cholate and dipalmitoyl phosphatidylcholine as the lipid component. Interestingly, the use of polar lipids from *H. halobium* led to vesicle preparations showing little or no pH response.

In all the vesicle preparations described so far the asymmetric orientation of the purple membrane has not been controlled by any specific manipulation. In a recent study Happe et al. (1977) showed that the direction of proton translocation and thus the orientation of the purple membrane in proteoliposomes depended on the pH during vesicle formation. Proteoliposomes containing cardiolipin as the sole lipid component when formed at a pH below 2.75 showed proton extrusion. When formed at a pH above this

value, proton influx occurred. Similar results were obtained by use of halobacterial lipid extracts. The conclusion was drawn that the degree of protona-
tion of the polar head group of the lipid molecules ($pK_{cardiolipin} \sim 3$) controls the direction of orientation of bacteriorhodopsin within the vesicle
membrane.

A further point of interest is the question of whether the function of
bacteriorhodopsin depends on the presence of the natural halobacterial
lipids. Happe and Overath (1976) have prepared lipid-depleted bacteriorhodopsin by treatment of the purple membrane with detergents (also see
Wildenauer and Khorana, 1977). Less than one lipid molecule was present
per molecule of protein corresponding to a ten fold decrease in lipid content.
These residual lipids were of the same composition as found in intact purple
membrane. When lipid-depleted bacteriorhodopsin was incorporated into
liposomes, proton pumping activity was observed. On the assumption that
the remaining lipids were evenly distributed over all protein molecules the
conclusion is justified that the functioning of bacteriorhodopsin is independent of the presence of any specific type of lipid.

6.3.3. Bacteriorhodopsin in a planar interphase

6.3.3.1. Octane—water interphase

Boguslavski et al. (1975) have reported the absorption of bacteriorhodopsin as well as of respiratory chain enzymes to an interphase between
octane and water. When illuminated, the system which contained bacteriorhodopsin showed positive charging of the octane phase of up to 500 mV
potential difference. Photopotential generation was dependant upon the
presence of the proton acceptor dinitrophenol (DNP) in the octane phase.
Interestingly, it was found that the addition of soluble mitochondrial
ATPase led to ATP synthesis when the interphase film was illuminated
(Yaguzhinsky et al., 1976). Apparently bacteriorhodopsin is oriented with
its extracellular face towards the octane phase. Thus protons pumped into
the organic phase are accepted by DNP and are then transported back via the
ATPase to the water phase.

The physiological function of bacteriorhodopsin is to mediate vectorial
transport of protons between two aqueous phases. The experiments utilizing
the octane—water system illustrate that both the orientation of the purple
membrane and vectorial proton transport are possible even if one of the two
phases separated by the film is non-aqueous.

6.3.3.2. Lipid films

Mueller and Rudin (1969) have devised a method for the preparation of
planar bilayer membranes which is based on the following technique: a
"membrane forming solution" of lipid in an organic solvent is applied to a
small hole of about 1 mm diameter in a partition between two compart-

ments. A lipid film forms over the hole and thins down spontaneously. During this process a series of interference colours are seen. When the film thins down to less than 300 Å it does no longer reflect light and thus appears black. Membranes of this type are therefore known as black lipid films. Black lipid films can further thin down to 50 Å, which then represents the dimension of a lipid bilayer. For a detailed and authorized review on the properties of black lipid films see Tien, chapter 4 of this volume.

Coloured films of >300 Å thickness containing bacteriorhodopsin were prepared from purple membranes and soya bean phospholipids by Drachev et al. (1974). Potential differences between the two chambers separated by the film were measured with Ag/AgCl electrodes. A stable and reproducible photopotential of 50 mV and a photocurrent of 5×10^{-13} A were recorded when the film was illuminated. The electric response observed was sensitive towards the effect of the uncoupler 2,4,6-trichlorocarbonyl cyanide phenyl hydrazone (CCCP). Both photopotential and photocurrent values were rather low, presumably due to random bidirectional orientation of the purple membrane sheets within the film. Skulachev and coworkers (Drachev et al., 1976a; Barskii et al., 1975) later improved the method of preparation of the films by fusing proteoliposomes containing bacteriorhodopsin to a preformed thick membrane. Exchange of the electrolyte solution did not alter the photoelectric response of the system, indicating that the fusion process was essentially irreversible. The membranes of this type had a resistance of 2×10^{-10} Ω and showed a maximum photopotential of 150 mV corresponding to a photocurrent of 7.5×10^{-12} Å. The action spectrum of the photoresponse indicated that the chromophore of bacteriorhodopsin was responsible for the electrical effects observed. The membrane was shunted by either CCCP, gramicidin S or by an externally applied resistance. It was further demonstrated that bacteriorhodopsin was able to work against an externally applied voltage (V_{ext}). The photovoltage increased with an increasing V_{ext} of the opposite sign, at a value of V_{ext} = 300 mV the photopotential was zero. The electromotive force developed by bacteriorhodopsin when illuminated was therefore calculated to be 300 mV.

The preparation of black lipid films containing bacteriorhodopsin was first reported by Dancshazy and Karvaly (1976) and Herrmann and Rayfield (1976). The former authors constructed a bilayer membrane containing bacteriorhodopsin over a 2-mm hole which was stable up to a breakdown voltage of 100—180 mV. Purple membrane was then fused to the film by addition to one of the two compartments. These membranes developed a maximum photocurrent of 2×10^{-12} Å and a photovoltage of 20 to 60 mV, the action spectrum of which indicated that bacteriorhodopsin was the species responsible for the photoelectric effect. The shape of the action spectrum further indicated that actinic light was partially absorbed by pigments present in the membrane which did not contribute to excitation of the chromophore via energy transfer.

Using the same technique Karvaly and Dancshazy (1977) demonstrated that the photovoltage induced by 560-nm light is quenched by blue light. They postulated that this effect was due to the photochemical reaction of the 412-nm intermediate of the photochemical cycle of bacteriorhodopsin which is known to result in a faster regeneration of the purple complex (see section 2.3). The effect of blue light points towards the possibility that in the bacterium the membrane potential might be modulated by alteration of the spectral composition of the incident light.

The main disadvantage of using lipid films is their instability, due to their large diameter to thickness ratio. By addition of polystyrene or polyacrylamide to the "membrane forming solution" the lifetime of the film can be increased up to 20-fold (Shieh and Packer, 1976). The breakdown voltage of the films stabilized by the addition of polymer are 1.30—1.80 V.

In all studies using lipid films containing bacteriorhodopsin there is still a dearth of specific information. The solvent content of films is unknown, the nature of the fusion process between the purple membrane and the film is also unknown nor is the role of Ca^{2+} ions in the fusion process understood. Of fundamental importance would be a knowledge as to the degree of bacteriorhodopsin orientation within any artificial lipid films.

6.3.3.3. Lipid impregnated millipore filters

The electrogenic nature of bacteriorhodopsin has been established by use of purple membranes incorporated into lipid films. The chief disadvantage of planar lipid films however is that the maximal diameter of a circular film is only 2 mm. Efforts to increase the total surface area of the artificial membrane have been most successful by attachment of liposomes containing bacteriorhodopsin to millipore filters (Skulachev, 1976). This can be done by allowing the liposomes to fuse to one side of a millipore filter which has previously been impregnated with lipids and fixed over an aperture of up to 12 mm diameter separating two compartments (Blok et al., 1977; Packer et al., 1977). The fusion process was found to be facilitated by the presence of Ca^{2+} ions. A photopotential of up to 215 mV could be generated which was sensitive to the effect of uncouplers. The most interesting property of the millipore filter system is its stability; membrane preparations obtained by Packer et al. (1977) were reported to retain their photovoltaic response for over a month without significant loss of activity. For this very reason bacteriorhodopsin contained within millipore filter systems appears to be the most useful technique for studying the suitability of the purple membrane with regard to the advancement of biotechnological techniques.

6.3.3.4. Air—water interphase

When amphiphilic molecules are spread on a water surface they tend to orientate in such a manner that the hydrophilic parts of the molecules face the water and the hydrophobic parts stack into the air. Using this method

James and Augenstein (1966) have studied membrane bound enzymes in an air—water interphase.

If a suspension of purple membrane in a hexane solution of soya phosphatidylcholine is applied to a water surface in a Langmuir trough, a lipid monolayer containing purple membrane sheets forms spontaneously (Hwang et al., 1977a and 1977b). Both surface pressure and surface potential of the film could easily be measured. Furthermore, the film could be transferred as a monolayer onto a solid support of glass or mica and by repeated insertion of the support into the interphase a multilayer could be formed containing up to 160 layers of membrane. These artificial membranes were then accessible to analysis. Absorption spectroscopy showed that the chromophore of bacteriorhodopsin retained its spectroscopic characteristics. Electron microscopic analysis revealed that the purple membrane fragments occupied about 35% of the area of the artificial film and were non overlapping. When the film was covered with a decane layer containing the proton acceptor FCCP a photovoltage of about 70 mV was recorded. The decane phase charged positively indicating that the extracellular face of the purple membrane was oriented towards the air and thus the cytoplasmic face oriented towards the water phase.

Of all the artificial systems at present available for the study of bacteriorhodopsin the most well characterized are the air—water interphase films. These films seem to be worth further study since rather large membrane areas can be formed which contain purple membrane sheets in a highly oriented manner.

6.4. CONCLUSIONS

Much progress has been made in recent years towards a better understanding of the structure and function of bacteriorhodopsin. However, questions concerning the molecular mechanism of light driven proton translocation and the mechanism by which the energy of the electrochemical proton gradient is converted into the energy of a chemical bond via an enzymatic reaction remain unanswered. This latter problem is of fundamental importance since it represents one of the central problems in bioenergetics. Elucidation of the molecular events during the photochemical cycle of bacteriorhodopsin would help to clarify some of the problems associated with the photochemistry of other retinal-protein complexes. An understanding of the molecular mechanism of membrane vectorial proton translocation would certainly help to reveal key features of the mechanism of other active transport processes.

6.5. ACKNOWLEDGEMENTS

The author thanks Prof. D. Oesterhelt for helpful discussions and Dr. P. Towner for correcting the English.
The help of H. Petrásek, B. Walckhoff and E. Dinkl during preparation of the manuscript is gratefully acknowledged.

REFERENCES

Bakker, E., Rottenberg, H. and Caplan, S.R. (1976) Biochim. Biophys. Acta, 440, 557—572.

Barskii, E.L., Drachev, L.A., Kaulen, A.D., Kondrashin, A.A., Liberman, E.A., Ostroumov, S.A., Samnilov, V.D., Semenov, A.Y., Skulachev, V.P. and Yasaitis, A.A. (1975) Bioorg. Chem., 1, 85—95.

Becher, B. and Ebrey, T.G. (1976) Biochem. Biophys. Res. Commun., 69, 1—6.

Belyakova, T.N., Kadzyauskas, Y.P., Skulachev, V.P., Smirnova, I.A., Chekulayeva, L.N. and Yasaitis, A.A. (1975) Dokl. Akad. Nauk SSSR, 223, 483—486.

Blaurock, A.E. (1975) J. Mol. Biol., 93, 139—158.

Blaurock, A.G. and Stoeckenius, W. (1971) Nat. New Biol., 233, 152—154.

Blaurock, A.E., Stoeckenius, W., Oesterhelt, D. and Scherphof, G.L. (1976) J. Cell Biol., 71, 1—22.

Blok, M.C., Hellingwerf, K.J. and van Dam, K. (1977) FEBS Lett., 76, 45—50.

Bogomolni, R.A. (1977) Fed. Proc., 36, 1833—1839.

Bogomolni, R., Hwang, S.B., Tseng, Y.W., King, G.I. and Stoeckenius, W. (1977) Biophys. J., 17, 98 a W-Pos-C 13.

Boguslavski, L.I., Kondrashin, A.A., Kozlov, I.A., Metelsky, S.T., Skulachev, V.P. and Volkov, A.G. (1975) FEBS Lett., 50, 223—226.

Bridgen, J. and Walker, I.D. (1976) Biochemistry, 15, 792—798.

Cherry, R.J., Heyn, M.P. and Oesterhelt, D. (1977) FEBS Lett., 78, 25—30.

Christoffel, V. (1976) Diplomarbeit. University of Würzburg.

Dancshazy, Z. and Karvaly, B. (1976) FEBS Lett., 72, 136—138.

Danon, A. and Stoeckenius, W. (1974) Proc. Natl. Acad. Sci. U.S.A., 71, 1234—1238.

Drachev, L.A., Kaulen, A.D., Ostroumov, S.A. and Skulachev, V.P. (1974) FEBS Lett., 39, 43—45.

Drachev, L.A., Frolov, V.N., Kaulen, A.D., Liberman, E.A., Ostroumov, S.A., Plakunova, V.G., Semenov, A.Y. and Skulachev, V.P. (1976a) J. Biol. Chem., 251, 7059—7065.

Drachev, L.A., Jasaitis, A.A., Kaulen, A.D., Kondrashin, A.A., Van Chu, L., Semenov, A.Y., Severina, I.I. and Skulachev, V.P. (1976b) J. Biol. Chem., 251, 7072—7076.

Dundas, I.E.D. (1977) Adv. Microb. Physiol., 15, 85—120.

Ebrey, T.G., Becher, B., Mao, B., Kilbride, P. and Honig, B. (1977) J. Mol. Biol., 112, 377—397.

Fischer, U. (1978) Ph.D. Thesis, University of Würzburg.

Happe, M. and Overath, P. (1976) Biochem. Biophys. Res. Commun., 72, 1504—1511.

Happe, M., Teather, R.M., Overath, P., Knobling, A. and Oesterhelt, D. (1977) Biochim. Biophys. Acta, 465, 415—420.

Hartmann, R. and Oesterhelt, D. (1977) Eur. J. Biochem., 77, 325—335.

Hartmann, R., Sickinger, H.-D. and Oesterhelt, D. (1977) FEBS Lett., 82, 1—6.

Hellingwerf, K.J., Arents, J.C. and van Dam, K. (1976) FEBS Lett., 67, 164—166.

Henderson, R. (1975) J. Mol. Biol., 93, 123—138.

Henderson, R. (1977) Annu. Rev. Biophys. Bioeng., 6, 87—109.

Henderson, R. and Unwin, P.N.T. (1975) Nature, 257, 28—32.

Herrmann, T.R. and Rayfield, G.W. (1976) Biochim. Biophys. Acta, 443, 623—628.

Hess, B. and Kuschmitz, D. (1977) FEBS Lett., 74, 20—24.

Heyn, M.P., Bauer, P.J. and Dencher, N.A. (1975) Biochem. Biophys. Res. Commun., 76, 897.

Hildebrand, E. and Dencher, N. (1975) Nature, 257, 46—48.

Hwang, S.-B. and Stoeckenius, W. (1977) J. Membr. Biol., 33, 325—350.

Hwang, S.-B., Korenbrot, J.I. and Stoeckenius, W. (1977a) J. Membr. Biol., 36, 115—135.

Hwang, S.-B., Korenbrot, J.I. and Stoeckenius, W. (1977b) J. Membr. Biol., 36, 137—158.

James, L.K. and Augenstein, L.G. (1966) Adv. Enzymol., 28, 1—40.

Kagawa, Y., Ohno, K., Yoshida, M., Takeuchi, Y. and Sone, N. (1977) Fed. Proc., 36, 1815—1818.

Karvaly, B. and Dancshazy, Z. (1976) FEBS Lett., 76, 36—40.

Kayushin, L.P. and Skulachev, V.P. (1974) FEBS Lett., 39, 39—42.

Keefer, L.M. and Bradshaw, R.A. (1977) Fed. Proc., 36, 1799—1804.

King, G.J. (1977) Fed. Proc., 36, 362, Abstr. EE3.

Korenstein, R. and Hess, B. (1977) Nature, 270, 184—186.

Kushwaha, S.C., Kates, M. and Martin, W.G. (1975) Can. J. Biochem., 53, 284—292.

Kushwaha, S.C., Kates, M. and Stoeckenius, W. (1976) Biochim. Biophys. Acta, 426, 703—710.

Lanyi, J.K. (1974) Bacteriol. Rev., 38, 272.

Lanyi, J.K. and MacDonald, R.E. (1976) Biochemistry, 15, 4608—4614.

Lanyi, J.K. and MacDonald, R.E. (1977) Fed. Proc., 36, 1824—1827.

Larsen, H. (1967) Adv. Microbiol. Phys., 1, 97—132.

Lewis, A., Spoonhower, J., Bogomolni, R., Lozier, R. and Stoeckenius, W. (1974) Proc. Natl. Acad. Sci. U.S.A., 71, 4462—4466.

Lozier, R.H. and Niederberger, W. (1977) Fed. Proc., 36, 1805—1809.

MacDonald, R.E. and Lanyi, J.K. (1975) Biochemistry, 14, 2882—2889.

MacDonald, R. and Lanyi, J.K. (1977) Fed. Proc., 36, 1828—1832.

MacDonald, R.E., Greene, R.V. and Lanyi, J.K. (1977) Biochemistry, 16, 3227—3235.

Martinosi, A. (ed.) (1976) The Enzymes of Biological Membranes, Vol. 1. John Wiley and Sons, London, New York, Sydney and Toronto.

Mao, B., Becher, B., Kilbride, P., Ebrey, T.G. and Honig, B. (1977) Biophys. Soc. (Abstr.) 76a.

Marcus, M.A., Lewis, E., Racker, E. and Crespi, H. (1977) Biochem. Biophys. Res. Commun., 78, 669—675.

McClare, C.W.F. (1967) Nature, 216, 766—771.

Michel, H. and Oesterhelt, D. (1976) FEBS Lett., 65, 175—178.

Mitchell, P. (1968) Chemiosmotic Coupling and Energy Transduction. Glynn Research Ltd., Bodmir, Cornwall.

Montal, M. (1976) Annu. Rev. Biophys. Bioeng., 5, 119—175.

Mueller, P. and Rudin, D.O. (1969) Curr. Top. Bioenerg., 3, 157—249.

Neugebauer, D.-Ch. and Zingsheim, H.P. (1978) J. Mol. Biol., 123, 235—246.

Neugebauer, D.-Ch., Blaurock, A.E. and Worcester, D.L. (1977) FEBS Lett., 78, 31—35.

Oesterhelt, D. (1975) CIBA Found. Symp., 31, 147—167.

Oesterhelt, D. (1976) Angew. Chem. Int. Ed. Engl., 15, 17—24.

Oesterhelt, D. and Christoffel, V. (1976) Biochem. Soc. Trans., 4, 556—559.

Oesterhelt, D. and Hess, B. (1973) Eur. J. Biochem., 37, 316—326.

Oesterhelt, D. and Krippahl, G. (1973) FEBS Lett., 36, 72—76.

Oesterhelt, D. and Schuhmann, L. (1974) FEBS Lett., 44, 262—265.

Oesterhelt, D. and Stoeckenius, W. (1971) Nat. New Biol., 233, 149—152.

Oesterhelt, D. and Stoeckenius, W. (1973) Proc. Natl. Acad. Sci. U.S.A., 70, 2853—2857.

Oesterhelt, D. and Stoeckenius, W. (1974) Methods Enzymol., 31, 667—678.

Oesterhelt, D., Gottschlich, R., Hartmann, R., Michel, H. and Wagner, G. (1977) Symposia of the Society for General Microbiology, Vol. XXVII, pp. 333—349.

Oesterhelt, D., Hartmann, R., Fischer, U., Michel, H. and Schreckenbach, T. (1975) 10th Meet. Eur. Biochem. Soc., 40, 239—251.

Oesterhelt, D., Meentzen, M. and Schuhmann, L. (1973) Eur. J. Biochem., 40, 453—463.

Ovchinnikov, Y.A., Abdulaev, N.G., Feigina, M.Y., Kiselev, A.V. and Lobanov, N.A. (1977) FEBS Lett., 84, 1—4.

Packer, L., Konishi, T. and Shieh, P. (1977) in Living Systems as Energy Converters (R. Buvet et al., eds.), pp. 119—134, Elsevier Amsterdam.

Pettei, M., Yudd, A.P., Nakanishi, K., Henselman, R. and Stoeckenius, W. (1977) Biochemistry, 16, 1955—1959.

Racker, E. (1973) Biochem. Biophys. Res. Commun., 55, 224—230.

Racker, E. and Stoeckenius, W. (1974) J. Biol. Chem., 249, 662—663.

Razi Nagvi, K., Gonzalez-Rodriguez, J., Cherry, R.J. and Chapman, D. (1973) Nat. New Biol., 245, 249—251.

Renthal, R. and Lanyi, J.K. (1976) Biochemistry, 15, 2136—2143.

Ryrie, I.J. and Blackmoore, P.F. (1976) Arch. Biochem. Biophys., 176, 127—135.

Schreckenbach, T. (1976) in Microbiology in Energy Conversion (H.G. Schlegel, J. Barnea, eds.), pp. 245—266. Goltze KG, Göttingen.

Schreckenbach, T. and Oesterhelt, D. (1977) Fed. Proc., 36, 1810—1814.

Schreckenbach, T., Walckhoff, B. and Oesterhelt, D. (1977) Eur. J. Biochem., 76, 499—511.

Schreckenbach, T., Walckhoff, B. and Oesterhelt, D. (1978) Photochem. Photobiol. 28, 205—211.

Schreckenbach, T., Walckhoff, B. and Oesterhelt, D. (1978) Biochemistry (in press).

Sherman, W.V., Slifkin, M.A. and Caplan, S.R. (1976) Biochim. Biophys. Acta, 423, 238—248.

Shieh, P. and Packer, L. (1976) Biochem. Biophys. Res. Commun., 71, 603—609.

Skulachev, V.P. (1976) FEBS Lett., 64, 23—25.

Skulachev, V.P. (1977) FEBS Lett., 74, 1—9.

Stoeckenius, W. (1976) Sci. Am., 234, 38—46.

Stoeckenius, W. and Rowen, R. (1967) J. Cell Biol., 1, 365—393.

Sumper, M. and Herrmann, G. (1976a) FEBS Lett., 69, 149—152.

Sumper, M. and Herrmann, G. (1976b) FEBS Lett., 71, 333—336.

Sumper, M., Reitmeier, H. and Oesterhelt, D. (1976) Angew. Chem., Int. Ed. Engl., 15, 187—194.

Tokunaga, F., Govindjee, R., Ebrey, T.G. and Crouch, R. (1977) Biophys. J., 19, 191—198.

Wagner, G. and Oesterhelt, D. (1976) Ber. Dtsch. Bot. Ges., 89, 289—292.

Wagner, G., Hartmann, R. and Oesterhelt, D. (1978) Eur. J. Biochem., 89, 169—179.

Wildenauer, D. and Khorana, H.G. (1977a) Fed. Proc., 36, 896, Abstr. 3290.

Wildenauer, D. and Khorana, H.G. (1977b) Biochim. Biophys. Acta, 466, 315—324.

Winget, G.D., Kanner, N. and Racker, E. (1977) Biochim. Biophys. Acta, 460, 490—499.

Yaguzhinsky, L.S., Boguslavsky, L.I., Volkov, A.G. and Rakhimaninova, A.B. (1976) Nature, 259, 494—495.

Yoshida, M., Sone, N., Hirata, H. and Kagawa, Y. (1975) Biochem. Biophys. Res. Commun., 67, 1295—1300.

Photosynthesis in relation to model systems, edited by J. Barber
© Elsevier/North-Holland Biomedical Press 1979

Chapter 7

Molecular and Functional Aspects of Immobilized Chloroplast Membranes

GEORGE C. PAPAGEORGIOU

Nuclear Research Center Demokritos, Department of Biology, Aghia Paraskevi, Athens, Greece

CONTENTS

7.1. INTRODUCTION

The rapidly growing interest in the scientific and technological aspects of immobilized enzymes is reflected in the numerous research papers, review articles, and edited volumes on this subject that appeared in the recent years (see for example Zaborsky, 1973; Kestner, 1974; Olson and Cooney, 1974; Pye and Wingard, 1974; Weetal, 1974; Mosbach, 1976; and references cited therein). The great majority of this literature concerns soluble enzymes, which can be immobilized, although remaining active, by a number of physical and chemical techniques. Comparatively little is known with regard to the immobilization of membrane-integrated enzyme functions such as those encountered in the photosynthetic membranes of higher plant chloroplasts, algae, and photosynthetic bacteria, in spite of their undisputed technological future as solar energy converters (see various chapters of this volume).

Admittedly, whereas in the case of soluble proteins immobilization is semantically equivalent to insolubilization without total loss of activity, the terms "immobilized membrane" and "immobilized organelle or cell" have noprecise meaning. We may, nevertheless, operationally define an immobilized membrane-bounded structure (e.g. a chloroplast thylakoid) as one whose volume no longer responds to the osmotic pressure of the surrounding medium. Immobilization, as defined can be achieved by covalent crosslinking of membrane proteins with bifunctional reagents, and it can be attributed to the inability of membrane proteins to move laterally, in order to allow for area expansion, as a result of the established crosslinking bridges. There is no direct evidence to support this, except that the proteins of osmotically insensitive membranes resist solubilization by detergents, and that the protein molecular sizes of the solubilized fraction greatly exceed the molecular sizes of native membrane proteins, suggesting extensive intermolecular crosslinking.

What is the objective in immobilizing biological entities artificially? The most frequently offered justification is the technological superiority of the immobilized preparations; in particular, their resistance to denaturation (and the associated with it functional inactivation), and the facility in industrial handling. The practicality approach, however, tends to obscure the fact that, often, significant scientific gains can be made by such manipulations. Protein crosslinking, for example, which can be classed under the more general theme of chemical modification of proteins, has provided insight to such fundamental problems as the 3-dimensional structure of proteins, the subunit structure of complex proteins, the arrangement of proteins in a biological membrane, and the necessity of molecular flexibility, and molecular mobility for functional expression. (Wold, 1967 and 1972: Fasold et al., 1971; Peters and Richards, 1977). Although more distant in the future, the technological importance of such knowledge, either in optimizing the use of biological preparations, or in employing them as models for synthetic systems, is beyond question.

In the present article, I shall review the currently existing knowledge about photosynthetic membranes that have been immobilized (or fixed) by covalent crosslinking of membrane proteins. Glutaraldehyde (GA; pentane-dial) is by far the most extensively used crosslinking agent, yielding fixed preparations of superior properties. Consequently, emphasis will be laid on the structure of its aqueous solutions, the chemistry of its reaction with proteins, and the structural and functional effects of its interaction with photosynthetic membranes. Other crosslinking agents that have been employed with photosynthetic membranes include formaldehyde (FA), malondialdehyde (MA), and more recently bifunctional imidoesters, such as dimethylmalonimidate (DMM), dimethyladipimidate (DMA), and dimethylsuberimidate (DMS). Finally, we shall also consider the few isolated attempts to immobilize chloroplasts by encapsulation either within a modified chloroplast envelope, or within synthetic membrane sacs.

7.2. BIFUNCTIONAL CROSSLINKING COMPOUNDS

7.2.1. Glutaraldehyde as a protein crosslinking agent

Since its introduction as a tissue fixative for electron microscopy by Sabatini et al., in 1963, GA has been the most widely employed compound for the chemical immobilization of soluble and membrane-integrated proteins. Its extensive use appears to be dictated not only by its superiority as protein crosslinker, relative to other aldehydes, such as MA (West and Packer, 1970), FA (Hopewood, 1969; West and Mangan, 1973; Zilinskas and Govindjee, 1976), and α-hydroxyadipaldehyde (Hopewood, 1973), but primarily because it allows rather large fractions of the original enzyme activities to survive. In this respect, GA is also clearly superior to bifunctional imidoesters, which inhibit the H_2O-splitting reaction of chloroplasts (Papageorgiou and Isaakidou, 1977). An informative review on the biological applications of GA was recently published by Russel and Hopewood (1976).

7.2.1.1. The structure of aqueous GA solutions

GA is commercially available in the form of aqueous solutions of varying content and degree of purity. Since the knowledge of the actual composition of the GA reagent is essential in the understanding of the chemistry of protein fixation, the structure of aqueous GA solutions has been studied in several laboratories. Richards and Knowles (1968), who were the first to apply ^{1}H-NMR spectroscopy to this problem, assigned the main observed chemical shifts (relative to tetramethyl silane) as follows:

$\tau = 4.77$	Broad singlet	$-\underline{CH}=C$	[~2H]
$\tau = 6.6$	Multiplet	$-CH=CH\underline{CH_2}C(CHO)=C$	[~3H]
$\tau = 8.3$	Broad multiplet	$-CH_2\underline{CH_2}CH_2-$	
		$-\underline{CH_2}CH=C$	[~18H]
		$-\underline{CH_2}CH(OH)_2$	

The protons which generate these absorptions are underlined, and their approximate abundance relative to CHO (obtained by peak integration) is given in the brackets. On the basis of these results, Richards and Knowles (1968) concluded that GA exists as a mixture of oligomeric and polymeric forms, which derive from a chain of successive aldol condensations of the monomer. The relative abundances of the different proton types do not indicate preponderance of one particular form, although the oligomeric α,β-unsaturated aldehydes appear to predominate. Some of the oligomers supposed to exist in commercial GA solutions are shown below:

$OHC{-}CH_2CH_2CH_2{-}CHO$

I. Monomer

$OHC{-}(CH_2)_3{-}CH{=}C(CHO){-}(CH_2)_2CHO$

II. Dimer

$OHC{-}(CH_2)_3CH{=}C(CHO){-}(CH_2)_2{-}CH{=}C(CHO){-}(CH_2)_2{-}CHO$

III. Trimer

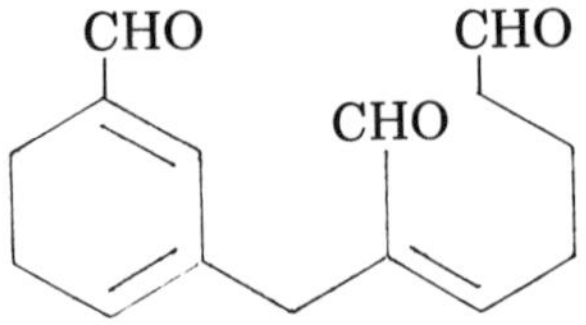

IV. Cyclic trimer

These interpretations, however, did not encounter general acceptance, and in fact the present concensus of opinion is that the structure of aqueous GA solutions, at acidic or neutral pH, is a lot simpler. On the basis of the weak absorption of commercial GA at 235 nm ($E^{1\%}_{1\,cm} = 1.5$ and 6.2), Hardy et al. (1969) concluded that α,β-unsaturated aldehydes are only minor components. Purification of the commercial product gave a solution of the monomer that was devoid of A235. ^{1}H-NMR spectra of the latter showed peaks at approx. the same positions as the spectra of commercial GA reported by Richards and Knowles (1968). Under the constraint of insignificant unsaturation they were interpreted in terms of the following hydration equilibria:

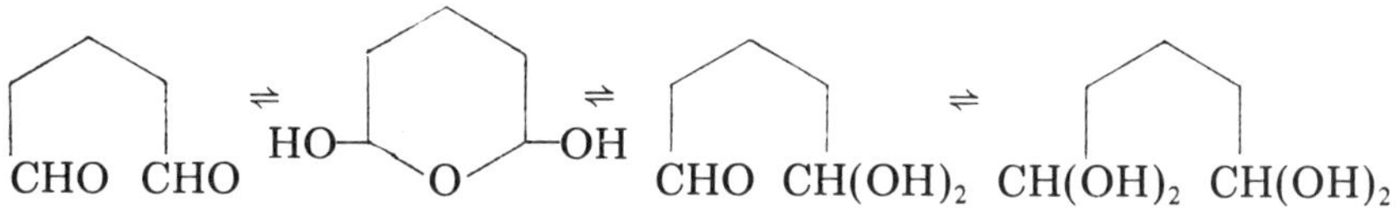

$$\text{I} \qquad\qquad \text{V} \qquad\qquad \text{VI} \qquad\qquad \text{VII}$$

At alkaline pH (pH = 8), GA was found to polymerize into a solid with an extinction coefficient at 235 nm of $E_{1cm}^{1\%}$ equal to 31.5, which is consistent with a high content in α,β-unsaturated polyaldehydes.

The composition of GA solutions of various commercial origins, examined by Korn et al. (1972), were found to be be the same except for the presence of CH_3OH in some of them. The observed chemical shifts in the ^1H-NMR spectrum were attributed to the following structures:

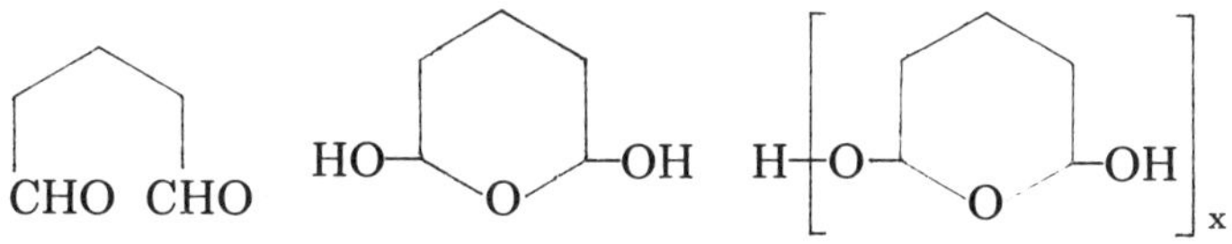

$$\text{I} \qquad\qquad \text{V} \qquad\qquad \text{VIII}$$

At 25°C, the monomer (I) is 15% and the hydrated forms V and VIII 85% of the total. In general, elevated temperatures favor dissociation to the monomer. In contrast to the suggestion of Hardy et al. (1969), the open chain hydrated forms VI and VII are not considered as significant components. The principal member of the heterocyclic polymer VIII appears to be the dimer (x = 2), as supported by the ^1H-NMR spectrum of acetylated aqueous GA, which corresponds to the structure:

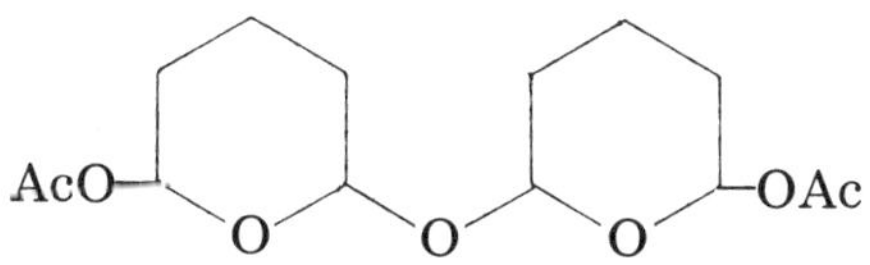

$$\text{IX}$$

The absence of polymeric forms was further confirmed by means of proton-decoupled ^{13}C-NMR spectra of commercial and purified GA solutions (Whipple and Ruta, 1974). The composition of aqueous GA solutions was found to vary slightly with pH and strongly with the temperature. At neutral pH, and 23°C, the monomer (I) makes 4% of the total, the monohydrate (VI) 71%. The latter exists in two stereochemical configurations, in approximate equal amounts.

V, cis   V, trans

A question of significant operational value for the experimentalist is how stable are aqueous solutions of GA on prolonged standing. Between −10°C and 4°C little polymerization occurs, as detected in terms of the absorbance at 235 nm (A235), but above this range the rate of polymerization increases exponentially (Rasmussen and Albrechtsen, 1974). Storage of GA under an inert gas atmosphere is unnecessary; glutaric acid, a product of GA oxidation absorbing at 207 nm, is not among the impurities (Gillet and Gull, 1972). Thin layer chromatography of monomeric GA on silica gel gives a single spot, even after a 7-day standing at pH 7.4, ambient temperature, and ambient light. Commercial grades give, in addition, slow moving spots, but these are much smaller than the respective monomer spots (Blass et al., 1975). Over a period of 6 months, we have noticed in our laboratory very little increase in A235 in purified commercial GA stored at −15°C, even though the GA and been frequently thawed for experimentation.

In summary, we may draw the following conclusions about the structure of aqueous GA solutions. At neutral or acidic environment, GA reagents consist primarily of the cyclic V (with a possible contribution by the dimer of VIII) and to a lesser extent of the monomer I, and the open chain hydrates VI and VII. At elevated temperatures, the equilibrium shifts in the direction of the monomer. Acid solutions of GA are quite stable, especially at low temperature, but they polymerize irreversibly at alkaline pH (Boucher, 1972; 1974), or in the presence of cationic catalysts (Aso and Aito, 1962a, b) and Al catalyst (Moyer and Grev, 1963). Polymerization is accompanied by a loss of CHO groups.

GA absorbs maximally at 280 nm ($\epsilon = 6.5$ M^{-1} cm^{-1} at 23°C; Blauer et al., 1975). Its purity is assessed in terms of the absorbance ratio A235/A280, whose acceptable range is between 0.15 and 0.25 (Anderson, 1967). For more stringent requirements, and in view of some variability in the magnitude of A235/A280 with temperature and solute content, the standardization of GA solutions should include, also, chemical titration of CHO groups, and measurements of the osmolality. (Hesse, 1973; Stibenz, 1973).

Techniques for GA purification include fractional distillation (Fahimi and Drochmans, 1968), adsorption of impurities on activated charcoal (Pease, 1964; Anderson, 1967), and passage through Sephadex G-10 (Hopewood, 1967). Chemical determinations of GA have been described by Frigerio and Shaw (1968), and by Hesse (1973).

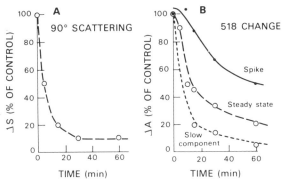

EFFECT OF GLUTARALDEHYDE
ON LIGHT—INDUCED 518 CHANGE AND 90° SCATTERING
pH 6.5

Fig. 7.1. Effect of 0.1% glutaraldehyde (7 μmol GA/mg Chl) on (A) the light-induced 90° light scattering by spinach chloroplasts at 518 nm, and (B) the light-induced absorbance change at 518 nm (ΔA518). ●———●, spike; ○———○, steady-state signal; ○- - - - -○, slow components, obtained as the difference spike minus steady state signal. Reproduced by permission from Thorne et al. (1975). For explanation see p. 225.

7.2.1.2. Reaction of glutaraldehyde with proteins

The exact chemistry of protein crosslinking by GA is not yet fully understood in spite of the significant advances made in the recent years. Given the tendency of aldehydes to condense to Schiff bases with primary amines, the expected site of attack is the free NH_2 of the lysyl residues, which disappears when proteins are reacted with GA (Fig. 7.1). (Papageorgiou and Isaakidou, 1977).

The presence of Schiff bases in the end products of the reaction of GA with proteins has been strongly disputed by Richards and Knowles (1968). Their main arguments were: (1) Schiff bases are acid labile, (except when a double bond is conjugated to the azomethine linkage) whereas in the acid hydrolysates of GA-reacted proteins partial loss of lysyl is observed. (2) GA modifies proteins irreversibly, since the products withstand treatments with urea, semicarbazide, as well as wide ranges of pH, ionic strength and temperature. (3) The pKa of GA-modified lysyl is about 8—8.5, whereas the pKa of a Schiff base does not exceed 5.

To reconcile these observations, Richards and Knowles (1968) postulated that crosslinking proceeds via the addition of side chain NH_2 to the double bonds of the unsaturated polymers, which were supposed to form as a result of consecutive aldol condensations of the monomer. A typical crosslinking bridge produced by the trimer III, would be

$$\begin{array}{ccc} & \text{CHO} & & \text{CHO} \\ & | & & | \\ -\text{CH}-\text{CH}-\text{CH}_2\text{CH}_2\text{CH}-\text{CH}_2- \\ & | & & | \\ \text{Enzyme}-\text{NH} & & \text{NH}-\text{Enzyme} \end{array}$$

X

The presence of unsaturated polymers in acidic and neutral solutions of GA was not borne out by subsequent studies. Hardy et al. (1976) found that GA reacts very slowly with the model compounds ω-aminohexanoic acid and a-N-acetyl lysine, with an optimal stoichiometry of 2 aldehydes/1 amino-acid. The product was a mixture of closely related compounds, absorbing at 265 nm (but bleached after reduction with borohydride), and containing one titratable group with pKa = 4.7 which was assigned to —COOH. The IR spectrum indicated the presence of OH, CH_2, C=O (carboxy, and carbonyl), C=C, and C=N, and the ^{3}H-NMR spectrum gave no definite evidence for CHO. The proposed structures of the product contained pyridinium rings, formed by condensation of GA with side chain NH_2. Such forms were actually iso-lated from the acid hydrolysis products of GA-reacted ovalbumin (Hardy et al., 1976b), and were assigned the structures shown below on the basis of ^{1}H-NMR spectra and chemical synthesis (Hardy et al., 1977).

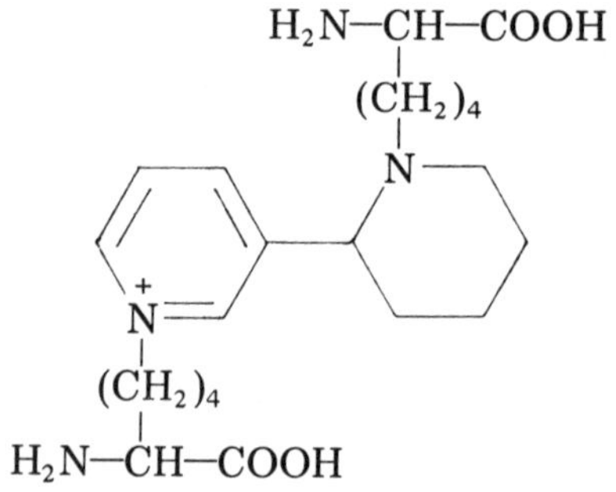

H₂N—CH—COOH
|
(CH₂)₄
|
N—

(pyridinium ring structure)

N⁺
|
(CH₂)₄
|
H₂N—CH—COOH

XI, (Anabilysine)

H₂N—CH—COOH
|
(CH₂)₄
|
N—

(piperidine ring structure)

N
|
(CH₂)₄
|
H₂N—CH—COOH

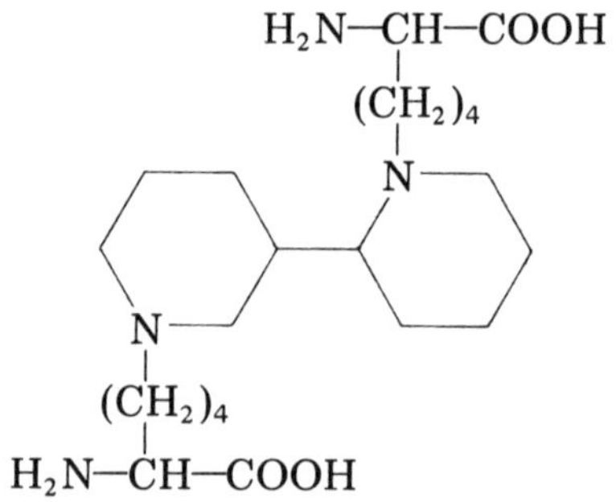

XII

Hardy et al. (1976a) speculated that by analogy to the plethora of protein crosslinking reactions with FA, the above mechanism is not the only one that may actually occur. Indeed, Habeeb and Hiramoto (1968) had estab-lished earlier that GA reacts also with SH, aromatic rings and imidazole rings of amino acids, peptides, and proteins.

We may summarize the above evidence as follows: (1) In acidic or neutral environment, GA reacts with proteins irreversibly via its monomeric form I. (2) Protein modification by GA includes, among other effects, intra- and intermolecular crosslinking, as evidenced by their resistance to treatment with urea, and their increased molecular weight. (3) The GA monomer can alkylate several side chain functions, such as NH_2, SH, imidazole rings, and aromatic rings. The end products of these little known reactions can be quite complicated. (4) Due to their inherent instability, Schiff bases are not found among the acid-stable products of the reaction of GA with proteins.

At alkaline pH, the GA monomer reacts with protein NH_2 with an attendant release of H⁺ to give stable Schiff bases. An interesting side reac-tion was proposed by Blauer et al. (1975) in order to explain the observation that the disappearance of CHO exceeds stoichiometrically the release of H^+.

$$RNH_3^+ + R'CHO \rightleftharpoons R\overset{+}{N}H_2CH(OH)R' \rightleftharpoons RN=CHR' + H^+$$

$$\Big\updownarrow_{RCHO}$$

$$RNH\overset{+}{\diagdown}\begin{matrix} CH(OH)R' \\ CH(OH)R' \end{matrix}$$

GA polymerizes to a variety of products on standing at alkaline conditions and its reaction patterns with proteins are expected to become even more complex. Alkaline conditions are essential for the sporicidal activity of GA (Russel and Haque, 1975; Russell and Vernon, 1975), which is attributed to the prevention of metabolite uptake (Munton and Russell, 1973). On the other hand, Bowes and Cater (1966) reported that alkaline GA does not react well with proteins. These examples illustrate the paucity of existing information with regard to the molecular effects of alkaline GA.

7.2.2. Protein crosslinking with bifunctional imidoesters

Imidoesters are protein modifiers, which in the pH range 7 to 10 react specifically with the free NH_2 groups (Hunter and Ludwig, 1962; 1972; Ludwig and Hunter, 1967). The reaction between an imidoester (XIII) and a primary amine proceeds via the tetrahedral addition intermediate XVI (Hand and Jencks, 1962).

$$R-C=NH_2^+ + H_2N-R'' \rightleftharpoons \left[R-C-NH_2 \right] $$

with groups OR'', XIII; OR', NHR'', XIV; products:

$$R-C=\overset{+}{N}H_2 + R'OH, \quad NHR'' \quad \text{XV}$$

$$R-\overset{\shortparallel}{C} + NH_3, \quad \overset{+}{N}HR'' \quad \text{XVI}$$

(with OR' on the XVI branch)

Thesc reactions were studied recently by means of ^{1}H-NMR spectroscopy by Browne and Kent (1975; see also Peters and Richards, 1977, for a relevant discussion). The end products depend on the pH of the reaction medium. Alkaline environment (pH $\sim$ 10) favors the formation of N-alkyl amidines (XV), with $t_{1/2} \sim 3$ min, and a yield of $\sim$60%. On the other hand, at neutral or slightly alkaline environment, the reaction proceeds in the direction of N-alkyl imidates (XVI), with $t_{1/2} < 1$ min. Subsequently, two-thirds of this product is hydrolysed to the carbonic acid ester and the free amine, and one-third reacts back with the NH_3 produced to yield N-alkyl amidine (XV).

$$R-\overset{\overset{+}{N}HR''}{\underset{\|}{C}}-OR'-\begin{cases} \overset{H_2O}{\longrightarrow} RCOOR' + R''NH_2 + H^+ \\ \\ \overset{NH_3}{\longrightarrow} R-\underset{\underset{NH_2^+}{\|}}{C}-NHR'' + R'OH \end{cases}$$

Bifunctional imidoesters were first applied to protein crosslinking by Dutton et al. (1966), and Hartman and Wold (1967). At sufficiently alkaline pH, crosslinking proceeds via the amidination reaction.

$$P-NH_2 + RO-\underset{\underset{NH_2^+}{\|}}{C}-(CH_2)_n-\underset{\underset{NH_2^+}{\|}}{C}-OR + H_2N-P \rightleftharpoons PNH-\underset{\underset{NH_2^+}{\|}}{C}-(CH_2)_n-\underset{\underset{NH_2^+}{\|}}{C}-NHP + 2ROH$$

The crosslinking property of bifunctional imidoesters was introduced to the study of subunit structure of oligomeric enzymes (Davies and Stark, 1970), of the 30S (Bickle et al., 1972) and the 70S (Nakamura and Wada, 1973) ribosome particles, and as tissue fixatives for electron microscopy (Hassel and Hand, 1974); also, to probe the structure-function interdependence of various membrane-integrated enzyme systems, such as in erythrocyte membranes (Krinsky et al., 1974), sarcoplasmic reticulum vesicles (Yuthavong et al., 1975), mammalian mitochondria (Tinberg et al., 1976) higher plant chloroplasts (Packer et al., 1974), and blue-green algae (Papageorgiou, 1977).

The principal advantages of imidoesters as protein crosslinkers are their specificity for NH_2, and the conservation of the protein-bound positive charge in the vicinity of the original NH_2 group. The first property is essential in arriving to meaningful conclusions about the tertiary and quaternary structure of proteins; the second, in correlating electrophoretic mobility to molecular size. Additional advantages are: their relative simple synthesis (Pinner, 1892); the possibility of introducing "cleavable structures" in the aliphatic chain (e.g. —SS— bridges, Traut et al., 1973; glycol groups, Coggins et al., 1976) which allow further manipulations of the crosslinked products; their high solubility in H_2O; and the not so intolerable conditions (pH ~ 10; Browne and Kent, 1975) at which the amidination reaction is carried out. Their main disadvantages are their tendency for hydrolysis and for the formation of N-alkyl imidates. The first confuses the stoichiometry between NH_2 reacted and crosslinking bridges formed, by allowing for monofunctional attack by the imidoester. The second permits a single imidoester function to crosslink two NH_2 groups, by a further reaction of the N-alkyl imidate (XVI) with an unreacted amine. Reviews on protein crosslinking with bifunctional imidoesters have been published by Wold (1967; 1972),

Fasold et al. (1971), Malcolm and Coggins (1977), and Peters and Richards (1977).

7.2.3. Quantitation of chloroplast crosslinking

Quantitative disagreements in the reported experimental results with chemically crosslinked photosynthetic membranes probably result from different fixation protocolls employed by different laboratories. Table 7.1 illustrates this in the case of GA-treated chloroplasts. An important consideration, therefore, when comparing the functional consequences of membrane protein crosslinking is the extent to which the fixation reaction has proceeded. Relative estimates of it can be obtained in terms of several physical and chemical parameters of chloroplast suspensions.

A convenient way is to measure light or salt-induced, changes in the light scattering properties of the suspension. Chloroplast thylakoids contract in the presence of hyperosmotic salt (Tolberg and Macey, 1965), or under strong illumination in the presence of weak acid anions (Crofts et al., 1967) with an attendant increase in their light scattering ability. Treatment with GA (Packer et al., 1968), or with DMS (Papageorgiou and Isaakidou, 1977) has been shown to obliterate the capacity of thylakoids for osmotic volume changes. In the case of GA, the concentration threshold of osmotic inactivation was found to be a function of the duration of the fixation reaction. For 5-min treatment, it is approximately 30 μmol GA/mg Chl, but for 30-min treatment it drops to approximately 10—15 μmol GA/mg Chl (Papageorgiou and Isaakidou, 1977). The progressive suppression of chloroplasts ability for light-induced scattering changes during incubation with 7 μmol GA/mg Chl is illustrated in Fig. 7.1A.

Light-induced changes of thylakoid membrane conformation, occurring within intact algal cells, can be visualized in terms of the slow change in the yield of Chl a fluorescence (Papageorgiou and Govindjee, 1968a, b; Govindjee and Papageorgiou, 1971). In one species, the red alga *Porphyridium cruentum*, it was shown that GA suppresses the slow fluorescence changes (Mohanty et al., 1972). However, although not explicitly understood in molecular terms, the osmotic insensitization of thylakoids is the most direct manifestation of membrane immobilization as a result of protein cross-linking.

In principle, intermolecular crosslinking of proteins can be quantitated by reference to a spectrum of molecular sizes. In the case of GA-treated chloroplasts the determination of molecular size by means of gel electrophoresis is complicated by the loss of ionizable protein groups (presumably $-NH_3^+$; Sane and Park, 1970). The main drawback, however, to the application of gel filtration techniques is that chloroplast proteins become highly insoluble after chemical crosslinking with GA (Hallier and Park, 1969b; Sane and Park, 1970) and DMS (Isaakidou and Papageorgiou, unpublished results), resisting quantitative extraction by detergents.

TABLE 7.1.

GLUTARALDEHYDE FIXATION OF HIGHER PLANT CHLOROPLASTS (REACTION CONDITIONS AS REPORTED BY SELECTED AUTHORS).

Authors	Plant species	Isolation and fixation media (Conc in mM)	Level of glutaraldehyde		Incubation time (min)	Typical surviving PSII activity (%)
			(% (v/v) or mM)	(μmol/mg Chl)		
Park et al. (1966)	Spinacea oleracea	Chloroplasts isolated from leaves which were infiltrated with 6% glutaraldehyde.	6%	Unspecified	20–180	25
Packer et al. (1968)	S. oleracea	Isolation: Choline—Cl⁻ 100; Tris 10; pH 8. Fixation: Same.	5%	Unspecified	30	Not assayed
Hallier and Park (1969b)	S. oleracea	Isolation: Sucrose 400; Tricine 50; pH 7.4. Fixation: Sucrose 450; Tricine 25; pH 7.4.	6%	Unspecified	20	6
West and Packer (1970)	Pisum satirium	Isolation: Choline-Cl⁻ 100; Tris 0.5; pH 7.8. Fixation: Same.	0–200 mM	0–150 [a]	6	30–35
Sane and Park (1970)	S. oleracea	Isolation: Borate 50; EDTA-washed chloroplasts; pH 8. Fixation: Sucrose 3000; Borate 50; pH 8.	5%	Unspecified	20	Not assayed
West and Mangan (1970)	Brassica oleracea	Isolation: Sucrose 400; Na-phosphate 40; EDTA 10; Na-isoascorbate 10; pH 7.4. Fixation: Sucrose 500.	—	0–160 [a]	7	60

West and Mangan (1973)	P. sativum	Isolation: Sucrose 400; NaCl 10; Na$_2$HPO$_4$—KH$_2$PO$_4$ 80; isoascorbate 10; pH 7.4.	0 50 mM	0—33	7	60
Oku et al. (1973)	S. oleracea	Isolation: Sucrose 400; NaCl 10; Tris 50; pH 7.8. Fixation: Sucrose 20: NaCl 0.5; Tris 2.5; pH 7.8.	6	Unspecified	5	46
Zilinskas and Govindjee (1976)	S. oleracea	Isolation: Sucrose 400; Tris—HCl 50, pH 7.2. Fixation: NaCl 10; phosphate 50; pH 6.8.	0—2%	0—120 [a]	5	23
Hardt and Kok (1976, 1977)	S. oleracea	Isolation: Sucrose 400; NaCl 10; Na-ascorbate 10; Tris · HCl 50; pH 7.5 Fixation: Sucrose 400; NaCl 50; Tris or Tricine 50; pH 7.4.	0— 12.5%	0—300 [a]	5	20—80 [c]
Boardman and Thorne (1977)	S. oleracea	Isolation: Sorbitol 400; NaCl 10; MgCl$_2$ 5; MnCl$_2$ 1; EDTA 2; iso-ascorbate 2; BSA 4%; MES 50: pH 6.5 (Intact chloroplasts) Fixation: Sorbitol 100; NaCl 20; MgCl$_2$ 4; K$_2$HPO$_4$ 10; EDTA 2; Tricine 0.5; pH 8.0.	10 mM	6.7	0—5	Not assayed
Papageorgiou and Issakidou (1977)	S. oleracea	Isolation: NaCl 175; TAPS, 50 Fixation: Same.	—	0—300	5	45

[a] Estimates.

[b] NH$_4^+$-uncoupled rates.

[c] Higher value assayed with PDox as electron acceptor without uncoupler: lower value with FeCN as electron acceptor in the presence of 30 mM CH$_3$NH$_3$Cl.

Another way to estimate the extent of chemical crosslinking is to monitor the disappearance of protein NH_2, which is most conveniently done by the rapid TNBS (trinitrobenzene sulfonic acid) assay of Fields (1972). The method is only approximate, in view of the ability of GA to alkylate several side chain functions besides lysyl ϵ-NH_2, and the possibility for monofunctional attack by the diimidoesters. For example, chloroplasts that are osmotically insensitized by treatment with 100 μmol GA/mg Chl for 5 min, and whose proteins resist quantitative solubilization by detergents, have about 90% of the TNBS detectable NH_2 still free (Isaakidou and Papageorgiou, unpublished experiments). Also, treatment of chloroplasts with 240 mM GA for 7 min, does not affect their susceptibility to tryptic digestion, indicating that all the necessary lysyl residues are still unmodified (West and Mangan, 1970). Although the establishment of crosslinking bridges is not stoichiometrically related to the disappearance of NH_2, the latter observable affords a molecule insight to the events that are associated with the immobilization of the thylakoid membrane.

7.3. GLUTARALDEHYDE-TREATED CHLOROPLASTS AND ALGAE

7.3.1. Biophysical properties

Following treatment with GA, chloroplast membranes become more refractive and scatter a larger fraction of incident light than before (Packer et al., 1968). This may be due either to a GA-effected contraction of the thylakoids, or it may represent a true membrane modification. The second alternative appears more probable, in view of the fact that hyperosmotically contracted chloroplast, scatter appreciably less strongly than the GA-fixed preparations. The light scattering ability of the latter is completely oblivious to osmolarity changes.

At room temperature, higher plants and algae emit Chl a fluorescence, which is asymmetrically distributed in a band centered at 684 nm. At very low temperature, as for example at the boiling point of N_2 ($77°$K), the single band is resolved into three distinct bands with maxima at 684 nm (F684), 696 nm (F696), and 720—735 mm (F735). F684 originated from Chl a of both photosystem II (PSII) and photosystem I (PSI), whereas F696 from Chl a associated only with PSII, and F735 from Chl a associated only with PSI (see chapter 2 and earlier volumes of this series; also Papageorgiou, 1975b). Compared to unfixed chloroplasts, the GA-fixed preparations have the F696 and F735 bands suppressed, while the F684 band is not affected (Oku et al., 1973).

Another manifestation of GA-effected modification of the membrane is the extent of fluorescence quenching by N-methyl phenazonium methosulfate (PMS). As an organic cation having the charge delocalized over the entire molecule, PMS is an ideal quencher for Chl a in solution, because all collisions with excited Chl a — whose excitation is also delocalized in the π-electron system of the porphyrin ring — are effective (Papageorgiou, 1975a). In a chloroplast suspension, PMS quenches by two mechanisms: indirectly by catalyzing cyclic electron transport through PSI, which results in H^+ uptake by the thylakoids (Wraight and Crofts, 1971), and directly by partitioning into the pigment domains of the membrane, where it traps and dissipates the migrating Chl a excitation (Papageorgiou, 1975a). Since direct quenching is independent of the relative PMS—Chl a orientation, its extent should reflect the PMS fraction that partitions to the pigment domain. Zilinskas and Govindjee (1976) employed this method to establish that GA-fixation modifies the membrane in such a way as to allow easier penetration of PMS into its hydrophobic domains.

When light is turned on a darkened sample of chloroplasts, a transient electric field (approx. $1-2 \cdot 10^5$ V/cm) develops across the membrane, as a result of the vectorial electron transport toward the outer face of thylakoids (for recent reviews, see Witt, 1975; Junge, 1977a, b). The macroscopic manifestations of the field are absorbance changes over the entire spectrum of the photosynthetic pigments, whose most pronounced expressions are an absorbance increase at about 510—520 nm ($\Delta A518$), and a decrease at 470—480 nm ($\Delta A480$, Junge and Witt, 1968). The absorbance changes are due to the interaction of the electric field with the transition moments of chlorophylls and carotenoids (analogous to the Stark effect in atomic spectra). The field rises within 20 ns, and decays to a lower steady-state value, as a result of transmembrane ion fluxes, in 10—200 ms, depending on the state of chloroplasts. Channel forming antibiotics (gramicidin D) and ionophores (K^+-valinomycin) accelerate the dissipation of the field.

The light-induced $\Delta A518$ signal, recorded by an ordinary dual wavelength spectrophotometer (equipped with a fast recorder) consists of two parts: a spike which is the vestige of the decaying true electrochemical signal, and an ensuing slow absorbance rise, which is also maximal at about 518 nm. The latter effect is related to the light scattering changes of osmotic origin, brought about by the transmembrane ion fluxes, and has nothing to do with the field-induced electrochromic band shifts. (Witt, 1975; Thorne et al., 1975). Fig. 7.1B illustrates that the true electrochromic change (the "spike") is more resistant to low levels of GA than the slow components, whose decay along the incubation time with the fixative, parallels the suppression of the light-induced light scattering changes (cf. Fig. 7.1A, p. 217). A similar result was also reported by Oku et al. (1973).

Several species of microalgae were found to become permeable to dichlorophenol indophenol (DCIP; an exogenous cofactor for photosynthetic electron transport) following treatment with FA, or GA. (Park et al., 1969a; Ludlow and Park, 1969). Optimal photosynthetic reduction of DCIP required short exposures (~ 4 min) to the fixative, at levels characteristic for each examined species. The extent of inner membrane modification is unknown, but at least for *Porphyridium cruentum* we know that GA penetrates to the stroma regions of the chloroplast, since it inhibits CO_2 fixation, while at the same time it allows the photoreduction of DCIP to proceed (Mohanty et al., 1972).

7.3.2. Primary photoreactions and photosynthetic electron transport

Several of the photosynthetic electron transport intermediates of plants are proteins. Crosslinking is expected, therefore, to have profound effects, particularly if translation, rotation, or intramolecular rearrangements of the carriers are involved in the electron transport process. The surprising result is that osmotically rigid chloroplasts, whose membrane proteins are no longer extractable by detergents, preserve substantial levels of electron transport activity. This is especially noteworthy with regard to the extremely sensitive H_2O-splitting function of chloroplasts.

Park and coworkers were the first to demonstrate that chloroplasts, which were either isolated from GA infiltrated spinach leaves (Park et al., 1966), or were fixed after isolation (Hallier and Park, 1969b), photoevolve O_2 by reducing Hill oxidants, such as ferricyanide (FeCN) and DCIP. These observations were subsequently confirmed in many laboratories (see Table 7.1) and the following regularities were established with regard to the fraction of surviving activity.

In general, the main determinants appear to be the conditions for the fixation and the assay reactions. Low GA levels (20—50 μmol GA/mg Chl) and short periods of incubation (5—10 min) result in high activities, while at the same time they suffice to immobilize the chloroplasts. High fractions are, also, indicated when the activities of control and GA-fixed chloroplasts are compared under weak illumination (Park et al., 1966; Hardt and Kok, 1976), and when lipophilic oxidants, such as PDox and DADox (respectively, mixtures of p-phenylene diamine and diaminodurene with FeCN) are employed as electron acceptors. Nearly 80% of the control O_2 evolution can be recovered with these oxidants (Hardt and Kok, 1976).

High activity levels are observed in the absence of photophosphorylation cofactors (ADP and phosphate, Pi), as well as in the absence of photophosphorylation uncouplers (West and Packer, 1970). The reason is that phosphorylating conditions stimulate the electron transport rate of control chloroplasts, while they fail to do so in the case of the GA-fixed ones (i.e. absence of "photosynthetic control"). Proton transporting uncouplers, such as NH_4Cl and carbonyl cyanide-p-trifluoromethoxy-phenylhydrazone (FCCP) stimulate the electron transport rate of fixed preparations, but to a lesser degree than the rate of the unfixed controls, thereby suppressing the fraction of surviving activity. GA-concentration curves of FeCN-supported O_2 evolution by isolated pea, chloroplasts are shown in Fig. 7.2.

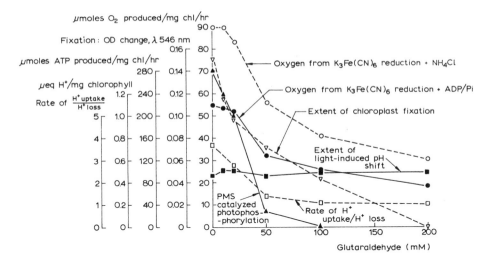

Fig. 7.2. Effect of glutaraldehyde concentration on chloroplast properties. Pea chloroplasts were treated at the indicated glutaraldehyde concentration for 6 min at 0°C. 200 mM glutaraldehyde corresponds to a fixation level of 70 to 130 μmol GA/mg Chl. Ferricyanide reduction was uncoupled with 1 mM NH₄Cl. The rates of active H⁺ uptake, and passive H⁺ efflux, were determined from the initial slopes following illumination and return of chloroplasts to darkness. For further experimental details, see Table I. Reproduced by permission from West and Packer (1970).

Recently, Hardt and Kok (1977) reported that treatment of spinach chloroplasts with an estimated 250–300 μmol GA/mg Chl for 5 min at 0°C has no effect on the primary oxidoreduction reactions at the PSII and PSI reaction centers, while it may inhibit electron transport from H_2O to FeCN by more than 90%. Their evidence is as follows: (a) GA-fixed chloroplasts can handle low fluxes of exciting light equally well as the unfixed chloroplasts; therefore electron transport should be limited at a dark reaction step, rather than a photochemical step. (b) The pattern of O_2 signals, obtained by illuminating the chloroplasts with a series of equally spaced flashes, shows the same periodicity in both preparations. (c) GA-fixed chloroplasts show the fast rise of Chl a fluorescence upon illumination. This is known to reflect the reduction of the electron transport intermediates that link PSII to PSI (Duysens and Sweers, 1963; Malkin and Kok, 1966; also chapter 6 of volume 2 of this series).

According to Hardt and Kok (1977), GA inhibits by interacting specifically with plastocyanin (PC), an electron transporting protein operating between Cyt f and the PS I reaction center (P700).

$$H_2O \rightarrow \ldots (PSII)P680 \rightarrow \ldots Cyt\, f \rightarrow PC \rightarrow P700(PSI) \rightarrow$$

228

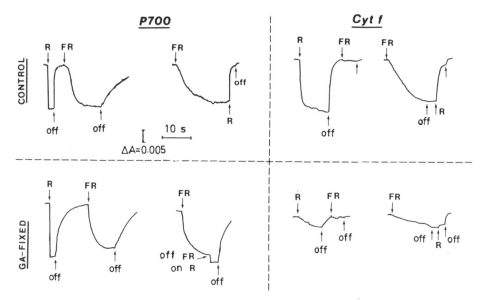

Fig. 7.3. Light-induced P700 and Cyt f absorption changes in control and glutaraldehyde-fixed chloroplasts. R, red actinic light, $\lambda > 650$ nm; FR, far red actinic light, $\lambda = 720$ nm. Chloroplasts were treated with 5% glutaraldehyde for 5 min at 0°C. (A) P700: measuring and reference beams were 703 and 720 nm, respectively; the reaction mixture contained 100 μg Chl/ml and 20 μM methyl viologen. (B) Cyt f: measuring and reference beams were 554 and 540 nm, respectively; the reaction mixture contained 136—160 μg Chl/ml, 100 μM methyl viologen, 10 μM dichlorophyl dimethylurea (DCMU), 5 mM ascorbate, and 100 μM dichlorophenol indophenol. For further experimental details, see Table I. Adapted by permission from Hardt and Kok (1977).

Their evidence is as follows (cf. Fig. 7.3): Red light (λ 650 nm) causes a rapid bleaching (oxidation) of P700 in the control and the fixed preparations by exciting both photosystems. After the red light is removed, P700 is reduced by the pool of intersystem intermediates, which itself was reduced by PSII during the light period, but the recovery is slower in the fixed sample indicating the presence of inhibition. Far red light ($\lambda = 720$ nm), driving PSI only, is capable of reoxidizing P700 in both preparations. When it is turned off, P700 is again reduced by the pool intermediates. However, if far red light is given after a long period of dark incubation (during which the pool intermediates are autoxidized), the P700 signal cannot recover in the subsequent dark period. At this state, red illumination causes a rapid reduction of P700 in the control chloroplasts, but has no such effect in the GA fixed chloroplasts where the intervening inhibition prevents the reducing power of PSII to be communicated to P700 readily, in order to compete with the oxidizing power of PSI. In the latter case, P700 is reduced with electrons leaking from the pool only when the red light is turned off.

Complementary to the above studies were investigations into the properties of the light-induced absorbance changes of Cyt f. Both red and far red light photooxidize Cyt f in the control sample, and the signal recovers in the dark either when sufficient pool reductants were produced (case of red excitation), or by a subsequent red illumination when the pool is empty of reductants (case of far red excitation given after long dark incubation). On the other hand, the two kinds of illumination result in small Cyt f signals only, in the case of GA-fixed chloroplasts, indicating that the oxidizing power of P700 is not readily communicated to Cyt f.

These experiments, therefore, indicate that GA inhibits the communication of the reducing power of PSII to P700, and the communication of the oxidizing power of P700 to Cyt f. Since the only intervening electron carrier is PC, Hardt and Kok (1977) postulated that it is the site where GA inhibits the photosynthetic transport. Additional support for this was provided by the observation that GA precipitates PC from its solutions and inhibits its chemical oxidation and reduction, whereas GA only partially inhibits oxido-reductions of Cyt c (an analogue of Cyt f). As a PC inhibitor, GA resembles KCN (Izawa et al., 1973), $HgCl_2$ (Kimimura and Katoh, 1972), and poly-cations (Brand et al., 1972; Ort et al., 1973). However, it has not been determined whether GA inhibits by blocking a specific group on the PC molecule, or by blocking it in an unfavorable conformation by crosslinking bridges.

Similar results were reported by Boardman and Thorne (1977). Fixation of chloroplasts at approx. 7 μmol GA/mg Chl, has no appreciable effect on the light-induced fast rise of Chl a fluorescence, nor on the reoxidation of Q by PSI, while it suffices to immobilize them (cf. Fig. 7.1). When the GA level was raised to approx. 70 mol GA/mg Chl, the reoxidation of Q by PSI was 50% inhibited. More severe GA treatments were found to suppress the fast rise of Chl a fluorescence in *Porphyridium cruentum* (Mohanty et al., 1972) and in isolated spinach chloroplasts (Oku et al., 1973).

There are several literature reports which do not agree with a specific inhibition of photosynthetic electron transport by GA near the reaction center of PSI. According to Hallier and Park (1969b), chloroplasts treated with 5% GA (v/v) for 20 min were capable of photoreducing methyl viologen (MV) with ascorbate/DCIP as the electron donating system, whereas chloroplasts obtained from spinach leaves that were infiltrated with 6% GA were unable to photoreduce Cyt c with reduced quinone as the electron donor (a typical PSI reaction, Park et al., 1966). Oku et al. (1973) found that chloroplasts preserve 30—40% of the ascorbate/DCIP to MV electron transport activity after treatment with 6% GA (v/v) for 20 min. The coincidence of this fraction with the fraction of P700 that could be titrated in the fixed preparations, both by chemical and by photochemical oxidation, led them to suggest that GA interacts specifically with the PSI reaction center.

Perhaps, the clue to these discrepancies should be sought in the organizational state of isolated chloroplasts prior to fixation with GA. Chloroplasts incubated in low salt media (especially low salt Tricine) disintegrate into unstacked thylakoids, whereas in the presence of sufficient monovalent, or divalent, metal salt the grana stacks are preserved. The process is reversible; stacking is observed on adding salts to the suspension of disorganized thylakoids, and grana disintegrate on lowering the salt concentration of the medium (see chapter 3 of Volume 2 of this series and also Izawa and Good, 1966; Arntzen and Briantais, 1975). Incubation media which give rise to grana containing chloroplasts gave GA-fixed products with strongly inhibited PSI electron transport (Hardt and Kok, 1976; 1977; Papageorgiou and Isaakidou, 1977; cf. also Table 7.1). Also, inhibition of PSI activity was reported when the grana structures were preserved by infiltrating spinach leaves with GA before chloroplast isolation (Park et al., 1966). On the other hand, when the incubation conditions provided for disorganized grana (e.g. Hallier and Park, 1969b; Oku et al., 1973; cf. also Table 7.1) substantial levels of PSI activity survived fixation with GA.

7.3.3. Ion transport and photophosphorylation

Several literature reports maintain that chloroplasts preserve the capacity for light-induced H^+ uptake when fixed with GA. The fraction of surviving activity depends on experimental variables, such as the duration of the fixation reaction, the chemical composition of the fixation and the assay media, and the regime of illumination. Chloroplasts fixed for 30 min with 250 mM GA have 40—60% of the PMS-catalysed H^+ uptake of the respective controls (Packer et al., 1968). With lighter fixation, Boardman and Thorne (1977; cf. Table 7.1) observed less than 20% inhibition of the light-induced H^+ uptake, and nearly no effect at all on the transmembrane pH gradient. More severe fixation, on the other hand, and perhaps the different ionic composition of the fixation medium, were found to suppress H^+ uptake more than 90% (Oku et al., 1973).

The magnitude of H^+ translocation to the thylakoid interior depends on the rate of photosynthetic electron transport which the fixed preparation can sustain. Since GA inhibits primarily a dark step in the electron transport sequence (viz. plastocyanin), but has no effect on the reaction center photochemistry, fixed chloroplasts are capable of utilizing weak light fluxes nearly as well as the unfixed controls. (Hardt and Kok, 1976). A corollary to this is the expectation that the fraction of surviving H^+ uptake will be higher when control and GA-fixed chloroplasts are compared at light fluxes sufficiently low as not to impose a strain on the dark oxidoreduction steps. This is indeed, what has been observed. With continuous illumination, the surviving H^+ transport activity, which is coupled to the photoreduction of FeCN, or of MV, is 20—30%, whereas with flash illumination, GA-fixed chloroplasts show H^+ uptake activities up to 75—90% of that of the control. A lower inhibition of H^+ uptake is also detected in the presence of PMS, which catalyses a cyclic electron transport around PSI that bypasses the GA inhibition site (Hardt and Kok, 1977).

The light-induced transmembrane pH gradient can be collapsed by the monocarboxylic antibiotic nigericin in exchange for monovalent cations (Na^+, K^+, NH_4^+), suggesting that GA fixation does not eliminate cation movements across the membrane (Packer et al., 1968). H^+ transporting uncouplers, such as NH_4Cl, $CH_3NH_2 \cdot HCl$, and FCCP were found to stimulate the rate of noncyclic electron transport in GA-fixed chloroplasts, but to a lesser degree than in the unfixed controls (West and Packer, 1970; West and Mangan, 1970). As a consequence, the percent surviving electron transport activity after treatment with GA is lower when control and fixed chloroplasts are tested in the presence of H^+-transporting uncouplers, than when tested in their absence. Phosphorylation cofactors (ADP and inorganic phosphate) fail to stimulate electron transport in GA-fixed chloroplasts, which is consonant with the observed complete inhibition of the noncyclic and the cyclic photophosphorylation (West and Packer, 1970; see also Fig. 7.2). Photophosphorylation is, indeed, very easily damaged by covalent cross-linking, indicating its requirement for structural flexibility (cf. Jagendorf, 1975) and perhaps for protein mobility. Two minute exposure, to about 7 μmol GA/mg Chl is all that is required to stop photophosphorylation (Boardman and Thorne, 1977).

7.4. IMIDOESTER-TREATED CHLOROPLASTS AND ALGAE

Bifunctional imidoesters were introduced as chloroplast modifiers by Packer and coworkers (1974; see also Papageorgiou et al., 1974). Several structural and functional expressions of isolated chloroplasts are altered after treatment with these compounds. The structural effects are related, more or less, to the multiple interlinking of protein subunits into supermolecules. The functional effects are more complex and subtle and they can be related to such causes as the amidination of specific NH_2 groups, and the molecular rigidity and immobility imposed by inter- and intramolecular crosslinking. In contrast to GA, imidoesters react with biological substrates rapidly, causing extensive protein modification within a few minutes. Also, whereas GA is a polar nonelectrolyte, which can easily penetrate membrane barriers, imidoesters are strongly dissociated electrolytes whose ability to permeate into hydrophobic domains is not yet sufficiently evaluated.

Chloroplasts reacted with 10—20 μmol DMS/mg Chl can no longer contract in response to increases in the osmotic pressure of the surrounding medium (Papageorgiou et al., 1974). At the DMS concentration threshold for osmotic immobilization of the chloroplasts, about 60% of the TNBS-detectable NH_2 are still free, and 50—60% of the SDS-solubilized membrane proteins can enter polyacrylamide gels excluding molecular sizes above 300 000 daltons. When the fixation level is raised to 50 μmol DMS/mg Chl (50% free NH_2), the SDS-solubilized proteins are no longer admitted into such gels (Isaakidou and Papageorgiou, unpublished results). Since chloroplast proteins range from 10 000 to 100 000 daltons (Anderson, 1974), it appears that several protein subunits become interconnected after treatment of chloroplasts with DMS. These phenomena parallel similar observations concerning the inhibition of osmotic swelling of rat liver mitochondria after DMS treatment (but significantly not after ethyl acetimidate treatment), which is also accompanied by the failure of SDS-solubilized proteins to enter polyacrylamide gels with a molecular exclusion threshold of 500 000 daltons (Tinberg and Packer, 1976; Tinberg et al., 1976). DMS has been reported also to effectively insolubilize liver tissue proteins (Hassel and Hand, 1974).

Isolated chloroplasts, immobilized by DMA or DMS, are capable of FeCN-supported O_2 evolution (PSII activity), provided that fixation is carried out at about $0°C$. Fixation at room temperature, at the osmotic immobilization threshold (i.e. 10—20 μmol DMS/mg Chl), yields photosynthetically inactive preparations. Inactive preparations are obtained, also, by low temperature fixation with 50 μmol DMS/mg Chl, or more (Isaakidou and Papageorgiou, 1977). The inhibition of the PSII activity appears to be the result of NH_2 amidination rather than of protein crosslinking because: (a) Methyl acetimidate (MA), a monofunctional imidoester is completely inhibitory when administered at 250 μmol/mg Chl (a level sufficient for the exhaustive amidination; of chloroplast NH_2 groups; Oliver and Jagendorf, 1976); and (b) GA-immobilized chloroplasts are capable of substantial levels of Hill activity (cf. section 7.2.2).

Unlike GA-fixed chloroplasts, the O_2 evolution performance of DMS-fixed chloroplasts is not improved in the presence of lipophilic Hill oxidants (e.g. PDox). When diphenylcarbazide (DPC) is used to substitute for H_2O as the electron donor to PSII, the inhibition is removed. This, then, may suggest the amidination of the H_2O-splitting enzyme complex as the major cause of the inhibition of photosynthetic O_2 evolution by DMS (Isaakidou and Papageorgiou, unpublished results). It should be remarked here that DPC does not relieve the partial inhibition of O_2 evolution of GA-fixed chloroplasts, except in the case when an aged chloroplast preparation was subjected to fixation (Oku et al., 1973).

In contrast to the specific inhibition of PSI activity by GA (cf. section 7.2.2), neither monofunctional imidoesters (MA; Oliver and Jagendorf, 1976), nor bifunctional imidoesters (DMS; Papageorgiou and Isaakidou, 1977) have any effect on the electron transport through PSI. Table 7.2 presents a comparison of the electron transport properties of GA and DMS-fixed chloroplasts.

DMS and/or its hydrolysis products uncouple sarcoplasmic reticulum vesicles (Yuthavong et al., 1974), as well as isolated spinach chloroplasts (Isaakidou and Papageorgiou, unpublished results). Crosslinking with bifunctional imidoesters does not impair the light-induced H^+ uptake by chloroplasts, as deduced by their ability to quench the fluorescence of acridine

TABLE 7.2.

RATES OF FERRICYANIDE-DEPENDENT AND PDox-DEPENDENT OXYGEN EVO-
LUTION BY ISOLATED CHLOROPLASTS FOLLOWING FIXATION WITH GLUTA-
RALDEHYDE (GA) AND DIMETHYLSUBERIMIDATE (DMS)

Type of sampe	Rate (μeq/mg Chl $\cdot$ h)	
	FeCN	PDox
Control chloroplasts	362	361
Fixed chloroplasts, 50 μmole GA/mg Chl	120	262
Control chloroplasts	30$\bar{5}$	436
Fixed chloroplasts, 10 μmol/DMS mg/Chl	135	148
Fixed chloroplasts, 50 μmol/DMS mg/Chl	0	0

dyes upon light energization (Papageorgiou et al., 1974). Once formed, the
transmembrane pH gradient is held longer by the imidoester-fixed preparations than by the controls (Packer et al., 1974). According to Oliver and
Jagendorf (1976), after extensive amidination with the monofunctional
imidoester MA, chloroplasts retain about 80% of the H^+ uptake activity, 75%
of cyclic photophosphorylation, and 60% of the Ca^{2+}-dependent ATPase
activity.

The only alga which has been subjected to crosslinking with a bifunctional
imidoester is the unicellular cyanophyte *Anacystis nidulans* (Papageorgiou,
1977). DMS was shown to amidinate the lamellar proteins, which indicates
that it passes through the cell envelope. Protein components of the latter are
also crosslinked by DMS, since permeaplasts (i.e. permeable cells prepared
by partial digestion of the cell wall with lysozyme) are more resistant to
osmotic damage when they are derived from fixed cells than from the controls. Permeaplasts from DMS-fixed cells are capable of photoinduced O_2
evolution and electron transport (across both photosystems), but quite unable to photoreduce CO_2. Their O_2 evolution activity, however, is temperature sensitive and it is lost above 22°C, whereas in the unfixed cells it is
stable up to 43°C.

7.5. MICROENCAPSULATED CHLOROPLASTS

The encapsulation of biological materials within artificial semipermeable
membranes protects them from the proteolytic and the lipolytic enzymes
that may exist in the outer bulk phase, while it allows their activities to
operate on permeant solutes. The high density of the enclosed material
assists additionally in maintaining the native active structures. Technology,
here in a way, simple mimics nature by placing sensitive macromolecules

within enclosures of limited permeability in order to protect them from a hostile environment.

Preparations of photosynthetic membranes can be encapsulated either within natural membranes, such as the cell membrane and the outer chloroplast membrane, or within synthetic membranes. Natural membranes may require further chemical manipulation (e.g. intermolecular protein crosslinking) to increase their resistance to osmotic rupture and made them permeable to small solutes. Chloroplasts can be isolated with their envelopes intact by mild techniques. In general, they are more active than their naked counterparts, since their photosynthetic membranes have not been exposed to the destructive factors of the vacuolar sap during leaf homogenization (cf. Barber, 1976). In contrast, the reaction conditions for the formation of synthetic membranes can be intolerably harsh for biological materials, inasmuch as the latter must be emulsified in an organic phase.

In this area little relevant work has been reported. Algae permeable to Hill oxidants can be prepared by short exposures to GA and FA (see section 7.3.1), and permeable *Anacystis nidulans*, which is photosynthetically active and resists osmotic lysis can be prepared by treating with lysozyme DMS-fixed cells (see section 7.6). In spite of their advantages, intact chloroplasts have not been exposed to crosslinking by bifunctional reagents, but Morita and Kono (1972; 1973) succeeded in preparing naked chloroplasts, which retained the stroma proteins, by homogenizing rice leaves in sucrose medium containing 3% GA.

The only attempt for artificial microencapsulation of chloroplasts was made by Kitajima and Butler (1976). Their approach was to disperse a mixture of chloroplast fragments, gelatin, and protamine in organic phase (dibutylphthalate), and then crosslink the protamine at the micelle interface with toluene diisocyanate. The product, spheres $10-50 \mu m$ in diameter, could photoreduce MV with ascorbate /DCIP as the electron donor, at 33% of the control rate, but they could not photooxidize H_2O. When hydrogenase (isolated from *Chromatium binosum*) was also present in the capsules, H_2 evolution could be detected with ascorbate/DCIP as the reductant and MV as the electron mediator. Considering that the total microencapsulation process lasted for 15 h, and that the capsules had to be washed free of the organic solvent with petroleum and strong detergent solutions, the surviving activities are quite remarkable.

7.6. STRUCTURAL AND FUNCTIONAL STABILITY OF IMMOBILIZED CHLOROPLASTS AND ALGAE

Preparation of higher plant chloroplasts are notoriously unstable, even when they are maintained at $0°C$ and darkness. Aging appears to be caused by the release of lipoprotein particles and of free fatty acids from the membrane, through the action of lipolytic enzymes. Active chloroplast life can be prolonged by their isolation in strongly buffered alkaline media, containing antioxidants, followed by storage in weakly buffered media that are supplemented with nonionic osmotica and fatty acid scavengers (e.g. bovine serum albumin; see Papageorgiou, 1978, and references cited therein). The salient question, is whether covalent crosslinking can delay the structural and the functional deterioration of chloroplasts. Prolongation of chloroplast activity (especially the photooxidation of H_2O) is of potential technological value since chloroplasts can decompose H_2O into H_2 and O_2 when coupled to hydrogenase (see chapter 10 and Beneman et al., 1973), a process which at best can last for approximately 7 h in reaction mixtures containing O_2 and peroxide scavenging systems (Rao et al., 1976; Fry et al., 1976; see also chapters 10 and 11 of this volume).

Crosslinking of membrane proteins with GA is a positive factor in preserving structure, as it can be inferred from the stabilization of chloroplast morphology and the resistance of membrane components to solubilization. GA-fixed chloroplasts retain their characteristic morphology under the light microscope after extraction with 80% and 100% aqueous acetone (Park et al., 1966), and release less protein when treated with neutral and ionic detergents than the controls (Sane and Park, 1970). On the other hand, acetone extraction completely disorganizes the membrane structure of unfixed chloroplasts (Berkaloff and Deroche, 1975). On dilution of unfixed chloroplast suspensions with buffered medium, chlorophyll is lost from the precipitable particles (typical loss, ca. 35%) and soluble proteins are released. Since the latter are lost faster than Chl, the ratio of soluble N/Chl decreases on successive washings. Fixation with 3% GA prevents these losses (Bottrill and Possingham, 1968). Also, chloroplasts isolated by homogenizing rice leaves in sucrose medium containing 3% GA were found to preserve their morphology under the electron microscope, and retain the stroma proteins, although their envelope was removed (Morita and Kono, 1972; 1973).

Park et al. (1966) reported that GA-fixed chloroplasts, stored in darkness, could photoevolve O_2 in the presence of Hill oxidants, long after the respective unfixed controls were inactivated. Typical "shelf life" at $4°C$ extends to approximately 3 to 4 weeks, but it is shortened to approx. 1 week when the fixed chloroplasts are stored at 17 to $20°C$ (Hardt and Kok, 1976). The electron donor DPC, which bypasses the H_2O-splitting enzyme complex and donates electrons to PSII directly (see chapter 8), stimulates electron transport in old fixed chloroplasts, but not in young fixed chloroplasts; hence, it appears that GA cannot stop the aging of the most sensitive O_2 evolution mechanism, although it may delay it (Oku et al., 1973). GA, further, affords partial protection to chloroplasts against activity losses by such membrane-disruptive processes, as treatments with trypsin, detergents, and chaotropic compounds, as well as heating at $50°C$ for 5 min (Zilinskas and Govindjee, 1976). In contrast, protein crosslinking with DMS does not extend the life of isolated chloroplasts stored at low temperature and darkness (Isaakidou and Papageorgiou, unpublished results).

Covalent crosslinking of membrane proteins is expected to contribute to the structural and functional stability of isolated preparations, since it achieves the same objective as low temperature storage, namely it prevents molecular motions. The important question, however, from a technological viewpoint, is whether such manipulations actually prolong the activities of biological preparations under conditions of continuous work stress. Soluble enzymes show, generally, increased tolerance toward heat denaturation after immobilization, but they are also less active than native enzymes for reasons of steric hindrance (cf. Zaborsky, 1973; Kestner, 1974). On the other hand, when envelope-free spinach chloroplasts were tested for stability of the PSII and the PSI functions at 3°C, under alternating 1-min light—1-min dark cycles, they were found to deteriorate initially faster than GA-fixed chloroplasts, but then to become indistinguishable from them within 2—4 cycles. Eventually, after about 15 min of saturating light illumination, both preparations were rendered completely inactive (Papageorgiou and Isaakidou, 1977). Also, chloroplasts fixed with the bifunctional imidoester DMA, were inactivated earlier than the unfixed controls under continuous illumination (Papageorgiou, unpublished experiments).

We have, also, examined whether protein crosslinking with DMS stabilizes the photosynthetic functions of the unicellular blue-green alga *Anacystis nidulans* (Papageorgiou, 1977). Anacystis can photoevolve O_2 by reducing Hill oxidants after its cells are made permeable to them (permeaplasts Ward and Myers, 1972) by partial digestion of the cell wall with lysozyme. Intact cells and unfixed permeaplasts exhibited optimal H_2O-splitting activity at 43°C. Above this temperature, activity deteriorated rapidly, but below it the ability to evolve oxygen was fairly stable to warming and cooling cycles. In contrast, permeaplasts prepared from DMS-fixed cells (200 μmol/mg Chl) had a threshold of temperature inactivation at 22°C. Below this point O_2 evolution activity was more or less constant. Above it, it deteriorated irreversibly. At 40°C, i.e. at about the temperature optimum of unfixed Anacystis activity, the DMS-fixed cells were totally inactive.

A clue to this remarkable behavior may be found by considering the coincidence of the inactivation temperature of DMS-fixed Anacystis with the phase transition temperature of its membrane lipids (22—24°C; Murata et al., 1975). The unusually high transition temperature of this algal membrane is due to the absence of linolenic acid. (Hirayama, 1967), which is the major unsaturated fatty acid of photosynthetic lamellae in other organisms. As long as the DMS-fixed preparations are kept below the phase transition temperature of membrane lipids, the FeCN-dependent O_2 evolution is fairly stable. This, however, is not possible with isolated higher plant chloroplasts, since their membranes are fluid above 0°C (and probably even at lower temperatures).

We may, then, conclude that artificial crosslinking of membreane proteins into supermolecules destabilizes integral membrane functions by making lipid molecule motions irreversible. This applies primarily to PSII function since the PSI functions of control and DMS-fixed Anacystis were found to be equally resistant to heat denaturation.

7.7. CONCLUDING REMARKS

Comparatively little research has been done, until now, with respect to the preparation of immobilized photosynthetic membranes, and the understanding of their properties. Of the whole arsenal of diverse techniques that has been developed for the immobilization of soluble enzymes (cf. Mosbach, 1976), only the chemical crosslinking of proteins in the lamellar plane has been explored to some detail. Most of the relevant work was performed with GA, whose stretched end-to-end length is 7.5 Å, but recently the bifunctional imidoesters DMA (8.5 Å) and DMS (11.5 Å) were also tried.

At present, we cannot easily differentiate between the functional consequences of the intramolecular and of the intermolecular crosslinking of lamellar proteins. We may infer, however, that there is no strict requirement for protein mobility, and perhaps for protein flexibility, with regard to photosynthetic electron transport in the plane of the lamella. On the other hand, these properties appear to be absolutely necessary for the esterification of phosphate to ADP, and the photosynthetic reduction of CO_2.

With the prospect of some very innovative commercial applications of photosynthetic membranes in vitro (e.g. as transducers of solar energy into low potential reducing power that can be supplied to soluble, or to immobilized bacterial hydrogenases and nitrogenases; cf. chapters 9—12), the question of stability gains immediate importance. In vivo, structural and functional stability of chloroplasts is achieved through continuous regeneration of their chemical constituents. In vitro this is not possible. Nevertheless, the longer storage life of the GA-crosslinked preparations can be used to advantage in a system which continuously regenerates the "spent" membranes in a reactor.

Perhaps, the microencapsulation of photosynthetic membranes by means of a chemical modification of their natural nonphotosynthetic envelopes (e.g. algae, permeaplasts, intact chloroplasts) is the route to follow in order to achieve relatively stable partial photosynthetic capacity in vitro. At present, this area is quite unexplored.

7.8. ACKNOWLEDGMENTS

I would like to thank all authors who so promptly responded to my call for relevant reprints, and who some times provided me also with material

238

not yet published. I would like, also, to express by gratitude to Drs. N.K. Boardman, B. Kok, and L. Packer for their permission to reproduce illustrations from their publications. Finally, I acknowledge the assistance of Mrs. Thoula Lagoyanni in collecting most of the bibliographical material on which this review is based.

REFERENCES

Anderson, J.M. (1974) Biochim. Biophys. Acta, 416, 191—235.

Anderson, P.J. (1967) J. Histochem. Cytochem., 15, 652—661.

Arntzen, C.A. and Briantais, J.-M. (1975) in Bioenergetics of Photosynthesis (Govindjee, ed.), pp. 52—113. Academic Press, New York.

Aso, C. and Aito, Y. (1962a) Bull. Chem. Soc. Jpn., 35, 1246.

Aso, C. and Aito, Y. (1962b) Macromol. Chem., 58, 195.

Barber, J. (ed.) (1976) The Intact Chloroplast. Elsevier North-Holland Biomedical Press, Amsterdam.

Beneman, J.R., Berenson, J.A., Kaplan, N.O. and Kamen, M.D. (1973) Proc. Natl. Acad. Sci. U.S.A., 70, 2317—2320.

Berkaloff, C. and Deroche, M.E. (1975) C.R. Acad. Sci. Paris, Ser. D, 280, 439—442.

Bickle, J.A., Hershey, J.W.B. and Traut, R.R. (1972) Proc. Natl. Acad. Sci. U.S.A., 69, 1327—1331.

Blass, J., Verriest, C. and Leau, A. (1975) J. Chromatogr., 107, 383—387.

Blauer, G., Harmatz, D., Meir, E., Swenson, M.K. and Zvilichovsky, B. (1975) Biopolymers, 14, 2585—2598.

Boardman, N.K. and Thorne, S.W. (1977) Plant Cell Physiol., Special Issue: Photosynthetic Organelles, pp. 157—163.

Bottrill, D.E. and Possingham, J.V. (1969) Biochim. Biophys. Acta, 189, 74—79.

Boucher, R.M.G. (1972) Am. J. Hosp. Pharm., 29, 660.

Boucher, R.M.G. (1974) Am. J. Hosp. Pharm., 31, 546.

Bowes, J.H. and Cater, C.W. (1968) Biochim. Biophys. Acta, 168, 341—352.

Brand, J., Baszynski, J., Crane, F.L. and Krogmann, D.W. (1972) J. Biol. Chem., 247, 2814—2819.

Browne, D.T. and Kent, S.B.H. (1975) Biochem. Biophys. Res. Commun., 67, 126—132.

Coggins, J.R., Hooper, E.A. and Perham, R.N. (1976) Biochemistry, 15, 2527—2533.

Crofts, A.R., Deamer, D.W. and Packer, L. (1967) Biochim. Biophys. Acta, 131, 97—118.

Davies, G.E. and Stark, G.R. (1970) Proc. Natl. Acad. Sci. U.S.A., 66, 651—656.

Dutton, A., Adams, M. and Singer, S.J. (1966) Biochem. Biophys. Res. Commun., 23, 730—739.

Duysens, L.N.M. and Sweers, H.E. (1963) in Studies on Microalgae and Photosynthetic Bacteria (Japanese Society of Plant Physiologists, eds.), pp. 353—372. Univ. of Tokyo Press, Tokyo.

Fahimi, H.D. and Drochmans, P. (1968) J. Histochem. Cytochem., 16, 199—204.

Fasold, H., Klappenberger, J., Meyer, C. and Remold, H. (1971) Angew. Chem., Int. Ed. Engl., 10, 795—801.

Fields, R. (1972) Methods Enzymol., 25B, 464—468.

Frigerio, N.A. and Shaw, M.J. (1968) J. Histochem. Cytochem., 17, 176—181.

Fry, I., Papageorgiou, G.C., Tel-or, E. and Packer, L. (1976) Z. Natuforsch., 32c, 110—117.

Gillett, R. and Gull, K. (1972) Histochemie, 30, 162—167.

Govindjee, and Papageorgiou, G.C. (1971) in Photobiology (A.C. Giese, ed.) Vol. 6, pp. 1—46. Academic Press, New York.

Habeeb, A.F.S.A. and Hiramoto, R. (1968) Arch. Biochem. Biophys., 126, 16—26.

Hallier, U.W. and Park, R.B. (1969a) Plant Physiol., 44, 535—539.

Hallier, U.W. and Park, R.B. (1969b) Plant Physiol., 44, 544—546.

Hand, E.S. and Jencks, W.P. (1962) J. Am. Chem. Soc., 84, 3505—3514.

Hardt, H. and Kok, B. (1976) Biochim. Biophys. Acta, 449, 125—135.

Hardt, H. and Kok, B. (1977) Plant Physiol., 60, 225—229.

Hardy, P.M., Nicholls, A.C. and Rydon, H.N. (1969) Chem. Commun., 565—566.

Hardy, P.M., Nicholls, A.C. and Rydon, H.N. (1976a) J. Chem. Soc., 958—962.

Hardy, P.M., Hughes, G.J. and Rydon, H.N. (1976b) Chem. Commun., 157—158.

Hardy, P.M., Hughes, G.J. and Rydon, H.N. (1977) Chem. Commun., 759—760.

Hartmann, F.C. and Wold, F. (1967) Biochem. Biophys. Res. Commun., 23, 730—739.

Hassel, J. and Hand, A.R. (1974) J. Histochem. Cytochem., 22, 223—239.

Hesse, G. (1973) Acta Histochem., 46, 253—266.

Hirayama, O. (1967) J. Biochem. (Tokyo), 61, 179—185.

Hopewood, D. (1967) Histochemie, 11, 289—295.

Hopewood, D. (1969) Histochemie, 17, 151—161.

Hopewood, D. (1973) in Recent Advances in fixation of Tissues in Electron Microscopy and Cytochemistry (E. Wisse, W.T. Daems, I. Molenaar, P. van Duijn, eds.), pp. 367—381. North-Holland Publ. Co., Amsterdam.

Hunter, M.J. and Ludwig, M.L. (1962) J. Am. Chem. Soc., 84, 4160—4162.

Hunter, M.J. and Ludwig, M.L. (1972) Methods Enzymol., 25B, 585—596.

Izawa, S., Kraayenhof, R., Ruuge, E.K. and Devault, D. (1973) Biochim. Biophys. Acta, 314, 328—339.

Izawa, S. and Good, N.E. (1966) Plant Physiol., 41, 544—552.

Jagendorf, A.T. (1975) in Bioenergetics of Photosynthesis (Govindjee, ed.), pp. 414—492. Academic Press, New York.

Junge, W. (1977a) Annu. Rev. Plant Physiol., 28, 503—536.

Junge, W. (1977b) in Encyclopedia of Plant Physiology, New Series (A. Trebst, M. Avron, eds.), Vol. 5, pp. 59—93. Springer Verlag, Berlin.

Junge, W. and Witt, H.T. (1968) Z. Naturforsch., 23b, 244—254.

Kestner, A.I. (1974) Usp. Khim., Engl. Ed., 43, 690—707.

Kimimura, M. and Katoh, S. (1973) Biochim. Biophys. Acta, 325, 167—174.

Kitajima, M. and Butler, W.L. (1976) Plant Physiol., 57, 746—750.

Korn, A.H., Feairheller, S.H. and Filachione, E.M. (1972) J. Mol. Biol., 65, 525—529.

Krinsky, N.I., Bymum, E. and Packer, L. (1974) Arch. Biochem. Biophys., 160, 350—352.

Ludlow, C.J. and Park, R.B. (1969) Plant Physiol., 44, 540—545.

Ludwig, M.L. and Hunter, M.J. (1967) Methods Enzymol., 11, 595—604.

Malkin, S. and Kok, B. (1966) Biochim. Biophys. Acta, 126, 413—434.

Malcolm, A.D.B. and Coggins, J.R. (1977) Annu. Rep. Chem. Soc. (in press).

Mohanty, P.K., Papageorgiou, G.C. and Govindjee (1972) Photochem. Photobiol., 10, 667—682.

Morita, K. and Kono, M. (1972) Soil Sci. Plant Nutr., 18, 225—231.

Morita, K. and Kono, M. (1973) Soil Sci. Plant Nutr., 19, 317—320.

Mosbach, K. (ed.) (1976) Immobilized Enzymes Methods of Enzymology, Vol. 44. Academic Press, New York.

Moyer, W.W. and Grev, D.A. (1963) Polym. Lett., 1, 29.

Munton, J.J. and Russell, A.D. (1973) J. Appl. Bacteriol., 36, 211—217.

Murata, N., Troughton, J.H. and Fork, D.C. (1975) Plant Physiol., 50, 508—517.

Nakamura, V. and Wada, A. (1973) Biochem. Biophys. Res. Commun., 52, 35—42.

Oku, T., Sugahara, K. and Tomita, G. (1973) Plant Cell Physiol., 14, 385—396.

Oliver, D. and Jagendorf, A. (1976) J. Biol. Chem., 251, 7168—7175.

240

Olson, A.C. and Cooney, X.X. (ed.) (1974) Immobilized Enzymes in Food and Micro-biological Processes. Plenum Press, New York.

Ort, D.R., Izawa, S., Good, N.E. and Krogmann, D.W. (1973) FEBS Lett., 31, 119—122.

Packer, L., Allen, M.J. and Starks, M. (1968) Arch. Biochem. Biophys., 128, 142—152.

Packer, L., Torres-Pereira, J., Chang, P. and Hansen, S. (1974) Proc. 3rd Int. Congr. on Photosynthesis, Vol. 2, pp. 867—872. Elsevier Sci. Publ. Co., Amsterdam.

Papageorgiou, G.C. (1975a) Arch. Biochem. Biophys., 166, 390—399.

Papageorgiou, G.C. (1975b) in Bioenergetics of Photosynthesis (Govindjee, ed.), pp. 319—371. Academic Press, New York.

Papageorgiou, G.C. (1977) Biochim. Biophys. Acta, 461, 379—391.

Papageorgiou, G.C. (1978) Methods Enzymol. (in press).

Papageorgiou, G.C., Case, G.D., Hansen, S. and Packer, L. (1974) Lawrence Berkeley Laboratory Annual Report, pp. 30—31.

Papageorgiou, G.C. and Govindjee (1968a) Biophys. J., 8, 1299—1315.

Papageorgiou, G.C. and Govindjee (1968b) Biophys. J., 8, 1316—1328.

Papageorgiou, G.C. and Isaakidou, J. (1977) in Bioenergetics of Membranes (L. Packer, G.C. Papageorgiou, A. Trebst, eds.), pp. 257—268. Elsevier North-Holland, Amsterdam.

Park, R.B., Kelly, J., Drury, S. and Sauer, K. (1966) Proc. Natl. Acad. Sci. U.S.A., 55, 1056—1062.

Pease, D.C. (1964) Histological Techniques for Electron Microscopy. Academic Press, New York.

Peters, K. and Richards, F.M. (1977) Annu. Rev. Biochem., 46, 523—551.

Pinner, A. (1892) Die Imidoäther und Ihre Derivative. Oppenheim, Berlin.

Pye, E.K. and Wingard Jr., L.B. (eds.) (1974) Enzyme Engineering. Plenum Press, New York.

Rao, K.K., Rosa, L. and Hall, D.O. (1976) Biochem. Biophys. Res. Commun., 68, 21—28.

Rasmussen, K.E. and Albrechtsen, J. (1974) Histochemie, 38, 19—26.

Richards, P.M. and Knowles, J.R. (1968) J. Mol. Biol., 37, 231—233.

Russel, A.D. and Haque, H. (1975) Microbios, 13, 151—153.

Russell, A.D. and Hopewood, D. (1976) Prog. Med. Chem., 13, 271—301.

Russell, A.D. and Vernon, G.N. (1975) Microbios, 13, 147—149.

Sabatini, D.D., Bensch, K. and Barnett, R.J. (1963) J. Cell Biol., 17, 19—58.

Sane, P.V. and Park, R.B. (1970) Plant Physiol., 46, 852—854.

Stibenz, D. (1973) Acta Histochem., 47, 83—88.

Thorne, S.W., Horvath, G., Kahn, A. and Boardman, N.K. (1975) Proc. Natl. Acad. Sci. U.S.A., 72, 3858—3862.

Tinberg, H.M., Nayudu, P.R.V. and Packer, L. (1976) Arch. Biochem. Biophys., 172, 734—740.

Tinberg, H.M. and Packer, L. (1976) in Mitochondria: Bioenergetics, Biogenesis, and Membrane Structure, pp. 349—366. Academic Press, New York.

Tolberg, A.B. and Macey, R.I. (1965) Biochim. Biophys. Acta, 109, 424—430.

Traut, R.R., Bollen, A., Hershey, J.W.B., Sundberg, J. and Pierce, J.R. (1973) Bio-chemistry, 12, 3266—3273.

Ward, B. and Myers, J. (1972) Plant Physiol., 50, 547—550.

Weetall, H.H. (1974) Anal. Chem., 46, 602A—609A.

West, J. and Mangan, J.L. (1970) Nature, 228, 466—468.

West, J. and Mangan, J.L. (1973) J. Agric. Sci. Cambr., 80, 399—406.

West, J. and Packer, L. (1970) Bioenergetics, 1, 405—412.

Whipple, E.B. and Ruta, M. (1974) J. Org. Chem., 39, 1666—1668.

Witt, H.T. (1975) in Bioenergetics of Photosynthesis (Govindjee, ed.), pp. 493—554. Academic Press, New York.

Wold, F. (1967) Methods Enzymol., 11, 617—640.

Wold, F. (1972) Methods Enzymol., 25B, 623—651.

Wraight, C.A. and Crofts, A.R. (1970) Eur. J. Biochem., 17, 319—327.

Yuthavong, Y., Feldman, N. and Boyer, P.D. (1975) Biochim. Biophys. Acta, 382, 116—124.

Zaborsky, O.R. (1973) Immobilized Enzymes. CRC Press, Cleveland.

Zilinskas, B.A. and Govindjee (1976) Z. Pflanzenphysiol., 77, 302—314.

Photosynthesis in relation to model systems, edited by J. Barber
© Elsevier/North-Holland Biomedical Press 1979

Chapter 8

Photosynthetic Water-Splitting Process and Artificial Chemical Systems

A. HARRIMAN * and J. BARBER **

* The Royal Institution, 21 Albemarle Street, London W.1. and ** Department of
Botany, Imperial College of Science and Technology, London S.W.7, United Kingdom

CONTENTS

8.1. INTRODUCTION

The fact that photosynthetic organisms can give off oxygen has been known since 1771 when Joseph Priestley demonstrated that green plants could "purify" air made "foul" by animals. Soon after this the dutchman, Jan Ingenhousz showed that light was required for O_2 evolution to occur, while in 1804 Theodore de Saussure realized that water was involved in the process. However although Nurmser in 1930 had suggested that O_2 was derived from water itself the more popular view during the early part of the twentieth century was that the gas came from carbon dioxide the other main reactant of photosynthesis. For a more detailed historical account see Rabinowitch (1945) and Hill (1972). It was not until 1937 did Robert Hill demonstrate that O_2 evolution could occur from isolated chloroplasts in the absence of CO_2 fixation. Using artificial electron acceptors like ferrioxalate and ferricyanide he demonstrated that the photoreduction of these compounds was at the expense of the oxidation of water. Further support for the concept that O_2 was derived from water came from the experiments by Van Niel (1941) and Ruben et al. (1941).

Since these early experiments there has been considerable effort to understand how green plants can efficiently bring about the photooxidation of water. Progress has been hampered by the labile nature of the "water-splitting" enzyme so that isolation of this biochemically active complex has not as yet been possible. However useful information has been obtained from a wide range of different types of experiments particularly those involving kinetic studies.

In this chapter we give a brief survey of present day knowledge of photosynthetic oxygen evolution and then go on to review in some depth what progress is being made with artificial chemical systems designed to mimic this aspect of green plant functioning.

8.2. PHOTOSYNTHETIC O_2 EVOLUTION

8.2.1. General introduction

Photosynthetic O_2 evolution is associated with photosystem two (PSII) (see chapter 3 of this volume and chapters 3, 4 and 8 of volume 2 in this series). Here the energy of a red quantum (approx. 1.8 eV) is sufficient to induce charge separation in the PS2 reaction centre yielding a strong oxidant Z^+ and a weak reductant Q^- (see Barber, 1977).

$$\text{Z Chl Q} \xrightarrow{h\nu} \text{Z Chl}^+ \text{ Q}^- \rightarrow \text{Z}^+ \text{ Chl Q}^-$$

The reaction centre contains a special form of chlorophyll a (possibly a dimer) which undergoes an absorption decrease at about 680 nm when it is oxidised (Mathis et al., 1976). For this reason it is known as P680. As explained in chapter 3 of this volume, P680 has associated with it a light harvesting pigment system including several hundreds of chlorophyll *a* molecules. The "primary" acceptor Q is probably a quinone in a special environment (van Gorkom, 1974; Knaff et al., 1977) and normally it passes its electron to photosystem one (PS1) via a pool of plastoquinone molecules (see chapter 2 and Williams, 1977). The redox potential of the primary acceptor seems to be about zero volts but there is some evidence that it may be more reduced (Cramer and Butler, 1969; Malkin and Barber, 1978). Moreover recent experiments have suggested that perhaps there is an earlier electron acceptor than Q and pheophytin has been implicated (Klimov et al., 1977). The chemical nature of Z is unknown but it acts as a very efficient electron donor to P680$^+$ (donation time of 30 nsec; see Van Best and Mathis, 1978). However, what is known is that the PSII charge separation takes place across the thylakoid membrane with the oxidation process leading to O_2 evolution occurring on the inner side (see Trebst, 1974).

The redox state of Z$^+$ must be more positive than +0.8 V since this is the average potential necessary to oxidise water and there is evidence that Z$^+$ does not directly interact with water. Several types of experiments have indicated that the overall reaction time for O_2 evolution from photosynthetic tissue is 0.6 msec but this value reflects the rate determining step on the reducing side of PSII. According to Witt (1971) the average time for the donation of an electron from water to Z$^+$ is 0.2 msec.

8.2.2. Four quantum requirements

8.2.2.1. Oscillatory phenomenon

To form one molecule of oxygen it is necessary to remove four electrons from two molecules of water.

$$2 H_2O \rightarrow 4 e^- + 4 H^+ + O_2$$

As mentioned above, the average potential of the four oxidising equivalents must be higher than +0.8 V. Moreover at least four quanta are necessary to drive the four electron process. Therefore in green plants either four reaction centres cooperate together to bring about the evolution of one O_2 molecule or charge storage occurs within individual reaction centres. These two possibilities can be distinguished by monitoring O_2 evolution induced by short bright light flashes capable of bringing about single excitation of the PSII reaction centres. Thus, for a dark adapted system, if cooperation between reaction centres occurs then O_2 should be evolved on the first flash while further flashes would be required if charge accumulation was necessary. In fact years ago several workers demonstrated that there was

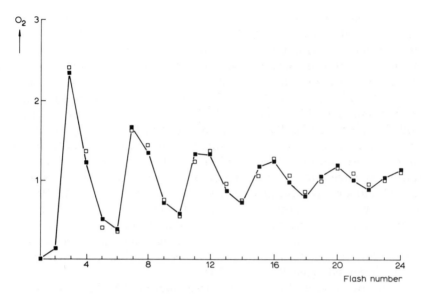

Fig. 8.1. Oxygen flash yield sequence observed with isolated chloroplasts with nicatin-amide adenine dinucleotide phosphate (NADP) as the terminal electron acceptor. Before experimenting, the chloroplasts were given a long dark pretreatment period (Data from Forbush et al., 1971).

no O_2 evolved on the first flash (Allen and Frank, 1955; Whittingham and Bishop, 1961). In more recent years, the availability of short duration flash sources, together with sensitive O_2 electrodes, has enabled a thorough study of this type to be made. Joliot et al. (1969) were the first to show that with successive flashes of light a characteristic pattern of oxygen evolution occurred. Fig. 8.1 shows an example of flash induced O_2 evolution as measured by Forbush et al. (1971) where it can be seen that there was essentially no O_2 evolved on the first and second flash while the third flash gave the maximum yield. Subsequent yields show a damped oscillation of four. Several hypotheses have been put forward to explain these results but the most widely accepted is that presented by Kok et al. (1970). In this hypothesis the O_2 evolving enzyme can exist in four photoactive charged states S.

In the "S-state" model it is envisaged that oxidising equivalents are accumulated by successive reduction of Z^+.

$$Z\ P680\ Q \xrightarrow{h\nu} Z\ P680^+\ Q^- \rightarrow Z^+\ P680\ Q^- \xrightarrow[\ \ \ \ \ S_n\ \ \ S_{n+1}\ \ \ \ ]{} Z\ P680\ Q^-$$

The sequence leading to O_2 evolution is thought to be:

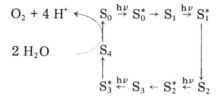

8.2.2.2. Properties of the "S-state" clock

Below are listed the main properties of the S-state model. We have not given details about the various experiments involved in determining these facts and for more information the reader is referred to excellent reviews by Joliot and Kok (1975) and Radmer and Cheniae (1977).

(i) The suffix represents the number of oxidised equivalents e.g. $S_1 = S^+$, $S_2 = S^{2+}$ etc.

(ii) Because the maximum yield of O_2 evolution is observed on the third flash then S_1 is the most abundant S species in the dark and that the S_0 state, although stable in the dark, can only be generated by the $S_4 \rightarrow S_0$ transition.

(iii) In continuous light, or after the oscillations are completely damped, all the S-states exist in equal concentrations so that on terminating the illumination the S_2 and S_3 states decay to S_1. Thus in the dark the S_1 to S_0 ratio is 3 : 1.

(iv) The damping of the oscillations can be explained in terms of the inability of the flash to activate some reaction centres (misses) while in some other reaction centres two photoacts occur during a single flash (doublet hits).

(v) Misses and double hits can also explain why the yield on the third flash is not quite three times the steady-state yield or the yield observed after a long series of flashes as would be expected.

(vi) The relaxation steps $S_n^* \rightarrow S_{n+1}$ could be due either to the reoxidation of Q^- (i.e. Q^- to Q) or to the turnover of the electron exchange reaction $S_n + Z^+ \rightarrow S_{n+1} + Z$. At room temperature these transitions take a few hundred μsec except for the $S_3^* \rightarrow S_4 \rightarrow S_0 + O_2$ which takes about 1 msec.

(vii) The times for deactivation of the S_2 and S_3 states to the S_1 state in the dark have been determined by studying the effect of varying the time between giving flashes (see Joliot and Kok, 1975). The back reactions vary considerably under different conditions ranging from 1 to 60 sec.

8.2.2.3. Proton release

According to the "S-state" scheme presented above the conversion of S_4 to S_0 should yield one oxygen molecule and four protons. Attempts have been made to prove that H^+ release follows O_2 evolution in the expected way. Using a sensitive pH electrode technique Fowler and Kok (1974) obtained results which suggested that O_2 and H^+ were released in a concerted reaction although the stoichiometry did not fit the model. The problem with this measurement is that water oxidation takes place on the inner side of the thylakoid membrane (see Trebst, 1974) and consequently external pH electrode measurements are hampered by the slow diffusion of protons across the membrane. More recently Auslander and Junge (1977) have used the pH indicator neutral red to monitor proton release on the inner side of the membrane and were unable to attribute the flash induced lowering of pH entirely to the $S_4 \rightarrow S_0$ transition. Apparently protons are released at other stages in the cycle, a conclusion also reached by Saphon and Crofts (1977).

8.3. CHEMICAL NATURE OF THE WATER-SPLITTING ENZYME

Although there have been many attempts to isolate a biochemically active complex from photosynthetic tissue containing the basic functional unit required for the photooxidation of water, none have been successful (see Barber, 1977). However despite this there are a number of observations which have given hints to its composition.

8.3.1. Involvement of manganese

In section 8.2 we have concluded that photosynthetic O_2 evolution occurs via a concerted four electron mechanism in which four oxidizing equivalents are sequentially stored. There is indirect evidence that the charge accumulator may involve manganese, a suitable candidate since it can exist in several different oxidation states. It is well known that when photosynthetic organisms are grown in the absence of manganese they loose the ability to evolve O_2 (Pirson et al., 1952) and also that this metal is required for PSII activity (Kessler et al., 1957; Cheniae, 1970). However it is possible to incorporate Mn into the depleted cells so as to restore normal O_2 evolution. The reincorporation process only occurs with illuminated intact living systems (see Radmer and Cheniae, 1977). For maximum activity it seems that one Mn atom is required per 50 to 100 chlorophyll a molecules in line with other evidence that there is about 4 Mn atoms per PSII reaction centre (Cheniae and Martin, 1970).

The involvement of Mn in the O_2 evolving enzyme system is also supported by experiments where isolated chloroplasts are subjected to various treatments which remove this metal from its binding site. For example, treatment with hydroxylamine, washing with alkaline Tris and gentle heating will release about two thirds of the bound Mn with a concomitant decrease in O_2 evolution (see Radmer and Cheniae, 1977). These treated chloroplasts do, however, maintain the ability to promote electron flow through PSII except they require the addition of an electron donor, such as tetraphenyl boron to the suspending medium. Thus it seems that the extractable Mn which controls the oxygen evolving process is not required for the photooxidation of primary electron donors (e.g., Z and P680). There is an intriguing report that O_2 evolution can occur from Tris treated chloroplasts after washing with reduced dichlorophenol indophenol or reduced hydroquinone (Yamashita et al., 1971) which at first sight suggests that the readily extractable Mn is not necessarily an intimate structural feature of the water splitting enzyme. However it is not clear whether these workers had removed Mn by their treatment.

Unlike free Mn, the bound Mn in the chloroplast does not seem to show a definite ESR signal although recently Siderer et al. (1977) have claimed to have observed an ESR signal due to bound Mn^{2+} which underwent light induced changes attributed to PSII activity. Their observations contrast with other workers, although all agree that an EPR signal indicative of Mn^{2+} $(H_2O)_6$ is readily seen when the bound Mn is released by the various treatments mentioned above (e.g. see Blankenship and Sauer, 1974, Blankenship et al. 1975).

Although it is still not absolutely clear which fraction of the chloroplast Mn pool is associated with the water splitting process, recent studies using the pulsed NMR technique to follow water proton relaxation do at least indicate the important role of this metal in photosynthetic O_2 evolution. If

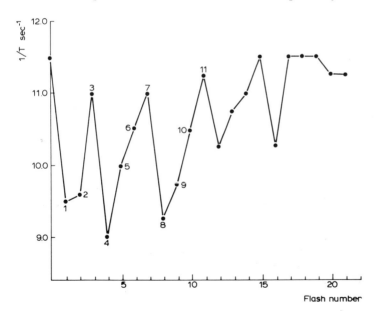

Fig. 8.2. The water—proton relation rate (1/T) as a function of flash number. The chloroplasts were dark pretreated and the relaxation rates following a particular flash determined at the beginning of the flash and followed until the next flash in the sequence (From Wydrzynski et al., 1976).

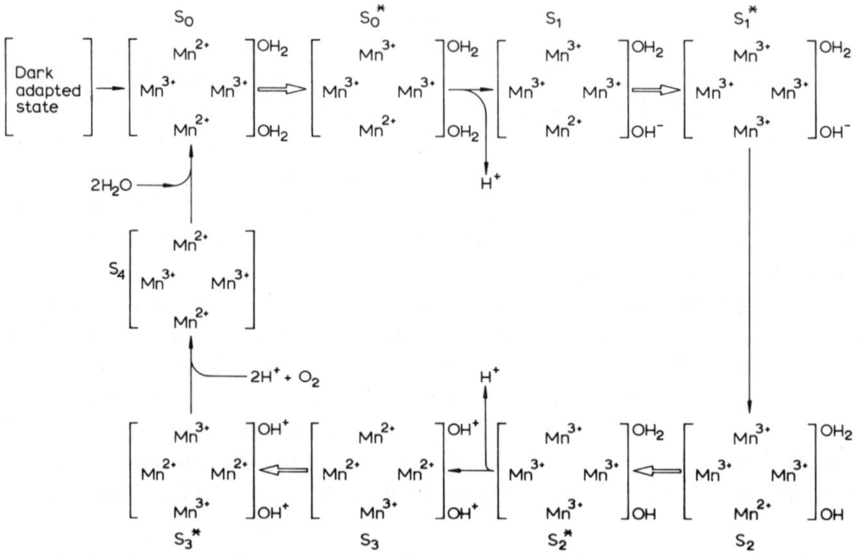

$$
\begin{bmatrix} \text{Dark} \\ \text{adapted} \\ \text{state} \end{bmatrix} \rightarrow
\overset{S_0}{\begin{bmatrix} & Mn^{2+} & \\ Mn^{3+} & & Mn^{3+} \\ & Mn^{2+} & \end{bmatrix}}\!\!\begin{bmatrix} OH_2 \\ OH_2 \end{bmatrix} \rightleftharpoons
\overset{S_0^*}{\begin{bmatrix} & Mn^{3+} & \\ Mn^{3+} & & Mn^{3+} \\ & Mn^{2+} & \end{bmatrix}}\!\!\begin{bmatrix} OH_2 \\ OH_2 \end{bmatrix} \leftarrow
\overset{S_1}{\begin{bmatrix} & Mn^{3+} & \\ Mn^{3+} & & Mn^{3+} \\ & Mn^{2+} & \end{bmatrix}}\!\!\begin{bmatrix} OH_2 \\ OH^- \end{bmatrix} \rightleftharpoons
\overset{S_1^*}{\begin{bmatrix} & Mn^{3+} & \\ Mn^{3+} & & Mn^{3+} \\ & Mn^{3+} & \end{bmatrix}}\!\!\begin{bmatrix} OH_2 \\ OH^- \end{bmatrix}
$$

$2H_2O \longrightarrow$ H^+

$$
\overset{S_4}{\begin{bmatrix} & Mn^{2+} & \\ Mn^{3+} & & Mn^{3+} \\ & Mn^{2+} & \end{bmatrix}}
$$

$\longrightarrow 2H^+ + O_2$ H^+

$$
\overset{S_3^*}{\begin{bmatrix} & Mn^{3+} & \\ Mn^{2+} & & Mn^{2+} \\ & Mn^{3+} & \end{bmatrix}}\!\!\begin{bmatrix} OH^+ \\ OH^+ \end{bmatrix} \Leftarrow
\overset{S_3}{\begin{bmatrix} & Mn^{2+} & \\ Mn^{2+} & & Mn^{2+} \\ & Mn^{3+} & \end{bmatrix}}\!\!\begin{bmatrix} OH^+ \\ OH^+ \end{bmatrix} \rightarrow
\overset{S_2^*}{\begin{bmatrix} & Mn^{3+} & \\ Mn^{3+} & & Mn^{3+} \\ & Mn^{3+} & \end{bmatrix}}\!\!\begin{bmatrix} OH_2 \\ OH \end{bmatrix} \Leftarrow
\overset{S_2}{\begin{bmatrix} & Mn^{3+} & \\ Mn^{3+} & & Mn^{3+} \\ & Mn^{2+} & \end{bmatrix}}\!\!\begin{bmatrix} OH_2 \\ OH \end{bmatrix}
$$

Fig. 8.3. A speculative model for photosynthetic oxygen evolution based on water—proton relaxation studies presented by Govindjee et al. (1977).

water is associated with bound Mn then this technique can indirectly follow changes in the redox state of Mn since the water proton relaxation rate can be changed by several orders of magnitude. Wydrzynski, Govindjee and colleagues (Wydrzynski et al., 1975, 1976; Govindjee et al., 1977) have adopted this approach and by using flash excitation have shown that water proton relaxation rates oscillate as a function of the flash number (see Fig. 8.2). In order to explain their observations and relate them with the flash induced O_2 evolution pattern they have suggested the speculative scheme shown in Fig. 8.3. This scheme takes into account the fact that protons are released at other stages in the cycle and not just at the $S_4 \rightarrow S_0$ transition (see also Renger, 1977). However in chemical terms the model can be criticized since, in practice, it would be expected that very short spacial distances must exist between the redox centres in order to allow the formation of a 0-0 band. It seems unlikely that the four coordinated manganese atoms in the proposed complex could be sufficiently close (only a few Angstroms) to facilitate the molecular bond of O_2 to be established without the involvement of free radicals.

8.3.2. Other effects

Except for those experiments which implicate the possible involvement of Mn in the water splitting process there are no other observations which give ideas as to the composition of the enzyme itself. However there are a number of findings which may ultimately be useful for elucidating the molecular process involved. For example, Hind and Whittingham (1963) and later Izawa et al. (1969) showed that NH_3 and methylamine can inhibit O_2 evolution. The mechanism of the inhibition does not involve Mn release but rather a direct interaction of these compounds with the water splitting enzyme itself (see Velthuys and Amesz, 1975). It is not difficult to imagine that the highly electronegative N-atom of these amines will compete effectively as an electrophile with the O atom of water, and indeed NH_3, like water will form complexes with Mn^{2+} (Angelici, 1973). In addition, ligand field stabilization effects in complexes of Mn^{3+} (but not Mn^{2+}) are greater with NH_3 than with H_2O (Orgel, 1966).

Some years ago Warburg and Lütggen (1946) suggested that chloride ions were required for photosynthetic O_2 evolution. Since then several groups of workers have found support for this suggestion (Arnon and Whatley, 1949; Punnet, 1959; Izawa et al., 1969). It seems that in some way Cl^- or other monovalent anions of highly dissociated inorganic acids are required for efficient transfer of electrons from water to $P680^+$ but not for the reduction of $P680^+$ when artificial electron donors are used.

In addition to Cl^-, bicarbonate ions have been reported to be necessary for O_2 evolution. Reports for this requirement stem back to the old ideas of Warburg (See Warburg and Krippahl, 1960) that O_2 evolution resulted from the photochemical decomposition of a chlorophyll-CO_2 complex. This involvement of bicarbonate in the photosynthetic oxygen evolution is still advocated by Metzner (1975), although Jursinic et al. (1976) have obtained convincing evidence that the HCO_3^- effect takes place on the reducing side of the PSII reaction centre.

8.4. IN VITRO WATER-SPLITTING SYSTEMS

Although knowledge of the photosynthetic water splitting process mainly relies on kinetic information with little known about the molecular processes involved, this has not restricted efforts to construct model systems. It is the purpose of the remaining part of the review to describe these various models, to compare their relative efficiency, and to show what progress has been made in this field.

8.4.1. Homogeneous solution

8.4.1.1. Formation of hydrogen

8.4.1.1.1. Redox cycle. Perhaps the simplest method for photogeneration of hydrogen from water involves irradiation of a low valence transition metal

TABLE 8.1

PHOTOGENERATION OF MOLECULAR HYDROGEN BY IRRADIATION OF METAL IONS IN DILUTE ACID SOLUTION

Metal	Φ_{H_2}	λ_{exct} (nm)	λ_{min} (nm)	Reference
Fe^{2+}	0.40	254	287	Heidt et al. (1962)
Cr^{2+}	0.15	254	380	Dainton et al. (1959)
Eu^{2+}	0.20	254	405	Davis et al. (1977)
Ce^{3+}	0.001	254	~270	Heidt and McMillan (1954)
$Mo_2(SO_4)_4^{4-}$	0.053	254	~560	Gray et al. (1977)
$(Rh^I)_2$	0.004	546	~600	Gray et al. (1977a)

ion at a wavelength corresponding to a charge transfer to solvent (CTTS) transition. This has the effect of transferring an electron from the metal ion on to a water molecule.

$$M^{n+} + H_2O \rightarrow M^{(n+1)+} + H^{\cdot} + OH^-$$

The early work in this field has been comprehensively reviewed by Marcus (1956) and by Dainton and James (1958) and it is well known that many transition metal ions give rise to hydrogen production by this method. A summary of much of the available data is given in Table 8.1 and it is seen that the quantum yield for hydrogen formation (Φ_{H_2}) can be very high. Unfortunately, the CTTS absorption band is normally found in the high energy UV region and most transition metal ions require 254-nm irradiation in order to bring about hydrogen formation from water. In fact the position of the CTTS band depends upon two terms, namely, the ionisation energy of the ion ($I_{M^{n+}}$) and a constant (A) which is characteristic of the water molecule and its derived ion and radical only.

$$hc/\lambda = A + E^0F$$

$$I_{M^{n+}} = E^0F + constant$$

(where E^0 is the standard redox potential, F is the Faraday and the other symbols have their usual meanings). Dainton and James (1958) have located the long wavelength edge of the CTTS absorption band for a wide series of transition metal ions and have shown that there does exist a linear relationship between hc/λ and E^0. The λ term represents the lowest energy excitation wavelength that can be used to generate hydrogen with a given metal ion. These values are included in Table 8.1 and, in general, show that the system cannot be used in the solar energy region.

TABLE 8.2.

WAVELENGTH DEPENDENCE FOR Φ_{H_2} FROM PHOTOLYSIS OF Eu^{2+} IN 1 M $HClO_4$ SOLUTION

λ_{exct} (nm)	Φ_{H_2}
436	0
405	0.053
365	0.065
334	0.072
313	0.079
254	0.20

From Davis et al. (1977).

However, for some of the systems, hydrogen formation can be achieved with visible light excitation. The threshold limit for Eu^{2+} is 405 nm where $\Phi_{H_2} = 0.053$ compared to a value of 0.20 at 254-nm irradiation. The absorption spectrum of Eu^{2+} consists of a long wavelength band, $\lambda_{max} \sim 325$ nm, with a barely discernible "staircase" structure on the high energy side and an overlapping band in the UV centred around 250 nm. These bands have been characterised as $4f \rightarrow 5d$ transitions, the two bands being due to ligand field splitting of the d-orbitals in a cubic field. The 5d-orbitals show appreciable CTTS character, thus giving rise to the observed photooxidation of Eu^{2+}. The two transitions have similar character as evidenced by the same type of chemistry, hydrogen formation occurs at all wavelengths, but reaction from the upper electronic level appears to occur with a larger dissociative rate constant and hence a higher quantum yield (Table 8.2). It is possible that the observed wavelength dependence can be explained in terms of the initially formed products being provided with more energy thus making it easier to escape from the solvent cage.

Davis et al. (1977) have found that the efficiency of hydrogen production is dependent upon the acidity of the medium. The observed quantum yield relative to that at 1 M $HClO_4$ follows the equation:

$$\Phi_{rel} = \Phi_0 + k[H^+]^{1/2}$$

with $\Phi_0 = 0.15$ and $k = 0.85$, for irradiation at 366 nm, as shown in Fig. 8.4. Ryason (1977) has reported that Φ_{H_2} remains constant for [HCl] = 0.48 and 4.8 M but this would seem to be in error. Certainly, the detailed data provided by Davis et al. show clearly that Φ_{H_2} is dependent upon acid concentration in a perchlorate medium. Possibly, an answer to this discrepancy may be found in terms of an anion effect since it has been found that Φ_{H_2} increased at higher ionic strength in chloride media but decreased in the presence of perchlorate. However, these ionic strength effects were slight and

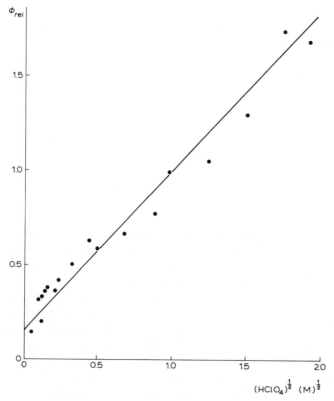

Fig. 8.4. The dependence of the relative quantum yield of hydrogen production (Φ rel) upon perchloric acid concentration (using 366-nm irradiation) (From Davis et al., 1977).

probably reflect small changes in activity coefficients rather than a primary salt effect.

The two binuclear complexes included in Table 8.1 have been reported only recently but both compounds are believed to be photoactive in the visible region. The molybdenum complex described by Gray et al. (1977) involves a strong metal to metal bond and irradiation results in a one electron transfer reaction.

$$[Mo_2(SO_4)_4]^{4-} + H^+ \xrightarrow{h\nu} H^{\cdot} + [Mo_2(SO_4)_4]^{3-}$$

Irradiation at 254 nm gives a poor stoichiometric balance but this is much improved when visible light (λ = 515 nm) is used. The authors do not give a Φ_{H_2} value for visible light irradiation but if the reaction follows the general pattern then this value will be much lower than that found with 254-nm irradiation (Φ_{H_2} = 0.053). Very recently, Gray et al. (1978) found that Φ_{H_2}

TABLE 8.3.

DISAPPEARANCE QUANTUM YIELDS FOR PHOTOLYSIS OF BINUCLEAR MOLYB-
DENUM(II) COMPLEXES IN AQUEOUS ACIDS (254-nm EXCITATION)

Complex	Solvent	Φ_{254}
$K_4Mo_2(SO_4)_4$	5 M H_2SO_4	0.19
	2.5 M H_2SO_4	0.17
	0.5 M H_2SO_4	0.16
	0.05 M H_2SO_4	0.15
	0.001 M H_2SO_4	0.18
$Mo_2(aq)^{4+}$	1 M HO_3SCF_3	0.035
	1 M HO_3SCF_3/	0.031
	0.15 M CH_3OH	
$K_4Mo_2Cl_8$	3 M HCl	0.14
	6 M HCl	0.092
$(NH_4)_4Mo_2Br_8$	3 M HBr	0.043

was independent of acid concentration and that the stoichiometric balance
became much better when the reaction was restricted to low percentage con-
versions. Other binuclear molybdenum complexes, such as $Mo_2(aq)^{4-}$ and
$Mo_2Cl_8^{4-}$, were also found to produce hydrogen gas upon 254-nm irradiation.
A table of the experimental results obtained in this study is given above
(Table 8.3) where Φ_{254} refers to the quantum yield for loss of compound
rather than formation of hydrogen. A particularly important observation
made during these experiments was that the reaction involved a hydride
intermediate (the significance of metal hydrides is discussed later). Thus,
irradiation of $Mo_2Cl_8^{4-}$ in 3 M HCl results in formation of $Mo_2Cl_8H^{3-}$ with a
quantum yield of 0.13. In a subsequent step, $Mo_2Cl_8H^{3-}$ thermally decom-
poses to yield one molecule of hydrogen gas and $Mo_2(\mu OH)_2aq^{4+}$. A similar
reaction was found with $Mo_2Br_8^{4-}$ although here the quantum yield was much
lower.

The binuclear rhodium complex, which was also described by Gray et al.
(1977a), consists of two Rh^I atoms bridged by four 1,3-dicyanopropane
linkages (Fig. 8.5). At the very high acid concentrations used, the complex
exists in a monoprotonated form. Irradiation of the complex in 12 M HCl at
546 nm gives clean conversion to the corresponding binuclear Rh^{II} complex
and formation of hydrogen (Fig. 8.6).

$$[Rh_2(bridge)_4H]^{3+} + H^+ \xrightarrow[2Cl^-]{h\nu} [Rh_2(bridge)_4Cl_2]^{2+} + H_2$$

A value for Φ_{H_2} of 0.004 was obtained but this is greatly reduced in the
presence of oxygen.

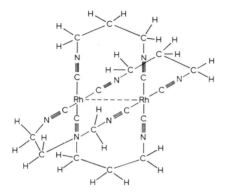

Fig. 8.5. Structure of the binuclear rhodium (I) complex.

The reaction mechanism has been well established for the mononuclear systems. It has been found that Φ_{H_2} depends upon several parameters, all associated with secondary thermal reactions of H˙ after it has escaped from the solvent cage. The reaction mechanism appears to be identical for each metal system and is well demonstrated by the chromium (II) photolysis. Here, Dainton et al. (1959) have shown that $\Phi_{H_2} = 0.5 \, \Phi_{Cr^{3+}}$ and is independent of the absorbed light intensity and concentration of Cr^{2+} ions. However, Φ_{H_2} increased slightly with the concentration of acid in the range 3×10^{-4} to 1 M and then increased steeply at higher acid concentrations.

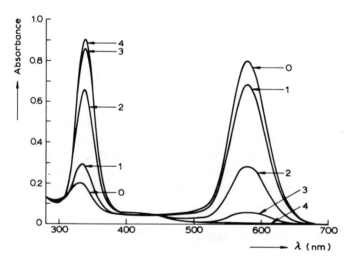

Fig. 8.6. Irradiation (546 nm) of $[Rh_2 \, (bridge)_4] \, (BF_4)_2$ in outgassed 12 M HCl. The four scans after t = 0 were taken at intervals of about 2 min (From Gray et al., 1977).

The dependence upon [H^+] was taken as evidence for reaction between protons and free hydrogen atoms;

$$H^. + H^+ \rightleftharpoons H_2^{.+}$$

whilst the absence of a light intensity effect precludes direct $H^.$ combination. Hydrogen gas is then formed by a thermal reaction with a second Cr^{2+} ion.

$$H_2^{.+} + Cr^{2+} \rightarrow H_2 + Cr^{3+}$$

It is known that the exact nature of both chromous and chromic ions depends upon the type and concentration of acid used. Thus, the different complex ions which exist in very strong acid solution were believed to account for the anomolous Φ_{H_2} values obtained under such conditions.

Therefore, many transition metal ions can be used for the photogeneration of hydrogen from water. However, in order to obtain a cyclic system it is necessary that the oxidised form of the metal can react with water to reform the original metal ion.

$$M^{(n+1)+} + H_2O \rightarrow M^{n+} + OH^. + H^+$$

For this process to proceed in the dark the $M^{(n+1)+}/M^{n+}$ redox couple must exceed 1.23 V. Of the systems listed in Table 8.1, only the Ce^{4+}/Ce^{3+} couple fulfills this basic requirement. However, although Ce^{4+} is thermodynamically unstable with respect to water, it was found by Heidt and Smith (1948) that aqueous solutions of Ce^{4+} were quite stable when left in the dark. Thus, none of the above systems complete the cycle by thermal reduction of $M^{(n+1)+}$ but both Ce^{4+} and Fe^{3+} are photoreduced upon UV irradiation and these systems are covered in a later section.

The photolysis of simple halide ions in aqueous solution also gives rise to hydrogen production. Again, irradiation corresponds to a CTTS absorption band and the reaction involves transfer of an electron from the anion to a water molecule which is intimately associated with the ion. The CTTS absorption bands occur in the UV region for all the halide ions and the reaction mechanism has been studied in detail for iodide ions only. Rigg and Weiss (1952) found that the Φ_{H_2} was dependent upon pH, [I^-], and [I_2] and they proposed a series of secondary reactions which accounted for all the experimental observations. The primary photochemical step involves dissociation of the reactants into radicals.

$$I^- + H_2O \overset{h\nu}{\rightarrow} I^. + H^. + OH^-$$

Secondary reactions proceed to give the observed products.

$$H^{\cdot} + H^{+} \rightleftharpoons H_2^{+}$$

$$I^{-} + H_2^{+} \rightarrow H_2 + I^{\cdot}$$

$$2\,I^{\cdot} \rightarrow I_2$$

$$I_2 + H^{\cdot} \rightarrow HI + I^{\cdot}$$

Even under the most favourable conditions, Φ_{H_2} is low (~ 0.025) and this is probably due to the low escape probability from the solvent cage.

$$H^{\cdot} + I^{\cdot} \rightarrow HI$$

Thus, although halide ions react in a similar manner to transition metal ions, the overall Φ_{H_2} values are very low and the reaction requires UV irradiation ($\lambda < 250$ nm). Such systems have no application for solar energy conversion or storage and little, if any, relevance to photosynthesis.

8.4.1.1.2. Dye-sensitised redox reactions. An obvious means of increasing the utility of the redox reaction is to add an electron donor. By choosing a suitable coloured compound for this purpose, it should be possible to drive the reaction by visible light. Thus, irradiation of a dye in the presence of a transition metal ion can lead to net electron transfer.

$$D^{*} + M^{n+} \rightarrow D^{+} + M^{(n-1)+}$$

Subsequently, the reduced metal ion can be used to reduce water.

$$M^{(n-1)+} + H_2O \rightarrow M^{n+} + H^{\cdot} + OH^{-}$$

There are several well known examples whereby a transition metal ion is used to quench an excited state of a dye. With simple transiton metal ions, the rate constants for quenching by both energy transfer and paramagnetic enhanced nonradiative decay processes are quite low ($<10^7\,M^{-1}\,s^{-1}$) whilst electron transfer often proceeds at a much higher rate. Of particular interest is the photooxidation of tris(2,2′-bipyridyl)ruthenium (II) by metal ions. Some of the important rate constants are collected in Table 8.4 where k_F refers to the rate constant for quenching the luminescent state of the Ru^{2+} complex and k_B is the rate constant for reverse electron transfer. All the systems shown in Table 8.4 involve net electron transfer.

$$^{*}(\text{bipy})_3Ru^{2+} + M^{n+} \rightarrow (\text{bipy})_3Ru^{3+} + M^{(n-1)+}$$

TABLE 8.4.

RATE CONSTANTS FOR QUENCHING THE LUMINESCENT STATE OF $(BIPY)_3 Ru^{2+}$
IN AQUEOUS SOLUTION (k_F) AND FOR REVERSE ELECTRON TRANSFER (k_B)

Quencher	k_F $(M^{-1} s^{-1})$	k_B $(M^{-1} s^{-1})$	Reference
Fe^{3+}	2.7×10^9	4.9×10^6	Sutin et al. (1976)
Eu^{3+}	6.0×10^4	4.0×10^8	Harriman (1978)
$(phen)_3 Co^{3+}$	2.3×10^9	$\geqslant 3 \times 10^7$	Lin and Sutin (1976)
$(NH_3)_6 Ru^{3+}$	2.7×10^9	3.7×10^9	Chou et al. (1977)

With the exception of Eu^{3+}, quenching is very efficient and k_F is close to the diffusion limited value. However, reverse electron transfer occurs, with high efficiency in most cases, so that the redox reaction is reversible.

$$(bipy)_3 Ru^{3+} + M^{(n-1)+} \rightarrow (bipy)_3 Ru^{2+} + M^{n+}$$

The reverse electron transfer process presents a great problem when photo-redox reactions are carried out in fluid solution. In all the above cases, the redox products are not sufficiently stable towards this reaction to enable reaction with the aqueous substrate. Lin and Sutin (1976), in particular, have carried out the reaction with Eu^{3+} as quencher using a high concentration of perchloric acid in the presence of a platinised Pt wire. This sytem was of interest because, on thermodynamic grounds, Ru^{3+} has the necessary potential to oxidise water whilst europium (II) is capable of reducing water to hydrogen. However, upon irradiation with light $\lambda > 400$ nm there was no observable formation of hydrogen, presumably because of the high efficiency for reverse electron transfer.

Several interesting systems have been proposed to minimise the importance of reverse electron transfer. Experiments have shown that useful results are obtained if a weak reductant is added to the system. Thus, upon irradiation with visible light, acridine orange will transfer an electron to methyl viologen.

$$^*AO^+ + MV^{2+} \rightarrow AO^{2+} + MV^+$$

Normally, reverse electron transfer proceeeds rapidly but in the presence of EDTA the oxidised form of the dye participates in a second redox reaction.

$$AO^{2+} + EDTA \rightarrow AO^+ + EDTA^+$$

The oxidised form of EDTA is a weak oxidant and does not oxidise the reduced methyl viologen. This allows high yields of MV^+ to be built-up. It is now well established that an enzyme such as a hydrogenase can be used to catalyse the reaction between MV^+ and water (see chapters 9 and 10 of this book).

$$2\ MV^+ + 2\ H^+ \xrightarrow{\text{hydrogenase}} 2\ MV^{2+} + H_2$$

In fact, it has been reported by Bolton et al. (1976) that this overall system allows a steady production of hydrogen using visible light although Φ_{H_2} data are not available.

Recent studies have used modifications of this reaction scheme. Acridine orange can be replaced by other dyes, such as proflavine, and the hydrogenase enzyme can be replaced with an active Pt catalyst. Thus, a recent publication by Shilov et al. (1977) presents a most interesting method for photogeneration of hydrogen from water at neutral pH. In this system, an acridine dye (A) is photoreduced by an electron donor such as cysteine or EDTA in water at pH 6.5. Addition of V^{3+} or Eu^{3+} salicylates causes reoxidation of the bleached dye. Molecular hydrogen appears in the gas phase after small amounts of Adams platinum have been added to the solution.

$$
\begin{array}{ccccc}
EDTA^+ & \curvearrowright A^- & \curvearrowright M^{3+} & \curvearrowright H_2 \\
EDTA & \curvearrowleft A & \curvearrowleft M^{2+} & \curvearrowleft H_2O
\end{array}
$$

Upon 450-nm irradiation, the primary photoredox reaction produces the semireduced dye (A^-) which is capable of reducing the metal ions to the divalent state. Oxidation of M^{2+} at the surface of the Pt catalyst leads to formation of hydrogen with an estimated quantum yield of about 1%. In fact, the semireduced dye has a potential suitable for hydrogen formation without involvement of the metal ions. In the presence of a homogeneous catalyst, it was found that water was reduced directly by the semireduced dye.

In a slight modification of this scheme, methyl viologen was used in place of the metal salts. The semi-reduced methyl viologen has a characteristic absorption spectrum showing strong bands around 600 nm. Using EDTA as oxidising agent, the quantum yield for formation of MV^+ was estimated by spectroscopy under a variety of experimental conditions.

$$^*A + EDTA \rightarrow A^- + EDTA^+$$

$$A^- + MV^{2+} \rightarrow A + MV^+$$

The maximum quantum yield of this reaction was 0.55, although at high concentrations of $^*A$ and MV^{2+} there exists the strong possibility of direct reaction.

$$^*A + MV^{2+} \rightarrow A^+ + MV^+$$

Semireduced methyl viologen has a redox potential close to that of hydrogen (E^0 = −0.44 V) and the addition of the Pt catalyst causes evolution of H_2. Hence, this overall reaction system shows some degree of versatility and the use of Pt and Rh catalysts allows photogeneration of hydrogen without involvement of enzymes.

A more recent development of this type of reaction sequence is the work of Lehn and Sauvage (1977). This concerns a very complex process which is reported to generate catalytic yields of hydrogen from aqueous solutions at pH 7.5 using visible light. The system comprises of a weak reductant, triethanolamine (TEA), an unknown rhodium bipyridyl complex, tris(2,2'-bipyridyl)ruthenium (II), and an active Pt catalyst in neutral aqueous solution. The reaction shows a very strong pH dependence suggesting that TEA/TEAH$^+$ acts as the source of protons. Light is absorbed by the ruthenium (II) complex whilst TEA acts as electron donor, the turnover for the ruthenium (II) complex is in the order of several hundred. It is suggested that the rhodium compound reacts to form a hydride which is subsequently decomposed to give H_2 by the Pt catalyst, possibly in a photochemical step. The authors have not estimated a quantum yield for H_2 formation and much more work is required before the mechanism can be clarified. Compared to the Shilov model, this system seems very complicated and it might be worthwhile replacing the unknown rhodium complex with a better classified compound.

The suggestion that hydrogen can be formed by photochemical decompotion of a metal hydride has been reviewed by Balzani et al. (1975). Several transition metal hydrides eliminate H_2 upon irradiation, often with quite high quantum yield.

$$\begin{array}{ccc} R & H & R \\ \diagdown \diagup & & | \\ M & \overset{h\nu}{\rightarrow} & M + H_2 \\ \diagup \diagdown & & | \\ R & H & R \end{array}$$

M = W, Mo, Ir, Ru, Co. However, the organometallic products (R_2M) arising from elimination of H_2 are normally highly reactive and insert into C-H bonds of the solvent or undergo polymerisation. In the presence of substrates such as CO, C_2H_2, and $P(CH_3)_3$ stable adducts can be trapped. It is, therefore, necessary to find a system whereby one of the adducts will react with water to give a photochemically labile hydride.

$$R_2M(CO)_2 + H_2O \rightarrow R_2MH_2 + 2\,CO + \tfrac{1}{2}\,O_2$$

As yet, there are no known systems capable of this reaction. That such a reaction may be possible is suggested by the work of Bino and Ardon (1977) who have described redox reactions of the complex ion $[Mo_2Cl_8H]^{3-}$ which has two bridging chlorides and one bridging hydride between the two molybdenum atoms. These authors confirmed the earlier work of Cotton et al. (1969, 1976) that the binuclear hydride reacted with water to liberate H_2. From isotopic studies it was found that the bridging hydride was oxidised by an aqueous proton. The molybdenum product was identified as a dichloromolybdenum (III) dimer.

$$[Mo_2Cl_8H]^{3-} + OH^- + H_2O \rightarrow H_2 + [Mo_2(OH)_2Cl_2]^{2+}$$

Thus, the initial Mo^{II}—Mo^{III} mixed valent complex is oxidised to Mo^{III}—Mo^{III}. However, in the presence of a carboxylic acid, $[Mo_2Cl_8H]^{3-}$ undergoes reduction to a Mo^{II}—Mo^{II} complex rather than oxidation as above.

$$[Mo_2Cl_8H]^{3-} + 4\ HOAc \rightarrow Mo_2(OAc)_4 + 8\ Cl^- + 5\ H^+$$

This reaction represents the reversal of the process in which the complex ion is prepared by the addition of HCl to the acetate salt. Thus, these complexes can undergo both oxidation and reduction reactions depending upon the type of anion added to the medium. This is believed to be due to the ability of the anion to stabilise very short Mo—Mo bonds. A better understanding of these anion effects may allow the construction of a suitable cyclic system for decomposition of water.

8.4.1.2. Formation of oxygen

8.4.1.2.1. Redox reactions. Photooxidation of water can be brought about by irradiation of a high valent transition metal ion at a wavelength corresponding to a CTTM absorption band. This is the reaction required to complete the cycle for the reductive case described earlier and leads to dissociation of a water molecule into a proton and a hydroxyl radical.

$$M^{(n+1)+} + H_2O \rightarrow M^{n+} + H^+ + OH^{\cdot}$$

In practice, the reaction is difficult to achieve and there are few transition metal ions which will photooxidise water. Probably the best known examples are Ce^{4+} and Fe^{3+}, both give rise to generation of oxygen when irradiation with UV light.

The photochemical production of oxygen from aqueous solutions of Ce^{4+} salts was first reported by Baur (1908) and then investigated in detail by later authors. One of the major problems associated with cerium (IV) is that the $Ce(H_2O)_n^{4+}$ ion probably only exists in concentrated $HClO_4$ solution. In other media, hydrolysis, complexation, and polymerisation result in formation of a variety of different species which makes a quantitative comparison of the photochemistry a difficult operation. However, many general features are common to all reported studies. Thus oxygen formation occurs only with UV irradiation ($\sim$254 nm), Φ_{O_2} does not depend upon light intensity or [H$^+$] but decreases with increasing concentration of Ce^{3+}. This latter property is due to an inner filter effect since the molar extinction coefficient of Ce^{3+} is some thousand times that of Ce^{4+} and also to secondary thermal reactions of Ce^{3+}. There is also an increase in Φ_{O_2} with increased concentration of Ce^{4+} which was originally explained by Heidt and Smith (1948) assuming that only Ce^{4+} dimers were photoactive, but this proposal was later questioned on the basis of spectroscopic data. Instead, Evans and Uri (1950) have suggested that the primary photochemical step involves electron transfer within an ion-pair.

$$Ce^{4+}OH^- \rightarrow Ce^{3+} + OH^{\cdot}$$

This step is followed by a series of thermal reactions that lead to the formation of hydrogen peroxide.

$$Ce^{3+} + OH^{\cdot} \rightarrow Ce^{4+}OH^-$$

$$Ce^{4+}OH^- + OH^{\cdot} \rightarrow Ce^{3+} + H_2O_2$$

$$2\,OH^{\cdot} \rightarrow H_2O_2$$

Subsequently, hydrogen peroxide is oxidised to oxygen by further reaction with cerium (IV).

$$2\,Ce^{4+}OH^- + H_2O_2 \rightarrow 2\,Ce^{3+} + O_2 + 2\,H_2O$$

The maximum quantum yield of oxygen observed is about 0.1 but the extent of reaction is restricted to low levels because of the build-up of the highly absorbing Ce^{3+} ions.

Iron (III) shows similar photochemical behaviour to Ce^{4+} and UV irradiation of aqueous solutions gives rise to a low yield of oxygen. In the presence of inorganic anions, iron (III) shows a great tendency to form ion-pairs and the exact nature of ferric species present in solution depends upon the background medium. In concentrated $HClO_4$ solution, the primary photochemical step can be expressed in terms of hydroxyl radical formation.

$$Fe^{3+} + H_2O \overset{h\nu}{\rightarrow} Fe^{2+} + H^+ + OH^{\cdot}$$

The hydroxyl radicals that escape geminate recombination can combine to give hydrogen peroxide. Oxygen is then formed by secondary reactions involving hydrogen peroxide as described by Buxton et al. (1962).

$$2\,Fe^{3+} + H_2O_2 \rightarrow 2\,Fe^{2+} + 2\,H^+ + O_2$$

Dain and Kachan (1948) report that due to competitive scavenging reactions and low escape probability from the solvent cage, the Φ_{O_2} is only about 10^{-3}. In the presence of halide ions, Uri (1952) suggests that the primary photoreaction involves formation of a halide atom.

$$Fe^{3+}Cl^- \xrightarrow{h\nu} Fe^{2+} + Cl^{\cdot}$$

Thermal reactions then lead to formation of hydroxyl radicals which react to give hydrogen peroxide and, ultimately, oxygen.

$$Cl^{\cdot} + H_2O \rightarrow HCl + OH^{\cdot}$$

Several other transition metal ions have been proposed for the photo-dissociation of water, although reduction of the metal only proceeds with UV light. Mostly, as with Eu^{3+}, reported by Haas et al. (1970), and V^{5+}, described by Shchegoleva and Kryukov (1977), reduction of the metal ion only takes place in the presence of a so-called hydrogen atom scavenger such as ethanol. In fact, the scavenger acts as the primary reductant so that the first step in the reaction involves oxidation of the organic substrate.

$$Eu^{3+} + EtOH \xrightarrow{h\nu} Eu^{2+} + EtOH^+$$

Subsequent irradiation of Eu^{2+} leads to formation of hydrogen but the overall process cannot be regarded as photooxidation of water.

One system in particular that has received great attention is the photolysis of the tris(2,2′-bipyridyl)ruthenium (II) ion. Photochemists have long been attracted by this compound, it shows intense CT absorption in the visible region, it luminesces strongly, it has a relatively long emission lifetime (680 ns), and it is substitution inert in water at ambient temperature. The excited state can act as both electron donor and acceptor and, with careful choice of reactant, quenching can lead to formation of the corresponding ruthenium (I) and (III) complexes. The formal redox potentials for the excited state have been determined and are remarkable in that, as shown by Meyer et al. (1977), the excited state is capable of both oxidising and reducing water at pH 7.

$$^*(bipy)_3Ru^{2+} + \tfrac{1}{2} H_2O \rightarrow (bipy)_3Ru^+ + H^+ + \tfrac{1}{4} O_2$$

$$\Delta G = -0.02 \text{ V}$$

$$^*(bipy)_3Ru^{2+} + H^+ \rightarrow (bipy)_3Ru^{3+} + H^{\cdot}$$

$$\Delta G = -0.43 \text{ V}$$

However, irradiation of the complex in aqueous solution does not lead to product formation. The complex does not sensitise the photolysis of water although it has been reported by Sutin and Creutz (1975) that the ruthenium (III) complex does oxidise hydroxyl ions to oxygen. This is a thermal reaction and the reported mechanism is complex, the very strong pH dependence has not been explained as yet.

The major difficulty associated with the photooxidation of water is that the formation of oxygen requires a four electron change.

$$2 H_2O \rightarrow 4 H^+ + 4 e + O_2$$

All photoredox reactions proceed in one electron steps so that it is necessary to devise some kind of charge storage system if formation of free radicals is to be prevented. With the simple transition metal ions there is no real need to avoid radicals such as $OH^.$. Although most of the initially formed radicals recombine within the solvent cage, some do escape and lead to oxygen formation by secondary reactions. However, in the presence of an organic molecule, such as a dye sensitiser, $OH^.$ radicals are trapped as adducts and result in formation of hydroxylated products rather than oxygen. Thus, both anthra-9,10-quinone (AQ) and N-methylphenazine sulphate are known to abstract an electron from water upon irradiation with visible light.

$$^*AQ + H_2O \rightarrow AQ^- + H^+ + OH^.$$

Hydroxyl radicals rapidly attack a second molecule of anthra-9,10-quinone forming the hydroxy derivative as a major product. It is possible to prevent loss of the chromophore by trapping $OH^.$ with sodium formate but this has the effect of stopping formation of oxygen. Thus, in the presence of a dye it is essential that the redox system accumulates more than one oxidation equivalent before the system is capable of liberating oxygen from water.

8.4.1.2.2. Binuclear complexes. A molecule of water contains only one oxygen atom whilst an oxygen molecule has two. Thus as pointed out earlier, the formation of an oxygen molecule requires that two molecules of water are brought together and four electrons abstracted from them. As discussed in section 8.3.1 it seems that the photosynthetic process uses an unidentified manganese complex for this purpose but it is unlikely that a single manganese ion could remove all four electrons. Calvin (1974) has proposed that a binuclear manganese complex is a more likely reactant since a binuclear manganese (IV) compound could act as an acceptor of all the necessary electrons without the need to involve free oxygen atoms or hydroxyl radicals. Thus arose the notion of a binuclear manganese complex which contains two water molecules as being the active component in the PSII process. However, as already mentioned in section 8.3.1 other arguments have been put forward for a complex containing four manganese ions (Govindjee et al., 1977).

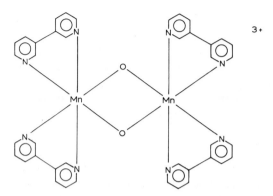

Fig. 8.7. μ-Di-oxotetrakis(2,2'-bipyridyl)dimanganese (III, IV).

Calvin (1974) attempted the construction of a model system using μ-di-oxotetrakis-(2,2'-bipyridyl)dimanganese (III, IV) perchlorate (Fig. 8.7). Irradiation of this complex at a CTTM absorption band was originally reported to result in reduction of the metal and formation of molecular oxygen but more recent work by Cooper and Calvin (1974) has shown that this result is wrong. In fact, the polarographic and photochemical reduction of the binuclear complex has now been reported in detail by Otsuji et al. (1977). Irradiation at 300 nm leads to reduction of the metal atoms, firstly from manganese (IV) to manganese (III) and then to manganese (II). During these reduction reactions, hydroxyl radicals are produced by oxidation of the bridging groups. The reaction is believed to proceed through a protonated excited state complex.

$$[L_2Mn^{III}\overset{O}{\underset{O}{\diagdown\diagup}}Mn^{IV}L_2]^{3+} \xrightarrow[H^+]{h\nu} {}^*[L_2Mn^{III}\overset{\overset{\overset{H}{|}}{O}}{\underset{O}{\diagdown\diagup}}Mn^{IV}L_2]^{4+}$$

$$\downarrow H_2O$$

$$2[L_2Mn^{II}]^{2+} + 2\ OH^{\cdot} \overset{h\nu}{\underset{}{\rightleftharpoons}} [L_2Mn^{III}\overset{\overset{\overset{H}{|}}{O}}{\underset{\underset{\underset{H}{|}}{O}}{\diagdown\diagup}}Mn^{III}L_2]^{4+} + OH^{\cdot}$$

In a separate study, Brown et al. (1977) have shown that the corresponding manganese (II) complexes could not be photooxidised by chlorophyll. Although a series of manganese (II) compounds were found to quench both singlet and triplet excited states of chlorophyll, reaction did not lead to formation of a redox pair. Instead, quenching was restricted to paramagnetic enhanced nonradiative decay of the excited state. Identical results were obtained by Ferreira and Harriman (1977) using thionine in place of chlorophyll and it is now concluded that the 2,2′-bipyridyl and 1,10-phenanthroline complexes of manganese do not represent a promising system for the photooxidation of water.

However, manganese (IV) forms many binuclear complexes with ligands such as oxalate, gluconate, Schiffs base, and phthalocyanine and it is possible that some of these complexes possess more interesting photochemical properties. In this context it is of interest to note the observed photoreduction of the decammine-μ-superoxodicobalt (III) ion in aqueous solution. The photochemistry of this complex was first reported by Sykes (1963) who found that irradiation with light λ 365 nm resulted in formation of oxygen. Later work by Barnes et al. (1968) established the stoichiometry of the reaction and showed that it was an intramolecular redox process.

$$[(NH_3)_5CoOOCo(NH_3)_5]^{5+} \xrightarrow[H^+]{h\nu} Co_{aq}^{2+} + O_2 + [Co(NH_3)_5(H_2O)]^{3+} + 5\,NH_4^+$$

The quantum yield for formation of oxygen is wavelength dependent (Table 8.5) but reaches a very high value for 313-nm irradiation. After some initial doubt, it now seems fairly well established that the bridging oxygen function is a superoxide ion, hence the preference for a one electron redox reaction. As yet, there have been no reported observations regarding the corresponding μ-peroxo complex, nor attempts to prepare the superoxide complex by photochemical methods. However, complexes involving an oxygen—oxygen bridge offer a more promising route for photogeneration of oxygen than the di-μ-oxo complexes of the Calvin model. This is particularly true of the very complex trinuclear ruthenium system that has been proposed by Earley (1973) since this system seems to offer no advantages whatsoever over the binuclear model.

TABLE 8.5.

QUANTUM YIELDS FOR OXYGEN FORMATION UPON IRRADIATION OF DECAMMINE-μ-SUPEROXO-DICOBALT(III) IN AQUEOUS SOLUTION.

λ_{exct} (nm)	Φ_{O_2}
750—600	0
800—470	0
365	0.65
313	0.85
254	0.30

From Barnes et al. (1968).

8.4.1.2.3. *Carbonate ions.* With regard to the photosynthetic process, Metzner (1975) has questioned whether the evolved oxygen comes directly from water and has suggested that bicarbonate ions may be involved in oxygen formation.

$$H_2O + CO_2 \rightleftharpoons H^+ + HCO_3^-$$

This controversy seems set to reign for some considerable time but it is worthwhile noting the few reported photoredox reactions involving carbonate anions.

Chibisov et al. (1972) have shown that carbonate ions quench the triplet state of anthra-9,10-quinone-2,6-disulphonate (AQS) in aqueous solution at pH 6.5. Flash photolysis shows the transient formation of semireduced quinone and oxidised carbonate.

$$^*AQS + CO_3^{2-} \rightarrow AQS^{\cdot-} + CO_3^{-}$$

The rate of reverse reaction is very fast but the decay of CO_3^- was enhanced in the presence of ethanol. This latter effect may indicate product formation arising from hydrogen abstraction from the alcohol.

Later, Scheerer and Gratzel (1976) found that the triplet state of duroquinone (DQ) is quenched by carbonate ions in both aqueous ethanol and micelle solutions. The use of the micellar system had a pronounced effect upon the efficiency of reverse and forward electron transfer processes. In aqueous ethanol, CO_3^{2-} quenches triplet duroquinone with a rate constant of $7 \times 10^7\,M^{-1}\,s^{-1}$ but in an aqueous micellar solution of dodecyltrimethylammonium chloride (DTAC) the quenching rate constant was extremely fast and could not be resolved by ns flash photolysis. In both systems, quenching resulted in net electron transfer.

$$^*DQ + CO_3^{2-} \rightarrow DQ^{\cdot-} + CO_3^{-}$$

The micellar system allows separation of the redox products and a kinetic analysis was used to propose that CO_3^- is able to reduce a second molecule of duroquinone.

$$DQ + CO_3^- \rightarrow DQ^{\cdot-} + CO_3$$

Subsequently, carbon peroxide is assumed to decompose by a reaction pathway that forms oxygen.

$$CO_3 \rightarrow CO_2 + O^{\cdot}$$

The significance of this report is limited because the authors do not provide experimental evidence for the formation of oxygen and the involvement of CO_3 is questionable.

8.4.2. Heterogeneous systems

8.4.2.1. Surfactants

Photoredox processes are of great interest because of the possibility of using light driven electron transfer reactions to store and convert solar radiation. The products of the redox processes are normally highly energetic and, in solution, undergo reverse electron transfer too rapidly to allow reaction with the substrate. Increasingly more attention is now being paid to systems which may inhibit these back reactions. In the previous section it was shown that competitive reduction with a weak reductant such as EDTA could lead to stabilisation of one of the initially formed products. A second important method is to carry out the redox reaction in a multiphase medium. In several cases it has been found that the controlled and highly condensed environment provided by an organised monolayer assembly can strongly modify photoreactivity and luminescence behaviour compared to fluid solution. Under such conditions, the reaction medium becomes more comparable to the chloroplast and it has been suggested that such conditions may enhance in vitro photodissociation of water.

Electron transfer reactions are known to proceed in many types of organised assemblies. Thus, quinones quench excited states of dyes in monomolecular layers, vesicles, micelles, lipid membranes, and thin films. So far, there are insufficient comparable data to estimate overall steady-state yields of redox products for identical systems conducted in both solution and organised layers. However, the oil/water interface, which is an integral part of most multiphase systems, allows separation of the redox products according to their solubility. This idea was first applied by Rabinowitch (1962, 1966) to the thionine—iron (II) system. Irradiation of thionine in the presence of iron (II) gives high yields of semireduced thionine and iron (III).

$$^*Th + Fe^{2+} \rightarrow Th^- + Fe^{3+}$$

The semireduced dye undergoes rapid dismutation to the fully reduced leuco dye which, in turn, is slowly oxidised by iron (III).

$$2\ Th^- \rightarrow Th + Th^{2-}$$

$$Th^{2-} + Fe^{3+} \rightarrow Th^- + Fe^{2+}$$

Although oxidation of the leucodye is relatively slow, the reaction is completely reversible and Rabinowitch converted the system to an irreversible redox process. This was achieved simply by carrying out the reaction in a water—ether emulsion. The leucodye is soluble in the ether layer whilst iron (III) resides in the aqueous phase thus the two redox products are separated and do not react together. More sophisticated versions of this type of reaction scheme use surfactants to separate the products rather than mixed solvents.

Recent work by Whitten et al. (1977a,b) has described the synthesis and photoredox properties of several hydrophobic ruthenium (II) complexes. It has been claimed that the shielding of the reactive core by the hydrophobic ligands could impart a selectivity to electron transfer processes. Luminescence quenching studies showed that amines, such as N,N-dimethylaniline and triethylamine, quenched the excited state of the hydrophobic ruthenium (II) complexes at close to the diffusion controlled limit. However, there was little reverse electron transfer and, instead, the ruthenium complex was permanently bleached.

$$^*Ru^{2+} + (C_2H_5)_3N \rightarrow Ru^+ + (C_2H_5)_3N^+$$

The ruthenium (I) product is stable over several days in dry acetonitrile but rapidly reverts to starting compound when air or water are introduced. Evidently with the hydrophobic complexes, the energy wasting reverse reaction is retarded sufficiently to allow one of the products to react by other processes. In this case, this involves oxidation of either the acetonitrile solvent or unreacted amine by $(C_2H_5)_3N^+$.

$$(C_2H_5)_3N^+ + CH_3CN \rightarrow (C_2H_5)_3^+NH + \dot{C}H_2CN$$

$$2 \, \dot{C}H_2CN \rightarrow NCCH_2CH_2CN$$

$$(C_2H_5)_3N^+ + (C_2H_5)_3N \rightarrow (C_2H_5)_3^+NH + (C_2H_5)_2N\dot{C}HCH_3$$

If a similar redox process can be brought about in aqueous solution then it may be possible to devise a system capable of the dissociation of water. It would be interesting to see if Sutin's experiment with $(bipy)_3Ru^{2+}/Eu^{3+}/Pt$ would work with the hydrophobic complexes.

Previous work by Whitten et al. (1977a) suggests that such a system may be achieved. Bearing in mind that excited state $(bipy)_3Ru^{2+}$ is thermodynamically capable of the dissociation of water, it was claimed that photolysis of monomolecular layers of surfactant ruthenium (II) complexes in contact with water resulted in formation of molecular oxygen and hydrogen. The complexes shown below (Fig. 8.8) were found to form stable monolayer films when deposited onto clean glass slides and slides coated with arachidate. The slides were prepared to give an outermost hydrophobic layer of the ruthenium (II) complex and were highly luminescent. After immersing the slide in water, the luminescence was considerably reduced and slightly red-

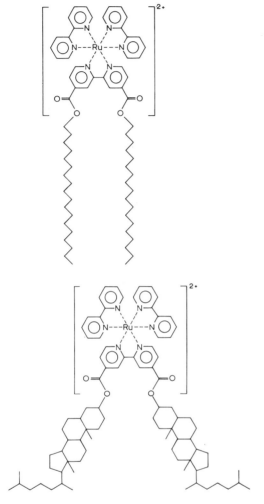

Fig. 8.8. Structures of the surfactant ruthenium (II) complexes used for monomolecular layer experiments.

shifted (Fig. 8.9). Gentle heating under vacuum restored the original luminescence properties. This effect was later explained by Harriman (1977) who reported that the photophysical properties of ruthenium (II) chelates were very solvent dependent. Emission yields and lifetimes were measured for several complexes and were found to be markedly lower in water than in less polar solvents. There was also a red shift in water exactly as found with the monomolecular assemblies. Hence, quenching by water is a characteristic of ruthenium (II) complexes and is not a unique feature associated with the surfactant compounds. The most intriguing aspect of Whitten's work was

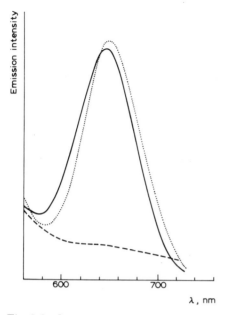

Fig. 8.9. Quenching and regeneration of the luminescence of the surfactant ruthenium (II) complex in monolayer assemblies; (· · · · · ·), freshly coated slide; (- - - - - -), after immersion in water; (————), after gentle heating in vacuum (Whitten et al., 1977).

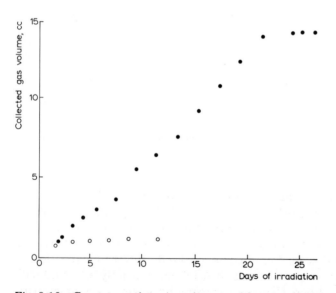

Fig. 8.10. Gas generation experiment with monolayer assemblies of the surfactant ruthenium complex irradiated in contact with water. Open circles are blank experiments; closed circles gas generated from irradiation of 20 slides. (Whitten et al., 1977).

that visible light irradiation of slides in contact with water was reported to give H_2 and O_2 with an approximate quantum yield of 0.1. Long irradiation times were required, because of the limited light absorption by monolayers, and a gas evolution profile is shown in Fig. 8.10. Gaines and Valenty (1977) and Harriman (1977) found that these experiments could not be repeated and it was concluded that pure surfactant complexes do not sensitise the photodissociation of water. Photolysis experiments were conducted in aqueous solution, micelles, and monolayer films but no experiment gave products arising from the decomposition of water. The absence of gas evolution by the pure surfactant complex may be due to its instability since it is photohydrolysed to the free acid. However, the experiments made by Harriman show that the simple ruthenium (II) complexes do not dissociate water suggesting that Whitten's original composition may have contained a discrete synergistic mixture of compounds.

One of the obvious limitations with models based on monomolecular layers is the small amount of light that can be absorbed by the system. As explained in detail in chapter 2, the plant overcomes this problem by using an array of chlorophyll molecules to harvest sunlight and channel the energy into a reaction centre. Knowledge of this light harvesting process has lead research workers to consider multilayers rather than a single monomolecular layer of pigment.

Wang (1969) claims to have constructed a multilayer model system that, like the photosynthetic process, can convert light to chemical free energy through pigment sensitised photooxidation of water. The system has two light harvesting subunits connected electrically in series. Each subunit consists of a multimolecular layer of zinc (II) tetraphenylporphine deposited onto clean aluminum surfaces and immersed into an aqueous mixture of potassium ferri- and ferrocyanides. Upon illumination with amber light, electron transfer is said to occur across about 70 molecular layers of the pigment with a photo-emf of 1.1 to 1.3 V per subunit. The photosensitive electrode is said to act as negative terminal and, by connecting two such units in series, a total cmf of 2.1 to 2.4 V can be obtained.

$$Pt \mid \begin{matrix} Fe(CN)_6^{4-} \\ Fe(CN)_6^{3-} \end{matrix} \mid Zn\ TPP \mid Al \mid Pt \mid Pt \mid \begin{matrix} Fe(CN)_6^{4-} \\ Fe(CN)_6^{3-} \end{matrix} \mid Zn\ TPP \mid Al \mid Pt$$

The emf produced by the above two stage cell is sufficient to oxidise water to oxygen. To demonstrate this possibility, a special microelectrolytic cell was constructed for use in conjunction with the photosensitive electrodes. An oxygen-sensitive membrane electrode was incorporated into the electrolytic cell which had two thin platinum coil electrodes connected, respectively, to the positive and negative electrodes of the photocell. In a typical experiment, a small aliquot of NADP/NAPD-reductase solution at pH 7.38 was placed in contact with the platinum foil electrodes. Illumina-

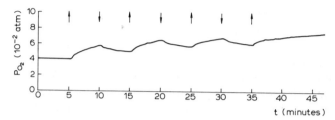

Fig. 8.11. Photooxidation of water by the model system. Arrow pointing upward indicates the time when the light was turned on; arrow pointing down indicates the time when the light was turned off. The partial pressure of oxygen was measured in atmospheres of O_2, and its decrease each time after the light was turned off is mainly due to the radial diffusion of O_2 molecules generated near the center of the thin filter-paper disk (From Wang, 1969).

tion of the pigment coated photosensitive electrode with amber light was reported to generate oxygen gas as measured by the membrane electrode (Fig. 8.11). Oxygen formation ceased when the light was turned off but there are no estimated quantum yield values. In fact, the data presented by Wang are not convincing with regard to the formation of oxygen. This is due mainly to the limited amount of information provided since there has not been a more detailed full paper. Therefore, it seems best to regard the Wang model as interesting but unproven.

8.4.2.2. Solid matrices

The photochemical properties of TiO_2 have received considerable attention (see also chapter 9). The photooxidation of organic substrates by TiO_2 has been known for a long time since it is responsible for the poor light stability of many paints. Similarly, TiO_2 sensitises photooxidation of many inorganic substrates such as Hg (to HgO) and CN^- (to CNO^-). Also, there are many known examples whereby olefins are photoreduced by TiO_2, for example the TiO_2 sensitised reduction of acetylene results in formation of CH_4, C_2H_4 and C_2H_6. The mechanism of the photooxidation reactions is fairly well understood but the reduction processes are less clear.

The photolysis of water on TiO_2 surfaces has been studied in conjunction with semiconductor electrodes. The Honda cell, and its many subsequent modifications, has been extensively reviewed and it is well established that irradiation of a TiO_2 crystalline electrode with UV light ($\lambda < 413$ nm) results in formation of oxygen and hydrogen.

$$H_2O + nh\nu \xrightarrow{TiO_2} H_2 + \tfrac{1}{2} O_2$$

$$\Delta G^{298} = 68.35 \text{ kcal} \cdot \text{mol}^{-1}$$

In reality, such systems represent a photoassisted electrolysis of water since it is necessary to apply a small potential to the electrode to overcome the inherent overpotential. Wrighton et al. (1975) have reported that a typical TiO_2 electrode which has its onset of response near to 400 nm gives an optical storage efficiency of about 1% as calculated by the yield of hydrogen gas. Much work is currently in progress to improve the range of available semiconductor electrodes and to increase the response to visible light. In particular, it is desirable to solve the difficulties associated with the preparation of suitable electrode surfaces.

Recent work by Schrauzer and Guth (1977) has shown that photolysis of chemisorbed water on incompletely outgassed TiO_2 powder yields H_2 and O_2 in a molar ratio of 2 : 1 if conducted under an argon atmosphere. In the presence of N_2, O_2 is still formed but the evolution of H_2 is inhibited as chemisorbed N_2 is reduced to NH_3 and N_2H_4. Illumination with near UV light transfers electrons from the valence band of TiO_2 into the lowest conduction band, the band gap corresponds to between 70 and 80 kcal and is sufficient to bring about dissociation of water. The electrons in the conductance band can be utilised for the reduction of substrates such as H^+, N_2, and C_2H_2. The positive holes generated in the valence band provide sites for oxygen production and other oxidation reactions. Heat treatment, sample history, and doping with other metals all contribute to the observed efficiency of the photosensitised reactions and the overall system seems to be of great potential for the production of reduced organic compounds (also see chapter 9 by Krasnovski).

Jacobs et al. (1977) have presented data to suggest that zeolites behave similarly to TiO_2 powders. It is well known that the colour of a Y-zeolite containing Ag^+ ions darkens upon ageing and, at the same time, hydroxy absorptions appear in the IR spectrum. In order to intensify this process, a sample of zeolite (AgY) was saturated with water vapour and exposed to sunlight for two hours. Mass spectrometry showed an increase in the partial pressure of oxygen in the vessel during irradiation and the sample turned grey. Analysis by IR spectroscopy indicated the build-up of hydroxyl groups and the decrease in concentration of Ag^+ ions. Overall, the effect of sunlight is to oxidise water to oxygen and reduce Ag^+ to silver metal.

$$2\ Ag^+ + 2\ Y^- + H_2O \xrightarrow{h\nu} 2\ Ag + 2\ YOH + \tfrac{1}{2}\ O_2$$

It was found that upon thermal treatment of the reduced silver zeolite at $873°K$ the hydroxy groups disappeared and Ag^+ ions were reformed. Mass spectrometry showed that hydrogen gas was desorbed and the zeolite recovered its original white colour.

$$Ag + YH \xrightarrow{873\ °K} Ag^+ + Y^- + \tfrac{1}{2}\ H_2$$

This cycle could be repeated several times and it was found that the yields could be improved by minor experimental modifications.

8.5. CONCLUSIONS

Advances made in the in-vitro photodissociation of water are of interest for two major reasons, understanding the in-vivo photosynthetic process and possible storage of solar energy. At the present time, there exists several methods for partial photodissociation of water using visible light, that is, photogeneration of hydrogen gas. This can be achieved simply by irradiation of a binuclear Rh^I complex in strongly acidic solution and by the more complicated dye sensitised redox reactions. However, such systems are not cyclic. With the Rh^I complex the product formed is the corresponding Rh^{II} complex which must be chemically reduced back to Rh^I before photogeneration of hydrogen can continue. The dye sensitised reactions proposed by Shilov et al. (1977) and Lehn and Sauvage (1977) do recycle the dye so that more hydrogen is formed than there is dye present in the system but the processes lead to formation of an oxidised amine, either EDTA or TEA. These amines are much easier to oxidise than water and therefore represent a much simplified photoredox system. The only way to build a completely cyclic system for the photodissociation of water is to oxidise water rather than to use an artificial electron donor. However, it is not sufficient to use a one electron redox reaction whereby water is oxidized to hydroxyl radicals since these products are highly reactive and lead to consumption of the chromophore if this is anything other than a simple metal cation. Thus, a prime requirement of the model system is that it must produce oxygen, only then will a truely catalytic process be achieved.

So far, there are no reported systems which can bring about oxygen formation from water in an homogeneous solution using visible light. Photolysis of Ce^{4+} and Fe^{3+} with UV light leads to formation of both hydrogen and oxygen and these systems are the best models attained yet. The use of semiconductor materials pushes the threshold wavelength further out into the visible region but is seems doubtful whether cheap semiconductors will be produced which work in the visible region proper. The use of μ-di-oxo binuclear complexes has given no advantage over simpler mononuclear compounds but this is because the structure of such complexes is wrong for oxygen production. The manganese complexes proposed by Calvin behave similarly to Ce^{4+} and Fe^{3+} in that they are photoreduced by a one electron step that leads to formation of hydroxyl radicals. Instead, it is only with the μ-superoxo cobalt complex, where there exists an O-O bond, that photolysis produces oxygen in high yield. Thus it seems reasonable that the system with the most potential for photoproduction of oxygen gas involves a highly oxidised μ-peroxo complex. In order to construct a cyclic system, the metal product formed by photoreduction of the μ-peroxo complex must be capable of undergoing oxidation to a higher valence complex, this step may be thermal or photochemical. Again, the intermediate oxidation state so formed should be capable of undergoing further oxidation to reform the μ-

peroxo complex so that the cycle can be completed. A further requirement of the model system is that there must be at least one water molecule coordinated to the transition metal centre, since the oxygen atoms from two water molecules are required to form the peroxo group. This gives an overall model system as outlined below.

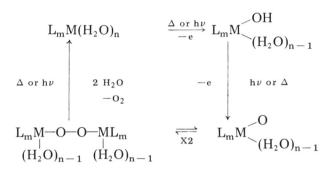

The ligand L must be chosen so that it is resistant to direct oxidation, substitution inert, and capable of stabilising the different oxidation states of M. The metal M must be capable of existing in three fairly stable oxidation states. Several first row transition metals have properties characteristic of M, such as Ti, V, and Mn, whilst most chelates fulfill the requirements expected of L. However, the formation of a μ-peroxo bridged species narrows the choice of compounds. Such species are poorly characterised and have received little study except in the case of cobalt. It is believed that the plant uses an unknown manganese complex to photooxidise water but little is known about the higher oxidation states of manganese. Certainly, manganese (IV) forms many μ-di-oxo binuclear complexes but no μ-peroxo complexes have been characterised as yet. We await developments in this field with great interest.

In addition to the pure chemical approach, it is hoped that, in the near future, careful biochemical separation techniques will yield a protein complex from the photosynthetic membrane which processes the basic properties of the in vivo water splitting enzyme. Only then will the chemist begin to know in molecular terms how green plants bring about the photooxidation of water, a reaction which is so vitally important for the existence of living organisms on our planet.

REFERENCES

Allen, F. and Frank, J. (1955) Arch. Biochem. Biophys., 58, 124—143.
Angelici, R.J. (1973) in Inorganic Biochemistry (G.L. Eichhor, ed.), Vol. 1, pp. 63—101. Elsevier, Amsterdam.
Arnon, D.I. and Whatley, F.R. (1949) Science, 110, 554—556.

278

Auslander, W. and Jung, W. (1977) Abstr. 4th Int. Congr. Photosynthesis, p. 12.

Balzani, V., Moggi, L., Manfrin, M.F., Bolletta, F. and Gleria, M. (1975) Science, 189, 852—856.

Barber, J. (1977) Proc. 4th Int. Congr. Photosynthesis, (D.O., Hall, J. Coombs, t.W. Goodwin, eds.), pp. 423—433. Biochem. Soc., London.

Barnes, J.E., Barrett, J., Brett, R.W. and Brown, J. (1968) J. Inorg. Nucl. Chem., 30, 2207—2210.

Baur, E. (1908) Z. Phys. Chem., 63, 683—685.

Bino, A. and Ardon, M. (1977) J. Am. Chem. Soc., 99, 6446—6447.

Blankenship, R.E. and Sauer, K. (1974) Biochim. Biophys. Acta, 357, 252—266.

Blankenship, R.E., Babcock, G.T. and Sauer, K. (1975) Biochim. Biophys. Acta, 387, 165—175.

Bolton, J.R., Markiewicz, S., Chan, M.S., Sparks, R.H. and Evans, C.A. (1976) Int. Conf. Photochemical Conversion and Storage of Solar Energy. Univ. of Western Ontario, Canada.

Brown, R.G., Harriman, A. and Porter, G. (1977) J. Chem. Soc., Faraday Trans. 2, 73, 113—119.

Buxton, G.V., Wilford, S.P. and Williams, R.J. (1962) J. Chem. Soc., 4957—4962.

Calvin, M. (1974) Science, 184, 375—381.

Cheniae, G.M. (1970) Annu. Rev. Plant Physiol., 21, 467—498.

Cheniae, G.M. and Martin, I.F. (1970) Biochim. Biophys. Acta, 197, 219—239.

Chibisov, A.K., Kuzmin, V.A. and Karyakin, A.V. (1972) Int. J. Chem. Kinet., 4, 639—644.

Chou, M., Creutz, C. and Sutin, N. (1977) J. Am. Chem. Soc., 99, 5615—5623.

Cooper, S.R. and Calvin, M. (1974) Science, 185, 376.

Cotton, F.A., Bennett, M.J. and Brencic, J.V. (1969) Inorg. Chem., 8, 1060—1065.

Cotton, F.A. and Kalbacher, B.J. (1976) Inorg. Chem., 15, 522—526.

Cramer, W.A. and Butler, W.L. (1969) Biochim. Biophys. Acta, 172, 503—510.

Dain, B.Y. and Kachan, A.A. (1948) Dokl. Acad. Nauk. SSSR, 61, 471—474.

Dainton, F.S., Collinson, E. and Malati, M.A. (1959) Trans. Faraday Sco., 55, 2096—2106.

Dainton, F.S. and James, D.G.L. (1958) Trans. Faraday Soc., 54, 649—663.

Davis, D.D., Stevenson, K.L. and King, G.K. (1977) Inorg. Chem., 16, 670—673.

Earley, J.E. (1973) Inorg. Nucl. Chem. Lett., 9, 487—490.

Evans, M.G. and Uri, N. (1950) Nature, 166, 602—603.

Ferreira, M.I.C. and Harriman, A. (1977) J. Chem. Soc., Faraday Trans. 1, 73, 1085—1092.

Forbush, B., Kok, B. and McGloin, M. (1971) Photochem. Photobiol., 14, 307—321.

Fowler, C.F. and Kok, B. (1974) Biochim. Biophys. Acta, 357, 299—307.

Gaines, G. and Valenty, S.J. (1977) J. Am. Chem. Soc., 99, 1285—1287.

Govindjee, Wydrzynski, T. and Marks, S.B. (1977) in Bioenergetics of Membranes (L. Packer, G.C. Papageorgiou, A. Trebst, eds.), pp. 305—316. Elsevier, Amsterdam.

Gray, H.B., Erwin, D.R., Geoffroy, G.L., Hammond, G.S., Solomon, E.L., Trogler, W.C. and Zagars, A.A. (1977) J. Am. Chem. Soc., 99, 3620—3621.

Gray, H.B., Kent, K.R., Lewis, N.S., Miskowski, V.M., Erwin, D.R. and Hammond, G.S. (1977a) J. Am. Chem. Soc., 99, 5525—5526.

Gray, H.B., Trogler, W.C., Erwin, D.K. and Geoffroy, G.L. (1978) J. Am. Chem. Soc., 100, 1160—1163.

Haas, Y., Stein, G. and Tomkiewicz, M. (1970) J. Phys. Chem., 74, 2558—2562.

Harriman, A. (1977) J. Chem. Soc., Chem. Commun., 777.

Harriman, A. (1978) (unpublished results).

Heidt, L.J. and Smith, M.E. (1948) J. Am. Chem. Soc., 70, 2476—2481.

Heidt, L.J. and McMillan, A.F. (1954) J. Am. Chem. Soc., 76, 2135—2139.

Heidt, L.J., Mullin, M.G. and Martin, W.B. (1962) J. Phys. Chem., 66, 336—341.

Hill, R. (1937) Nature 139, 881—882.

Hill, R. (1972) Proc. 2nd Int. Congr. Photosynthesis Res. (G. Forti, M. Avron, B.A. Melandri, eds.), Vol. I, pp. 1—18. Junk, The Hague.

Hind, G. and Whittingham, C.P. (1963) Biochim. Biophys. Acta, 75, 194—202.

Izawa, S., Heath, R.L. and Hind, G. (1969) Biochim. Biophys. Acta, 180, 388—398.

Jacobs, P.A., Uytterhoeven, J.B. and Beyer, H.K. (1977) J. Chem. Soc., Chem. Commun., 128—129.

Joliot, P., Barbieri, G. and Chabaud, R. (1969) Photochem. Photobiol., 10, 309—329.

Joliot, P. and Kok, B. (1975) in Bioenergetics of Photosynthesis (Govindjee, ed.), pp. 387—412. Academic Press, New York.

Jursinic, P., Warden, J. and Govindjee (1976) Biochim. Biophys. Acta, 440, 322—330.

Kessler, E., Arthur, W. and Brugger, J.E. (1957) Arch. Biochem. Biophys., 71, 326—335.

Klimov, V.V., Klevanik, A.V., Shuvalov, V.P. and Krasnovsky, A.A. (1977) FEBS Lett., 82, 183—186.

Knaff, D.B., Malkin, R., Myron, J.C. and Stoller, M. (1977) Biochim. Biophys. Acta, 459, 402—411.

Kok, B., Forbush, B. and McGloin, M. (1970) Photochem. Photobiol., 11, 457—475.

Lehn, J.M. and Sauvage, J.P. (1977) Nouv. J. Chim., 1, 449—451.

Lin, C.T. and Sutin, N. (1976) J. Phys. Chem., 80, 97—105.

Malkin, R. and Barber, J. (1978) J. Biochem. (Tokyo) (in press).

Marcus, R. (1956) Science, 123, 399—405.

Mathis, P., Haveman, J. and Yates, M. (1976) Brookhaven Symp. Biol., 26, 267—277.

Metzner, H. (1975) J. Theor. Biol., 51, 201—231.

Meyer, T.J., Nagle, J.K. and Young, R.C. (1977) Inorg. Chem., 16, 3366—3369.

Orgel, L.E. (1966) in Introduction to Transition-Metal Chemistry: Ligand Field Theory. Methven and Co. London.

Otsuji, Y., Sawada, K., Morishita, I., Taniguchi, Y. and Mizuno, K. (1977) Chem. Lett., 983—986.

Pirson, A., Ticky, C. and Wilhelmi, G. (1952) Planta, 40, 199—253.

Punnett, T. (1959) Plant Physiol., 34, 283—289.

Rabinowitch, E.I. (1945) Photosynthesis and Related Processes, Vol. I. Interscience, New York.

Rabinowitch, E. and Mathai, K.G. (1962) J. Phys. Chem., 66, 663—664.

Rabinowitch, E. and Frankowiak, D.J. (1966) Phys. Chem., 70, 3012—3014.

Radmer, R. and Cheniae, G.M. (1977) in Primary Processes of Photosynthesis, Vol. 2: Topics in Photosynthesis (J. Barber, ed.), pp. 303—348. Elsevier, Amsterdam.

Renger, G. (1977) FEBS Lett., 81, 223—228.

Rigg, T. and Weiss, J.J. (1952) J. Chem. Soc., 4198—4204.

Ruben, S., Randell, M., Kamen, M.D. and Hyde, J.L. (1941) J. Am. Chem. Soc., 63, 877—879.

Ryason, P.R. (1977) Sol. Energy, 19, 445—448.

Saphon, S. and Crofts, A.R. (1977) Abstr. 4th Int. Congr. Photosynthesis, p. 325.

Scheerer, R. and Gratzel, M. (1976) Ber. Bunsenges. Phys. Chem., 80, 979—982.

Schrauzer, G.N. and Guth, T.D. (1977) J. Am. Chem. Soc., 99, 7189—7193.

Shchegoleva, I.S. and Kryukov, A.I. (1977) Dopov Akad. Nauk Ikr. RSR, Ser. B, Khim. Biol. Nauki, 5, 429—432.

Shilov, A.E., Koryakin, B.V. and Dzhabier, T.S. (1977) Dokl. Akad. Nauk SSSR, 233, 620—622.

Siderer, Y., Malkin, S., Poupko, R. and Luz, Z. (1977) Arch. Biochem. Biophys., 179, 174—182.

Sutin, N. and Creutz, C. (1975) Proc. Natl. Acad. Sci. U.S.A., 72, 2858—2862.

Sutin, N., Creutz, C., Chou, M., Bottcher, W. and Lin, C.T. (1976) J. Am. Chem. Soc., 98, 6536—6544.

Sykes, A.G. (1963) Trans. Faraday Soc., 59, 1325—1347.

Trebst, A. (1974) Annu. Rev. Plant Physiol., 25, 423—458.

Uri, N. (1952) Chem. Rev., 50, 375—454.

Van Best, J.A. and Mathis, P. (1978) Biochim. Biophys. Acta (in press).

Van Gorkom, H.L. (1974) Biochim. Biophys. Acta, 347, 439—442.

Van Neil, C.B. (1941) Adv. Enzymol., 1, 263—328.

Velthuys, B.R. and Amesz, J. (1975) Biochim. Biophys. Acta, 376, 162—168.

Wang, J.H. (1969) Proc. Natl. Acad. Sci. U.S.A., 62, 653—660.

Warburg, O. and Lüttgens, W. (1946) Biokhimiya, 11, 303—322.

Warburg, O. and Krippahl, G. (1960) Z. Naturforsch., 15b, 367—369.

Whitten, D.G., Delaive, P.J., Lee, J.T., Sprintschnik, H.W., Abruna, H. and Meyer, T.J. (1977a) J. Am. Chem. Soc., 19, 7094—7097.

Whitten, D.G., Sprintschnik, H.W. and Kirsch, P.P. (1977b) J. Am. Chem. Soc., 99, 4947—4958.

Whittingham, C.P. and Bishop, P.M. (1961) Nature, 192, 426—427.

Williams, W.P. (1977) in Primary Processes of Photosynthesis, Vol. 2: Topics in Photosynthesis (J. Barber, ed.) pp. 99—147. Elsevier, Amsterdam.

Witt, H.T. (1971) Q. Rev. Biophys. 4, 365—477.

Wrighton, M.S., Ginley, D.S., Wolczanski, P.T., Ellis, A.B., Morse, D.L. and Linz, A. (1975) Proc. Natl. Acad. Sci. U.S.A., 72, 1518—1522.

Wydrzynski, T., Zumbulyadis, N., Schmidt, P.G. and Govindjee (1975) Biochim. Biophys. Acta, 408, 349—354.

Wydrzynski, T., Zumbulyadis, N., Schmidt, P.G., Gutowski, H.S. and Govindjee (1976) Proc. Natl. Acad. Sci. U.S.A., 73, 1196—1198.

Yamashita, T., Tsuji, J. and Tomita, G. (1971) Plant Cell Physiol., 12, 117—126.

Photosynthesis in relation to model systems, edited by J. Barber
© Elsevier/North-Holland Biomedical Press 1979

Chapter 9

Photoproduction of Hydrogen in Photosynthetic and Artificial Systems

A.A. KRASNOVSKY

A.N. Bakh Institute of Biochemistry of the U.S.S.R. Academy of Sciences, Moscow, U.S.S.R.

CONTENTS

Abbreviation

DCMU, 3-(3-dichlorophenyl-1,1-dimethylurea.

9.1. INTRODUCTION

As emphasised several times in this book the existence of mankind on the earth depends upon photosynthesis occurring in plants. The food and oxygen consumed by man and other organisms are produced by plants as a result of their photosynthetic activity. The energy which ensured the development of civilization on the earth was provided by the combustion of carbonaceous compounds resulting from photosynthesis (wood, coal, oil and natural gas). In addition to this, the photosynthetic process is responsible for maintaining significant levels of oxygen in the atmosphere.

Today some 95% of the energy consumed by mankind is derived from coal, oil and gas, and only about 5% comes from hydroelectric and nuclear power stations; according to some forecasts, by the year 2000 these stations will satisfy not more than 20% of the energy demand (Semenov, 1972), and most of the energy requirements will continue to be met by photosynthetic products. With the rates of energy consumption steadily growing, it is feared that the oil and natural gas reserves may be exhausted within the next few hundred years.

The practically unlimited production of energy in thermonuclear synthesis may be expected to lead to an economically profitable chemical synthesis of the basic food elements such as carbohydrates, proteins and lipids from carbonates, water and molecular nitrogen. Most authors, however, are very cautious in their forecasts of the possible time when thermonuclear reactions will be technically used.

The promising approach thus is to make a more efficient use of the solar energy reaching the earth's surface: the quantity of this energy exceeds by five orders the quantity of energy of all kinds produced on the earth. The question is how best to collect and utilize solar energy.

Possible ways of utilizing solar energy for heating and powering thermal generators of electricity have long been the subject of discussion, research and development. Much attention is given to the direct conversion of solar energy into electric energy in solar batteries (whose efficiency attains 20%). Although they are widely employed in earth satellites their broader applications in "terrestrial" energetics must await the solution of complicated technological and economic problems (see chapter 1).

I would like to dwell here on the traditional method of energy utilization habitual to man, namely, on prospects for the use of photosynthetic products not only as a source of food but also as an energy source.

The reaction between water and carbon dioxide which gives rise to sugars and oxygen proceeds with the increase of free energy of the system ($+\Delta F$). So, this reaction cannot be aided by catalysts, i.e. enzymes, because it proceeds "against" the thermodynamic potential gradient.

In order to bring about photosynthesis, a maximum of 120 kcal per gram molecule of evolved oxygen (or of consumed carbon dioxide) must be introduced in the system:

$$H_2O + CO_2 \rightarrow \tfrac{1}{6} C_6H_{12}O_6 + O_2 \ (+\Delta F = 120 \text{ kcal}) \ .$$

An Einstein of red light, $Nh\nu$, constitutes about 40 kcal. Thus, in order to gain free energy in the reaction of photosynthesis, a minimum of three quanta of red light must be put into the system.

Estimates of the minimum quantum requirement (as measured in different laboratories) range from 8 to 12 quanta. Hence the probable limit of absorbed light conversion to potential chemical energy in photosynthesis lies within 25% and 40%. The average estimate of 30% is usually accepted.

About half of all solar radiation energy reaching the earth's surface lies predominantly in the visible region of the spectrum (see chapter 1); the infrared region practically is not used by terrestrial plants; (but there are photosynthesizing bacteria *Rhodopseudomonas viridis* which can use infrared light with wavelengths down to 1 μs).

Assuming that all active solar radiation is absorbed the limit of utilization of incident solar radiation by plants is about 15%. Actually, on average green plants on the earth's surface use no more than 0.1% of the incident solar radiation.

Attainment of the maximum possible utilization of solar energy by plants (15%) presents a real challenge for the future: the vigorous development of plant physiology, biochemistry and genetics will undoubtedly permit an approach towards this limit. This problem is being investigated intensively and the reader is referred to relevant reviews (e.g. see Calvin, 1976).

The metabolism of animals and other non-photosynthetic organisms has been adapted through evolution to the utilization of various carbonaceous products of photosynthesis. However, one can ask whether it is reasonable to burn in furnaces complex organic compounds intended to use in specialized metabolism. Can photosynthesis be stopped at the stage of water decomposition to realise hydrogen as the product and thus avoid its normal function to reduce carbon dioxide?

As the preceeding chapters in this book emphasised, the possibility of using hydrogen in energetics is now attracting much attention. Hydrogen has several advantages as a fuel: it does not yield combustion products polluting the environment and with oxygen it can be used with a high efficiency (up to 80%) in fuel elements to produce electric power.

Green plants which acquired the capacity to use water molecules as a primary hydrogen donor combine two consecutive photochemical steps of electron transfer — one from water to cytochromes and quinones, and the other from reduced cytochromes to ferredoxin and pyridine nucleotides, resulting in their reduction. Electron transfer is coupled to the formation of adenosine triphosphate (ATP). The light energy-storing substances, reduced pyridine nucleotides and ATP, enter the carbon cycle. This reaction system involving electron transfer is located in membranes of chloroplasts and chromatophores in such a way that the water oxidation products (for example, oxygen) are spatially separated from the reduced products ($NADPH_2$). The range of the redox potential in which the photosynthetic electron transport chain operates, lies between the redox potentials of the oxygen and hydrogen electrodes. Therefore the energy model of photosynthesis represents a chlorophyll-sensitized decomposition of water into hydrogen and oxygen.

As discussed in some detail in the first three chapters of this book, primary light energy conversion in photosynthesis may be visualized as charge separation in reaction centers of photosystems. Electrons are transferred through a chain of intermediates with ferredoxin and NADP acting as the terminal electron acceptors. The electron vacant "hole" is also transported via suitable carriers finally to be used in the oxidation of water and bring about the evolution of molecular oxygen (see chapter 8).

In general, there is a possibility to stop an electron or its associated "hole" from reaching the terminal points by introducing suitable electron donors and electron acceptors.

For example, R. Hill in 1937 discovered that the introduction of electron acceptors to isolated chloroplasts switched off physiological electron transport to carbon dioxide but did not stop the path of the "positive hole" to the water splitting system so that oxygen was evolved as in the intact cell. Introduction of electron donors which can substitute for water molecules has also led to a possibility to switch off this branch of electron transfer. In the case where electron flow is stopped at the level of ferredoxin then in the presence of hydrogenase it is possible to observe hydrogen evolution (see chapter 10). Moreover it is possible to induce different types of hydrogen evolution by substitution of diverse physiological donors and acceptors of electron; so it is possible to construct simple systems of electron transfer leading to creation of non-physiological artificial model systems. The main purpose of this review is to consider the gradual simplification of the photosynthetic system but still maintaining the capability to evolve hydrogen. Firstly I will deal with hydrogen evolution by algal cells and then come to chloroplasts. Finally I will consider model systems using chlorophyll and inorganic photocatalysts; the case when charge separation is most pronounced. The review is mainly based on the experiments of my laboratory and the conditions used in the experiments are described in originaly papers cited.

9.2. PHOTOEVOLUTION OF HYDROGEN BY GREEN ALGAE

In 1942 Gaffron and Rubin discovered that illumination of unicellular algae under anaerobic conditions can lead to the formation of hydrogen gas (see review Gaffron, 1972). In 1949 Gest and Kamen found that photosynthesising bacteria were also capable of bringing about the photochemical evolution of hydrogen. In this case, however, hydrogen, is donated not by water but by various organic and inorganic substances used in bacterial metabolism. Under normal conditions molecular nitrogen inhibits the release of hydrogen so that this process is accompanied in bacteria by photosynthetic fixation of molecular nitrogen. The extensive literature dealing with the production of hydrogen by bacteria and algae has been surveyed in recent reviews by Kondratieva and Gogotov (1976) and Oshchepkov and Krasnovsky (1976) as well as by Hallenbeck and Benemann in chapter 11 of this book.

The evolution of hydrogen by unicellular algae has been studied in our laboratory (Oshchepkov and Krasnovsky, 1972, 1974). For this purpose, a device was constructed consisting of a gas chromatograph and a monochromator (Oshchepkov and Krasnovsky, 1974). A cuvette with a magnetic stirrer containing 1 ml of *Chlorella* suspension was connected to the gas chromatograph. Illumination of the algal suspension in air resulted in oxygen evolution. After bubbling of argon, hydrogen evolution occurred without any adaptation period.

In accordance with Gaffron, addition of glucose or other exogenous hydrogen donors sharply increased the production of hydrogen. Carbon dioxide was released simultaneously, and oxygen and hydrogen were evolved alternatively. Under steady-state conditions, no simultaneous stoichiometric release of oxygen and hydrogen was usually observed. However, Efimtzev et al. (1975), who used a sensitive amperometric method, recently recorded simultaneous evolution of hydrogen and oxygen in many photosynthesising organisms during the induction period. The evolution of hydrogen by

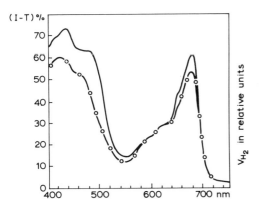

Fig. 9.1. Action spectrum of hydrogen photoproduction by *Chlorella* (○———○) as compared to the absorption spectrum.

Chlorella was measured as a function of wavelength of incident light (Oshchepkov and Krasnovsky, 1972, 1974). As Fig. 9.1 shows the action spectrum for the hydrogen evolved proved to be close to that of oxygen evolution. Some differences were noticed in the far-red region where hydrogen was evolved more effectively. These measurements have shown that no more than five quanta of red light are required to release 1 mol of hydrogen upon illumination of a *Chlorella* suspension. As 8—10 quanta are required to evolve 1 mol of oxygen during photosynthesis, the above data indicate that the photoevolution of hydrogen is more effective from the energetic point of view. However, the results of the above measurements should not be overestimated, since hydrogen is evolved as a result of photo-metabolism of the organic substances stored in the cell during ordinary photosynthesis. It is well known that this process may occur in general by way of enzymatic reactions without any light energy. Thus, a "dark" evolution of hydrogen is observable in *Chlorella* and, especially, in blue-green algae (Oshchepkov et al., 1973). In the case of the latter organisms, hetero-cysts can probably function by releasing hydrogen while normal cells can affect photosynthesis by giving off oxygen (Benemann and Weare, 1974).

Of significance is the fact that oxygen inhibits the photoevolution of hydrogen either by interacting with reduced products or by inhibiting the enzyme hydrogenase. The mechanism of hydrogen photoevolution could be visualized simply by a scheme in which oxygen is given off at one "end" of the photosynthetic electron transport chain while hydrogen is released at the other "end". Such a simple scheme, however, is not consistent with experi-mental evidence.

That hydrogen evolution is invariably accompanied by a release of carbon dioxide was already demonstrated in Gaffron's experiments. In the case of mutants deficient in Photosystem II (Bishop and Gaffron, 1963), hydrogen could still be evolved. Thus no rigid connection exist between photolysis of water (in Photosystem II) and evolution of hydrogen. This is consistent with the action of diuron which at a concentration of 10^{-6} M suppresses the release of oxygen without affecting that of hydrogen. It follows that the photoproduction of hydrogen is closely linked with the carbon metabolism of the cell (Gaffron, 1972; Gaffron and Rubin, 1942). Hydrogen is evolved by way of a number of enzymatic and photochemical intermediate reactions with participation of reduced compounds formed during the operation of carbon cycles of photosynthesis and respiration (see Fig. 9.2). However, in the overall process, the molecular hydrogen evolved is derived from water molecules since no other hydrogen source is present in the system. Most likely, in the course of carbon cycles of photosynthesis and respiration, active hydrogen donors are formed which enter Photosystem I of the elec-tron transport chain. Reduced pyridine nucleotides are formed in the Krebs cycle. These compounds may enter a locus of Photosystem I where chlorophyll sensitizes electron transport to ferredoxin. That such a mecha-

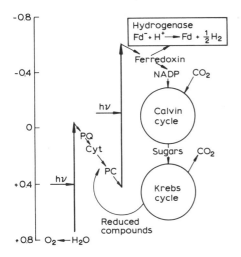

Fig. 9.2. A hypothetic diagram of metabolic pathways of hydrogen evolution (modified Gaffron's scheme).

nism is possible is indicated by model experiments described below.

The question arises as to whether photolysis of water can occur without the participation of carbon cycles. To answer this, the reactions must be done with isolated structures such as chloroplasts lamella where the photosynthetic electron transport chain is localized.

9.3. CHLOROPLASTS: PHOTOREDUCTION OF VIOLOGENS AND HYDROGEN EVOLUTION

To study the reducing branch of the photosynthetic electron transport chain, wide use is made of viologens, derivatives of γ-γ-dipyridyl, which were first applied in biochemical research by Michaelis more than four decades ago. The reversible reduction of viologens leads to the formation of blue-coloured ion radicals with an absorption maximum at about 600 nm. The redox potential of methyl viologen, $E_0' = -0.455$ V, is close to that of ferredoxin which is the final electron acceptor of Photosystem I, and is higher than the E_0' of the hydrogen electrode (-0.42 V). The reduced viologen is easily oxidized by oxygen to form hydrogen peroxide. To observe the accumulation of reduced viologens by chloroplasts, either oxygen production must be inactivated (i.e. Photosystem II), for example by heating, or an oxygen-consuming system must be introduced into the chloroplast suspension.

Arnon et al. (1961) reported photoreduction of methyl viologen by chloroplasts in the presence of cysteine and dichlorophenolindophenol, which acted as electron donors. Kok and his co-workers (1965) observed photoreduction of viologens by chloroplasts in the presence of glucose and glucose oxidase. Zweig and Avron (1965) and Black (1966) described the reduction of various viologens by chloroplasts in the presence of ethanol and catalase (see also chapter 10).

We have studied photoreduction of methyl viologen by chloroplasts using hydrazine as the hydrogen donor during the course of progressive disruption of chloroplast structure by organic solvents (Brin and Krasnovsky, 1972). This reaction proceeded most actively at pH close to 8.5. With 10% of solvent the process was activated, while with 50% solvent treatment complete inhibition occurred. With 70—80% of solvent, chlorophyll was extracted and active photoreduction of viologen was resumed. Illumination of chloroplasts in the presence of oxygen led to oxygen reduction with for-mation of hydrogen peroxide. Addition of methyl viologen resulted in a multifold increase of the amount of peroxide formed. Viologen competes with oxygen for electrons of Photosystem I and the cation radical of methyl viologen that is formed is reversibly oxidized by oxygen to form hydrogen peroxide (Shuvalov and Krasnovsky, 1975). Photoreduction of viologens by chloroplasts indicates that the system has reached the hydrogen electrode potential which is a precondition for release of molecular hydrogen if the system includes a catalyst (hydrogenase) promoting the reaction $2H + 2e \rightarrow H_2$.

Boichenko (1949) reported that isolated chloroplasts of higher plants were capable of releasing hydrogen in the presence of glucose. In Arnon's experiments (1961) photorelease of hydrogen by chloroplasts was observed in the presence of bacterial hydrogenase and with cysteine as the electron donor. ATP formation accompanied hydrogen production.

Benemann et al. (1973) have described photoproduction of hydrogen in a similar system (but without cysteine) consisting of isolated chloroplasts, ferredoxin and hydrogenase isolated from a *Clostridium pasteurianum* culture. They believed that the reaction is accompanied by oxygen produc-tion which, however, they were unable to measure. Oxygen is presumed to be used to oxidize exogenous electron donors such as glucose. However, as in Arnon's experiments, endogenous electron donors other than water may have been used. Hall and co-workers (1976) recently described a prolonged hydrogen evolution in similar systems and concluded that H_2O was the source of electrons (see chapter 10). Ben-Amotz and Gibbs (1975) showed hydrogen evolution in cell-free preparations from algae mixed with hydrogenase and dithioerythritol as electron donors.

In our laboratory photoproduction of hydrogen by chloroplasts was ob-served (see Table 9.1) in the presence of hydrogenase isolated by Gogotov et al. (1974) from photosynthesising bacteria, with $NADH_2$ acting as the electron donor (Krasnovsky et al., 1975). Also further experiments performed in our laboratory revealed that bubbling of argon or deoxygena-

TABLE 9.1.

HYDROGEN EVOLUTION IN RED AND WHITE LIGHT BY BEAN LEAVES CHLOROPLASTS

System	Hydrogen (μl/min)	
	Red light 600—750 nm	White light 400—700 nm
Without NADH$_2$	0.000	0.000
NADH$_2$	0.007	0.015
NADH$_2$ + MV^{2+}	0.072	0.150

Chlorophyll concentration 0.05 mg/ml; NADH$_2$ concentration, 10^{-3} M; methyl viologen (MV) concentration, 10^{-3} M. Light intensities: $7 \cdot 10^5$ (red light) and 10^6 (white light) erg cm^{-2} sec^{-1}.

tion in vacuo of the suspension led to inhibition of the ability of chloroplasts to photoreduce methyl viologen (Nikandrov et al., 1978a). The cause of this phenomenon is probably denaturation of functionally active membranes on the liquid—gas interphases. When glycerol (up to 50%) or bovine serum albumin was introduced into the system the activity of chloroplasts was preserved during the procedure of deoxygenation. It was revealed earlier in our laboratory that glycerol greatly activated the Hill reaction (Krasnovsky and Brin, 1968) as well as millisecond afterglow of chloroplasts (Chan-Van-Ni et al., 1977). For instance, without addition of exogenous electron donors chloroplast suspended in the presence of 50% glycerol were capable of photoreducing practically all the methyl viologen (10^{-4} M) introduced into the system.

The uncouplers studied (NH$_4$Cl, CH$_3$NH$_2$) greatly activated the reaction and DCMU inhibited the function of chloroplasts.

Probably water molecules serve as initial electron donors to photoreduce viologen as well as some endogenous electron donors probably stored inside the structures of chloroplast membranes. Nevertheless in this system it was impossible to measure evolution of oxygen which reacts rapidly with reduced viologen.

9.4. CHLOROPHYLL SOLUTIONS: PHOTOREDUCTION OF VIOLOGEN AND PHOTOEVOLUTION OF HYDROGEN

Studies carried out in our laboratory many years ago revealed that during the photoreduction of chlorophyll, an intermediate is formed with an E_0' close to that of the hydrogen electrode. In 1949, we revealed the possibility of chlorophyll-sensitized reduction of NAD (Krasnovsky and Brin, 1949).

In a reaction of this type, chlorophyll acts as a light-excited electron carrier from electron-donating molecules to electron acceptors (see for review Krasnovsky, 1974). More recently, we have investigated the photosensitized reduction of methyl viologen under the action of red light absorbed by chlorophyll in the presence of a number of electron donors. The reactions were done in organic solvents and aqeuous solutions of detergents where chlorophyll and other reaction components were dissolved (Brin et al., 1967; Krasnovsky, 1974; Krasnovsky and Brin, 1965). The most efficient photoreduction of methyl viologen under anaerobic conditions was observed when phenylhydrazine, cysteine and $NADH_2$ were used as electron donors; thiourea was inactive under these conditions. Efficient photosensitized reduction of methyl viologen in the presence of thiourea did, however, occur in experiments without preliminary evacuation of air (Luganskaya and Krasnovsky, 1970). The mechanism of this reaction was as follows. As a result of photosensitized oxidation of thiourea by oxygen, active long-lived reductants were formed which were capable of reducing methyl viologen. The anaerobiosis which had developed due to photosensitized reduction of oxygen, prevented reoxidation of reduced viologen. In pyridine solution, the "red" photoreduced form of chlorophyll is capable of a dark reaction with viologen. In this medium, a possible mechanism for the reaction consists of photoreduction of chlorophyll by electron donor followed by a reaction between the reduced chlorophyll and viologen (see Fig. 9.3). On the other hand, observations of fluorescence quenching by chlorophyll, and its analogs, by methyl viologen point to a possible primary photooxidation of the sensitising pigment (Krasnovsky and Drozdova, 1966). Thus, photoreduction of methyl viologen in a chlorophyll solution can be achieved at the expense of the light absorbed by chlorophyll. To have molecular hydrogen released in the reactions described above, a catalyst of the reaction $2H + 2e \rightarrow H_2$ should be introduced into the system. Indeed, addition of bacterial hydrogenase to an aqueous solution of Triton X-100 containing chlorophyll and cysteine (or $NADH_2$) did result in release of molecular hydrogen upon illumination (Krasnovsky et al., 1975a). Addition of methyl viologen considerably activated the reaction, as in the case of chloroplasts (see Table 9.2).

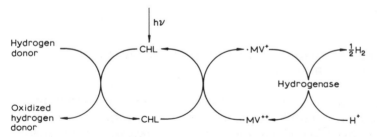

Fig. 9.3. Scheme of chlorophyll-photosensitized hydrogen evolution.

TABLE 2

CHLOROPHYLL-SENSITIZED PHOTOEVOLUTION OF HYDROGEN IN AQUEOUS SOLUTION OF TRITON X-100 UPON ILLUMINATION WITH RED AND WHITE LIGHT

System	Hydrogen (μl/min)	
	Red light 600—750 nm	White light 400—700 nm
$NADH_2$	0.070	0.100
$NADH_2 + MV^{2+}$	0.150	0.260
Cysteine	0.006	0.010
Cysteine + MV^{2+}	0.125	0.200

Hydrogenase was added in each system. $NADH_2$, cysteine and methyl viologen (MV) concentrations. $1.4 \cdot 10^{-3}$ M; Triton X-100 concentration, 0.5%. Light intensities, $5 \cdot 10^5$ (red light) and 10^6 (white light) erg cm^{-2} sec^{-1}.

9.5. PHOTOACTIVATION OF REDUCED PYRIDINE NUCLEOTIDES: REDUCTION OF VIOLOGEN AND HYDROGEN EVOLUTION

In the reactions described above, light-excited chlorophyll reacts which non-excited $NADH_2$. However, absorbing a light quantum in their own absorption region (at 340 nm), $NADH_2$ and $NADPH_2$ are activated and their redox potential becomes more negative than the E_0' of the hydrogen electrode. Thus excited $NADH_2$ ($NADH_2^*$) can reduce methyl viologen and ferredoxin (Krasnovsky et al., 1975):

$$NADH_2^* + 2\,MV^{2+} \rightarrow NAD + 2 \cdot MV^+ + H^+$$

$NADH_2$ probably reacts via a triplet excited state. The phosphorescence of $NADH_2$ was revealed in frozen solutions. The mechanisms of the reaction are

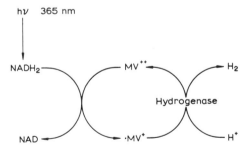

Fig. 9.4. Scheme of hydrogen evolution as a result of $NADH_2$ and $NADPH_2$ photoexcitation.

TABLE 9.3

HYDROGEN EVOLUTION UPON ILLUMINATION OF AQUEOUS SOLUTION OF PYRIDINE NUCLEOTIDE

System	Hydrogen (μl/min)
NADH$_2$	0.120
NADH$_2$ + MV^{2+}	0.400

Hydrogenase was added in each system. NADH$_2$ and methyl viologen (MV) concentrations, $1.4 \cdot 10^{-3}$ M. Light intensity, $2 \cdot 10^5$ erg cm^{-2} sec^{-1} at 365 nm.

considered in detail in our recent paper (Nikandrov et al., 1978). The activation of NADH$_2$ by light has been investigated in a series of studies undertaken in our laboratory. Just as in above-mentioned reactions, introduction of hydrogenase led to release of hydrogen (Krasnovsky et al., 1975a) upon illumination of aqueous solutions of NADH$_2$ with ultraviolet light at 365 nm (see Fig. 9.4 and Table 9.3).

9.6. USE OF INORGANIC PHOTOCATALYSTS FOR OXYGEN AND HYDROGEN EVOLUTION

It is generally assumed, that the primary reaction in photosynthesis is a charge separation process (see chapters 1 and 3). For this reason it is desirable to have a simple system to study so as to understand how light can generate electrons and "holes". Such a system exists with inorganic photoactive semiconductors like titanium dioxide, zinc oxide, etc. In this type of system the action of light is to excite electrons to the conduction band leaving behind "holes" in the valence band. Thus the light-induced charge separation which take place in these systems is ideal for model experiments.

About 50 years ago Emil Baur (1928) discovered that under the action of light zinc oxide gained the property to cause oxido-reductions. He described the phenomenon as "molecular electrolysis". Considering an excited sensitizer there were two sites, anodic and cathodic, so that oxido-reduction proceeded at the phase boundary. According to contemporary knowledge this may be regarded as a primitive point of view, but his basic idea is quite plausible. We tried to construct models, which in their action resembled oxido-reduction taking place in chloroplasts by using titanium dioxide, zinc oxide and tungsten trioxide (see for review Krasnovsky and Brin, 1970). Absorption bands of titanium and zinc oxides are situated near the border between visible light and UV. Titanium and zinc oxides are white powders, a sharp absorption begins at ~400 nm, so they are "black" in ultraviolet. In most cases we used excitation at 365 nm. Metzner's silver chloride models (Metzner, 1968) should be mentioned in this context.

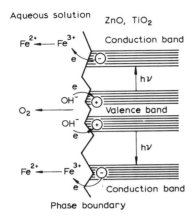

Fig. 9.5. Scheme of hydrogen evolution with inorganic photocatalysts (TiO_2 and ZnO).

We used fine powders of photocatalysts, having a specific surface up to ~ 20 m^2 g^{-1}. If we illuminate these TiO_2, ZnO or WO_3 suspensions in the presence of electron acceptors we observed an evolution of gas. In the first experiments sensitized photoevolution of oxygen was observed in solutions of some oxidants including potassium permanganate (Krasnovsky and Brin, 1961). Then we tried to simulate Hill reaction (Krasnovsky and Brin, 1962) by using as electron acceptors, ferricyanide or Fe^{3+}. The stoichiometry of the reactions that followed strictly corresponded to the equation:

$$2\ Fe^{3+} + H_2O \overset{h\nu}{\to} 2\ Fe^{2+} + 2\ H^+ + \tfrac{1}{2}\ O_2$$

Experiments of this kind were performed in water labelled with heavy oxygen. In this case the oxygen isotope ratio of the released O_2 was the same as in the water suspension used (Fomin et al., 1973) thus confirming that the oxygen had originated from water molecules (see Fig. 9.5).

It was possible to reduce not only ferric compounds and ferricyanides but also quinones, like p-benzoquinone (Krasnovsky et al., 1971). The quantum yield of the reactions studied did not exceed 1%.

Summarising these series of experiments we can conclude that under the excitation of photocatalysts charge separation proceeds if electrons can be donated to the "hole" from OH$^-$ anions or H_2O molecules and radicals can be produced which can recombine on the boundary to give molecular oxygen. The formation of OH $\cdot$ radicals under the described conditions had been found earlier (Korsunovsky, 1960). The electrons are accepted at the phase boundary by oxidants, that is by electron-accepting molecules.

A second type of photosynthetic reaction to attempt to simulate was the Mehler reaction of chloroplasts (see chapters 4 and 5 of Volume 1 of this series). In model systems studied oxygen may act as terminal electron

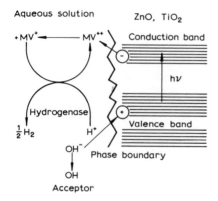

Fig. 9.6. Scheme of oxygen photoevolution with inorganic photocatalysts (TiO$_2$, ZnO, WO$_3$).

acceptor and water molecule (or OH$^-$) as a donor. The first observation of hydrogen peroxide formation with ZnO was made by Baur and Neuweiler (1927). We have used a sensitive measurement involving the chemiluminescence of luminol in the presence of peroxidase (Cormier and Prichard, 1968), in order to quantitatively determine the hydrogen peroxide produced, not only in the case of ZnO but also in water suspensions of TiO$_2$. If we added methyl viologen to this system, the reaction was greatly enhanced (Krasnovsky et al., 1976). Here methyl viologen probably functioned as intermediate electron carrier. So we could simulate a Mehler reaction in the photocatalysts suspension in the presence of oxygen.

There is a third type of reaction which is of interest. If we excluded oxygen from the suspension of photocatalysts so as to work under anaerobic conditions, a reduction of methyl viologen is observed (Krasnovsky and Brin, 1973). In this case methyl viologen acts as a terminal electron acceptor. The only possible electron donor in this system is H$_2$O or OH$^-$ radicals but we

TABLE 9.4

EVOLUTION OF HYDROGEN UPON ILLUMINATION OF AQUEOUS SUSPENSION OF INORGANIC PHOTOCATALYSTS

System	Hydrogen (μl/min)
Titanium dioxide	0.009
Titanium dioxide + MV^{2+}	0.094
Zink oxide	0.002
Zink oxide + MV^{2+}	0.005

Hydrogenase was added in each system. Methyl viologen (MV) concentration, 10^{-3} M. Light intensity, $2 \cdot 10^5$ erg cm^{-2} sec^{-1} at 365 nm. Illumination time, 1 min.

have not been able to observe oxygen evolution in this case. May be the OH ·
radicals formed react with viologen which may function as a scavenger of
these radicals. This, however, is not clear yet; the situation is rather similar
to viologen reduction in chloroplasts.

It is well known that the redox potential of methyl viologen slightly
exceeds that of hydrogen electrode. So by introduction of suitable catalysts
we may expect to observe hydrogen gas evolution. Using gas chromatography, a hydrogen photoproduction was indeed observed if the system was
supplied with a hydrogenase from photosynthetic bacteria (Krasnovsky et
al., 1975a).

One side of our system works like an oxygen electrode, releasing molecular oxygen while the other side may be compared to a hydrogen electrode.

In this connection I would like to mention studies of Fujishima and
Honda (1972). They have reported an electrochemical version of our experiments. In their case they demonstrate a charge separation on TiO_2 "membranes". By making these layers part of electrochemical cells it could be
demonstrated that these cells produce both hydrogen and oxygen. It is
remarkable that the authors obtained a great enhancement of oxygen evolution if they added Fe^{2+} to the system. This means that their set-up has a
striking similarity to that used in our experiments which we described in
1961.

It is important to emphasize that the semiconductors studied, function
not only as photosensitisers but also as catalysts. On the interfaces there may
be a recombination of the primary produced intermediates, which probably
makes these experiments successful. I suppose, that these simple models may
be of interest for those who are working on the problem of conversion of
light energy into chemical energy.

9.7. PROSPECTS FOR TECHNICAL USE

The current energy shortage has enhanced interest in the possible better
utilisation of solar energy. As a consequence some workshops have been
organized in the USA *. These workshops outlined possible lines of research
that could be developed and discussed various points of view regarding their
prospects. In particular, Gaffron, who discovered the phenomenon of
hydrogen photoevolution by algae, has pointed out that effective photolysis
of water by solar radiation absorbed in photosynthesis is a research problem
which is unlikely to receive practical outlets in the next few years to come.
He believes that this problem belongs to the domain of fundamental science
since a much better understanding of the mechanism of photosynthesis is

---

* The author is most grateful to Professor A. Hollander who has kindly sent me the
proceedings of these meetings for 1972 and 1973.

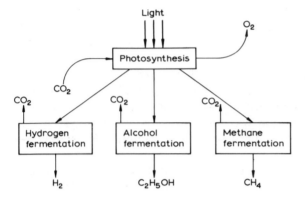

Fig. 9.7. Cooperation of photosynthetic and heterotrophic organisms producing hydrogen, methane or ethanol.

needed before it can be put to some form of technical use. Nevertheless, it is clear that unicellular algae are capable, under anaerobic conditions, of releasing molecular hydrogen by utilizing organic substances built up in the process of "normal" photosynthesis and therefore oxygen and hydrogen can be evolved through the use of the biochemical mechanisms operating in the same algal cell.

A further point to consider is whether it is better to use carbonaceous compounds within the same algal cell or to use a symbiosis of specialized organisms. One may conceive the following scheme of division of function between organisms of different types. "ordinary" photosynthesis leads to consumption of carbon dioxide, formation of organic substances and evolution of oxygen. In the next cycle, the organic substances formed (algal biomass) could be used by specialised heterotrophic organisms to form methane, hydrogen or alcohol (see Fig. 9.7). The carbon dioxide evolved in this process reenters the photosynthetic cycle. Such cooperation would probably require the introduction of intermediate bacterial cycles to prepare the organic substance of green cells for utilization by hydrogen or methane producers.

An important consideration is whether biochemical processes of this kind can use cellulose which constitutes the bulk of the photosynthetic product of forests and of the enormous amounts of agricultural wastes such as stems, straw, roots, etc. With such cooperation of autotrophic photosynthesizing organisms and heterotrophic organisms using photogenic organic matter, it would be possible to develop technological schemes that would probably be more efficient than a combination of processes within the green cell itself. Thus, for instance, a combination of algal photosynthesis and methane fermentation has been described where unicellular bacteria are cultivated together with methane-forming bacteria in pilot plants. Calvin (1974) proposed that addition of ethanol produced by fermentation to motor fuels may already be economically acceptable.

In contrast simpler photobiochemical reactions leading to a release of oxygen or hydrogen still remain to be studied. It is feasible, for example, to bring about photoreduction of viologens and, in a separate cycle, to achieve a release of molecular hydrogen with the aid of hydrogenase. The question is whether or not a reaction of this kind would be effective enough in a technological sense. The model photochemical systems described above have low quantum yields, and prospects for their technical use depend on the development of methods that would considerable enhance their efficiency.

Although many problems still remain unsettled the objectives and probable lines of research can be seen. One clear problem remaining for experimental solution is that of biophotolysis of water into hydrogen and oxygen with a sufficient high efficiency of solar energy utilization (up to 10%?) (see chapter 8).

In conclusion, some of the possible lines of research may be listed.

(i) A search for those algal strains which are capable of hydrogen evolution; of particular importance here are genetic studies aimed at finding appropriate mutants, including those that may block the carbon cycle.

(ii) Studies of various ways of combining photosynthetic systems with heterotrophic bacterial cultures to attain effective processing of photosynthetic products.

(iii) Development of effective photoenzymatic systems on the basis of chloroplasts and immobilized enzymes, and searches for a hydrogenase resistant to oxygen.

(iv) Creation of artificial photocatalytic systems releasing oxygen and hydrogen.

Clearly, research efforts along these lines are necessary not only in order that photosynthesis might be used to solve energy problems but also because they are important for intensifying photosynthesis as a source of food. The future will show which aspect of the problem will be more important.

REFERENCES

Arnon, D.I., Mitsui, A. and Paneque, A. (1961) Science, 134, 1425—1429.
Baur, E. (1928) Z. Phys. Chem., 131, 143—148.
Baur, E. and Neuweiler, C. (1927) Helv. Chim. Acta, 10, 901—906.
Ben-Amotz, A. and Gibbs, M. (1975) Biochem. Biophys. Res. Commun., 61, 355—359.
Benemann, J.R., Berenson, J.A., Kaplan, N.O. and Kamen, M.D. (1973) Proc. Natl. Acad. Sci. U.S.A., 70, 2317—2320.
Benemann, J.R. and Weare, N.M. (1974) Science, 184, 174—175.
Bishop, N.I. and Gaffron, H. (1963) in Photosynthetic Mechanisms in Green Plants, p. 441. N.A.S.-N.R.C., Publ.
Black Jr., C.C. (1966) Biochim. Biophys. Acta, 120, 332—340.
Boichenko, E.A. (1949) Dokl. Akad. Nauk SSSR, 64, 545—548.

298

Brin, G.P. and Krasnovsky, A.A. (1972) Dokl. Akad. Nauk SSSR, 204, 1253—1256.
Brin, G.P., Luganskaya, A.N. and Krasnovsky, A.A. (1967) Dokl. Akad. SSSR, 174, 221—224.
Calvin, M. (1974) Science, 184, 174—178.
Calvin, M. (1976) Photochem. Photobiol., 23, 425—444.
Chan-Van-Ni, Nikandrov, V.V. and Krasnovsky, A.A. (1977) Biofizika, 22, 1056—1061.
Cormier, M.J. and Prichard, P.M. (1968) J. Biol. Chem., 243, 4706—4714.
Efimtsev, E.I., Boichenko, E.A. and Litvin, F.F. (1975) Dokl. Akad. Nauk SSSR, 220, 986—989.
Fomin, G.V., Brin, G.P., Genkin, M.V., Liubimova, A.K., Blumenfeld, L.A. and Krasnovsky, A.A. (1973) Dok. Akad. Nauk SSSR, 212, 424—427.
Fujishima, A. and Honda, K. (1972) Nature, 238, 37—38.
Gaffron, H. (1972) Horizons of Bioenergetics. Academic Press Inc., New York.
Gaffron, H. and Rubin, J. (1942) J. Gen. Physiol., 26, 219—240.
Gest, G. and Kamen, M. (1949) Science, 109, 558—559.
Gogotov, I.N., Zorin, N.A. and Bogorov, L.V. (1974) Mikrobiologiya, 43, 5—10.
Kok, B., Rurainski, H. and Owens, O. (1965) Biochim. Biophys. Acta, 109, 347—356.
Kondratieva, E.N. and Gogotov, I.N. (1976) Izv. Akad. Nauk SSSR, Ser. Biol., 1, 69—86.
Korsunovsky, G.A. (1960) J. Fiz. Chim., 34, 510—517.
Krasnovsky, A.A. (1974) Transformation of light energy in photosynthesis. Nauka Publishing House, Moscow.
Krasnovsky, A.A. and Brin, G.P. (1949) Dokl. Akad. Nauk SSSR, 67, 325—328.
Krasnovsky, A.A. and Brin, G.P. (1961) Dokl. Akad. Nauk SSSR, 139, 142—145.
Krasnovsky, A.A. and Brin, G.P. (1962) Dokl. Akad. Nauk SSSR, 147, 656—659.
Krasnovsky, A.A. and Brin, G.P. (1968) Dokl. Akad. Nauk SSSR, 179, 726—729.
Krasnovsky, A.A. and Brin, G.P. (1970) in Molecular Photonics, pp. 161—178, Nauka Publishing House, Leningrad.
Krasnovsky, A.A. and Brin, G.P. (1973) Dokl. Akad. Nauk SSSR, 213, 1431—1434.
Krasnovsky, A.A., Brin, G.P. and Aliev, Z.Sh. (1971) Dokl. Aka. Nauk SSSR, 199, 952—955.
Krasnovsky, A.A., Brin, G.P. and Nikandrov, V.V. (1975a) Dokl. Akad. Nauk SSSR, 220, 1214—1217.
Krasnovsky, A.A., Brin, G.P. and Nikandrov, V.V. (1976) Dokl. Akad. Nauk SSSR, 229, 990—993.
Krasnovsky, A.A. and Drozdova, N.N. (1966) Dokl. Akad. Nauk SSSR, 167, 928—930.
Krasnovsky, A.A., Nikandrov, V.V., Brin, G.P., Gogotov, I.N. and Oshchepkov, V.P. (1975b) Dokl. Akad. Nak SSSR, 225, 711—713.
Luganskaya, A.N. and Krasnovsky, A.A. (1970) Mol. Biol. (Moscow), 4, 848—859.
Metzner, H. (1968) Hoppe-Seyler's Z. Physiol. Chem., 349, 1586—1590.
Nikandrov, V.V., Brin, G.P. and Krasnovsky, A.A. (1978) Biokhimiya 43, 636—645.
Nikandrov, V.V., Chan-Van-Ni, Brin, G.P. and Krasnovsky, A.A. (1978a) Mol. Biol. (Moscow), 12, 1278—1287.
Oshchepkov, V.P. and Krasnovsky, A.A. (1972) Fiziol. Rast. 19, 1090—1097; (1974) Fiziol. Rast. 21, 462—467.
Oshchepkov, V.P. and Krasnovsky, A.A. (1974) Prikl. Biokhim. Mikrobiol., 10, 760—764.
Oshchepkov, V.P. and Krasnovsky, A.A. (1976) Izv. Akad. Nauk SSSR, Ser. Biol., 1, 87—100.
Oshchepkov, V.P., Nikitina, K.A., Gusev, M.V. and Krasnovsky, A.A. (1973) Dokl. Akad. Nauk SSSR, 213, 739—742.
Rao, K.K., Rosa, L. and Hall, D.O. (1976) Biochem. Biophys. Res. Commun., 68, 21—28.
Semenov, N.N. (1972) Sci. Life (SSSR), No. 9—10.
Shuvalov, V.A. and Krasnovsky, A.A. (1975) Biokhimiya, 40, 358—366.
Zweig, G. and Avron, M. (1965) Biochem. Biophys. Res. Commun., 19, 397—400.

Photosynthesis in relation to model systems, edited by J. Barber
© Elsevier/North-Holland Biomedical Press 1979

Chapter 10

Hydrogen Production from Isolated Chloroplasts

K. KRISHNA RAO and DAVID O. HALL

Plant Sciences Department, King's College, 68 Half Moon Lane, London SE24 9JF, United Kingdom

CONTENTS

Abbreviations

BSA, bovine serum albumin; DBMIB, 2,5-dibromo-3-methyl-6-isopropyl-p-benzoquinone; DCPIP, 2,6-dichlorophenolindophenol; DCMU, 3-(3,4-dichlorophenyl)-1,1-dimethylurea; DMSO, dimethylsulfoxide; EDTA, ethylenediamine tetraacetate; FCCP, carbonyl cyanide p-trifluoromethoxyphenyl hydrazone; HEPES, N-2-hydroxyethylpiperazine-N'-2-ethanesulfonic acid; MV, methyl viologen; PS, photosystem; TMPD, N-tetramethyl-p-phenylenediamine.

10.1. INTRODUCTION

As emphasised by the theme of this book, those exploring alternative energy sources have recently turned their attention towards ways and means to harness solar radiation by processes other than the normal process of photosynthesis which is carried out by chlorophyll-containing organisms. One of the promising areas of research in this field already touched upon in chapter 9, is the utilization of illuminated photosynthetic organisms, either in whole or as cell free extracts, together with suitable catalysts, to split water into oxygen and hydrogen; the hydrogen thus produced can be used as fuel. In simple terms this photolytic reaction is

$$2 \, H_2O \xrightarrow[\text{catalysts}]{h\nu +} 2 \, H_2 + O_2 \, .$$

It may be thought of as the electrolysis of water but where sunlight is used as an energy source instead of electricity. In this chapter we will concentrate only on the use of isolated chloroplasts for the biophotolysis of water and discuss possible ways of synthesizing an artificial system based on our knowledge of the biological mechanism.

10.2. HISTORICAL BACKGROUND

The formation of H_2 gas by the coupling of the photosynthetic electron flow of isolated chloroplasts to bacterial hydrogenase was first demonstrated by Arnon et al. (1961), Arnon 1962, Paneque and Arnon (1962) and Mitsui and Arnon (1962) in the course of their study on the role of ferredoxin in photosynthetic phosphorylation. They showed that illuminated spinach chloroplasts in the presence of DCMU (which blocked oxygen evolution), cysteine (electron donor) DCPIP and methyl or benzyl viologen (electron carriers) and *Chromatium* hydrogenase gave rise to non cyclic photophosphorylation, the products of which were hydrogen gas and ATP. The electrons from cysteine were used to reduce hydrogenase through DCPIP and viologens; the reduced hydrogenase in turn converted protons to hydrogen gas. These authors also showed the photoproduction of H_2 by spinach chloroplasts, treated with DCMU or heated at 55° for 5 min, in the presence of ascorbate, DCPIP, and a crude hydrogenase from *Clostridium pasteurianum* — no addition of viologen was required and it was presumed the preparation contained ferredoxin. The central role of ferredoxin in H_2 evolution by a chloroplast-hydrogenase system was discussed by Tagawa and Arnon (1962). In all these experiments the water-splitting capacity of the chloroplasts was suppressed and organic compounds were added to the system as electron donors for H_2 production. A decade after the publication of these reports from Arnon's laboratory, Krampitz (1972) at a workshop on Biological Energy Conversion, reported that he had succeeded to produce

H_2 by coupling the reducing power created by the photolysis of water by washed spinach chloroplasts, with a crude hydrogenase from the bacterium *E. coli*, using viologen dyes as electron carriers. The proceedings of this workshop were not circulated widely. A year later Benemann et al. (1973) in a significant paper showed that spinach chloroplasts mixed with ferredoxin and *C. kluyveri* hydrogenase evolved H_2 in the light. The hydrogen production, though short lived, occurred in the absence of any added electron donors indicating that water was the electron donor in the reaction. Though the results indicated that H_2 evolution from sunlight and water by the photosynthetic process is possible, Benemann et al. cautioned that "even though the data presented indicate that O_2 is produced during the reaction, it remains to be established whether O_2 is actually evolved by the basic system". They also pointed out that the problems of ferredoxin autoxidation, hydrogenase inactivation and PSII instability must be resolved before the process could be considered for solar energy conversion. In the past three years Rao et al. (1976, 1978), Reeves et al. (1976), Hall et al. (1978) and Fry et al. (1977) have studied in detail the various factors which affect H_2 production and have improved considerably the efficiency of the system in terms of the rate and the total quantity of H_2 produced per given amount of chlorophyll. Ben Amotz and Gibbs (1975) were able to show light-dependent H_2 evolution from cell-free preparations of anaerobically adapted algae using dithiothreitol as electron donor to PSI. Krasnovsky (1976) used $NADH_2$ as an electron donor for the photoproduction of H_2 from a chloroplast-hydrogenase system with methyl viologen as mediator (see chapter 9). Hoffman et al. (1977) obtained very high rates of H_2 evolution by illuminating "closed" spinach thylakoid vesicles mixed with *Clostridium* hydrogenase, under anaerobiosis, in the presence of ascorbate and TMPD as electron donor and methyl viologen as electron carrier. Meanwhile many laboratories, including ours, are concentrating their efforts in attempts to stabilize the hydrogenase against oxygen inactivation and the chloroplast membranes against light inactivation and on finding suitable substitutes for ferredoxin and hydrogenase as electron and proton carriers in the system (For a brief review see Lien and San Pietro, 1975).

10.3. MATERIALS AND METHODS

A simplified scheme depicting the reaction pathway involved in the biophotolysis of water is shown in Fig. 10.1. The three essential catalytic components in the reaction are chloroplasts, ferredoxin and hydrogenase. Chloroplasts are normally isolated in a sorbitol medium according to Reeves and Hall (1973) yielding type B chloroplasts. Leaves of spinach (*Spinacia oleracea*), fat hen (*Chenopodium album* or *Ch. quinoa*), lettuce (*Latuca sativa*) and tobacco (*Nicotiana exelsior* or *N. sylvester*) have all been used as chloroplast sources depending upon the availability or the aim of the experi-

302

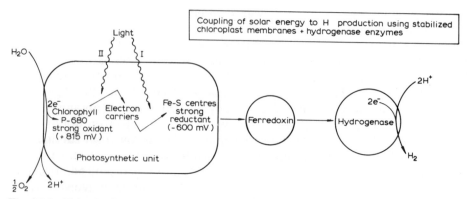

Fig. 10.1. Ferredoxin reduced by photosynthetic electron flow, donates electrons to hydrogenase which in turn converts the protons in solution to hydrogen gas.

ments. After isolation the chloroplasts are kept in ice in the dark before use or stored in liquid N_2 for use on a later day. Ferredoxins are isolated from bacteria, plants and algae by standard procedures (Buchanan and Arnon, 1970; Rao et al., 1971, Hall et al., 1972). Though ferredoxin is present in all algae and plants for a good yield of the protein we recommend using dry cells of *Spirulina maxima* available from Sosa Texcoco, Sullivan 51, Mexico 4, DF. Once isolated the ferredoxin can be stored for indefinite periods in liquid N_2 without much loss of activity. Hydrogenase can be extracted from many anaerobic and photosynthetic bacteria (Mortenson and Chen, 1974). We normally isolate the enzyme from cells of *C. pasteurianum* W5 purchased from the Microbiological Research Establishment, Porton, Salisbury, Wiltshire, England. The procedure of Chen and Mortenson (1974) with slight modifications is used for the isolation and purification of the hydrogenase. This enzyme is highly oxygen sensitive but can be stored anaerobically in liquid N_2. As long as the enzyme is not exposed to air, repeated thawing and freezing does not reduce the activity of the hydrogenase. Methyl viologen can be purchased from BDH Chemicals Ltd, Poole, Dorset, England.

Hydrogen evolution is assayed in 15 ml (4 dram) glass vials (C.E. Payne and Sons, Ltd., 6 Iveley Road, London, S.W.4) fitted with Suba-seal rubber stoppers (A. Gallenkamp and Co., P.O. Box 290, Christopher St., London, E.C.2). The basic reaction mixture usually contains chloroplasts equivalent to 100 μg chlorophyll, ferredoxin 50 nmol, BSA 5 mg, glucose 50 μmol, glucose oxidase 20 units, catalase 2000 units and ethanol 50 μl in a total volume of 2 ml made up with 50 mM HEPES—NaOH buffer, pH 7.5. All the reactants except chloroplast and hydrogenase are added to the vials, the vials closed with Suba-seal stoppers and flushed with oxygen-free N_2 using two $21G \times 1\frac{1}{2}$ in Gillett Scimitar hypodermic needles as inlet and outlet for the N_2. The vials become oxygen-free (as tested in the gas chromatograph) after 10 min flushing. An aqueous suspension of chloroplast is then injected into the vials which are then transferred to a shaking thermostatic circular Warburg water bath (Model V166, B. Braun Melsungen, Munich, Germany). Most of the experiments are done in the temperature range 20—30°C. The reaction is started by injecting hydrogenase (5—20 μl) into vials and illuminating the bath with an array of fourteen 40W incandescent lamps. The light falling on the vials is normally about 11 000 lux (about 46 W m^{-2}); the intensity can be varied to a certain extent with a rheostat connected to the circuit. At various intervals aliquots (100—10 μl— are withdrawn from the gaseous phase of the vials by means of a gas tight Hamilton syringe and injected into a gas chromatograph. The gas chromatograph (Taylor Servomex, Crowborough, Sussex, England) is fitted with a Poropak "Q", 80—100 mesh column heated to 55°C and a micro katharometer equipped with a thermal conductivity detection device. The gas chromatograph is connected to a chart recorder usually operated at 0.5 mV. With N_2 at a pressure of 20 lb/ inch^2. (1.5 kg/cm^2) flowing through as carrier gas, the H_2 and O_2 in the injected sample are easily resolved in the column; the H_2 peak being recorded 12 s after injection and the O_2 peak appearing after 14 s. The amounts of hydrogen in the vials are calculated from the recorder peak heights by comparison with peak heights obtained with calibration standards. The gas chromatograph can detect hydrogen gas in nanomolar quantities. Although the gas chromatographic technique is very convenient for measuring hydrogen, the hydrogen evolution can also be determined using a conventional Warburg manometer fitted with illumination devices. The simultaneous measurement of H_2 and O_2 can be monitored continuously using a modified Clark-type electrode (Jones and Bishop, 1976). Electron transport capacity of the chlorplasts is usually measured in a Clark oxygen electrode (Rank Brothers, Bottisham, Cambridge, England) using ferricyanide as electron acceptor.

10.4. CHARACTERISTICS OF THE H_2 EVOLUTION SYSTEM AND FACTORS INFLUENCING THE REACTION

10.4.1. Involvement of photosystems in hydrogen production

Due to the oxygen-sensitivity of *Clostridium* hydrogenase most of the early experiments testing the basic requirements of the H_2 evolution reaction were carried out under anaerobic conditions. Glucose and glucose oxidase were added to trap the O_2 evolved at PSII; catalase and ethanol can also be added to decompose any hydrogen peroxide formed in the reaction mixture (Benemann et al., 1973; Rao et al., 1976). Rao et al. (1976) showed conclusively that the H_2 evolution is dependent on light, the presence of hydrogenase and the presence of an active PSII and PSI in the chloroplasts (Fig. 10.2). Chloroplasts treated with PSII inhibitors such as DCMU or

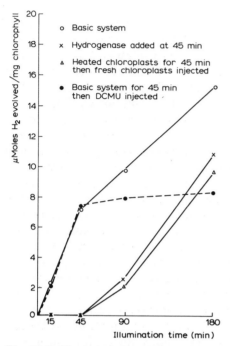

Fig. 10.2. Characteristics of H_2 evolution system. The components of the basic system are described in Methods section. Spinach chloroplasts, *C. pasteurianum* hydrogenase and spinach ferredoxin were used as catalysts. (From Rao et al., 1976).

DBMIB or chloroplasts heated at $55°C$ for 5 min did not catalyse H_2 evolution thus showing the need of PSII dependent electron transport for H_2 evolution. Such chloroplasts also catalyzed H_2 evolution when ascorbate and DCPIP were injected into the reaction mixture thus demonstrating the viability of PSI. This latter experiment also proved that the immediate source of electrons for hydrogenase is a PSI component. In these experiments average rates of 10 μmol H_2 per mg chlorophyll per h, measurable for 3 h, were obtained. Occasionally the reaction continued for more than 6 h at $20°C$. Fry et al. (1977) using additional stabilising agents in the reaction mixtures observed H_2 evolution for 20 h at $15°C$. At present we can produce on average 30 to 50 μmol of H_2 per mg chlorophyll per h by the addition of certain stabilising agents (discussed later) to the system. The maximum yield so far recorded is about 100 μmol H_2 per mg chlorophyll per h with *Chenopodium album* chloroplasts and *C. pasteurianum* hydrogenase in a reaction run at $25°C$ (Rao et al., 1978).

10.4.2. Chloroplasts

(i) Source and type. Chloroplasts isolated from different plants were found to have almost the same efficiency for H_2 production though there were marked variations in the rates of electron transport and H_2 evolution from chloroplasts isolated at different seasons. This was especially true of spinach, grown in the greenhouse or purchased from the market. The rate of decay of chloroplast activity on storage or illumination varied with the chloroplast source. Chloroplasts isolated from *Ch. quinoa* in sorbitol—HEPES medium and stored in the dark at 4°C were nine times more stable (as determined by their capacity to reduce ferricyanide) than spinach chloroplasts isolated and stored under the same conditions. Under continuous illumination at 20°C *Ch. quinoa* chloroplasts were three times more stable than spinach chloroplasts maintained identically (Morris and Hall, unpublished). The stability of *Chenopodium* chloroplasts in the H_2 evolution reaction paralleled their stability towards electron transport reactions measured in the oxygen electrode (our unpublished data). The factors which confer this unusual stability to chloroplast membranes from *Chenopodium* are being investigated. Whatever the source of chloroplasts their activity was always higher soon after isolation. The chloroplasts retained about 50% of their initial activity during storage in liquid N_2 for a month (Reeves et al., 1976).

The rate of hydrogen evolution mediated by ferredoxin and hydrogenase is not affected very much by the type of chloroplast used, i.e. whether *intact* (Type B), or *broken* (Type C). The chloroplasts are usually diluted to 20-fold their volume in 50 mM HEPES buffer or 20 mM phosphate buffer during the reaction and measurements are carried out at 30-min intervals for a period of three to six hours. Thus the chloroplasts are osmotically shocked prior to illumination. Hoffmann et al. (1977) have observed that H_2 evolution from fragmented spinach chloroplasts (sonicated or 0.4% digitonin-treated), mediated by anaerobically reduced MV, was only 35—40% of the rates obtained with thylakoid vesicles (Type C chloroplasts). In 1% digitonin-treated chloroplast particles no H_2 evolution was observed. The ruptured chloroplasts still retained PSI activity and the capacity for methyl viologen reduction. These authors found that the activity of the chloroplasts towards H_2 evolution ran parallel to their ATP forming capacity.

(ii) Stoichiometry of hydrogen evolution. Rao et al. (1976) compared the rates of ferricyanide reduction by chloroplasts using the oxygen electrode and in reaction vessels run in the Warburg bath parallel with H_2 evolution assays (Hall et al., 1978). The rate of ferricyanide reduction measured in the Warburg bath was only a third of the rate measured in the oxygen electrode. This may be due to the higher light intensity falling on the reaction mixture, the better conditions of stirring and the shorter measuring times employed in the oxygen electrode measurements. The rate of electron transport by chloroplasts in the uncoupled state (with 5 mM NH_4Cl) measured in the oxygen electrode varied between 200 to 400 μmol O_2 liberated per mg chlorophyll per h. Hydrogen evolution rates from such chloroplasts ranged between 30—50 μmol per mg chlorophyll per h. Thus only 10 to 20% of the total electron transport capacity of the chloroplasts is coupled to the H_2

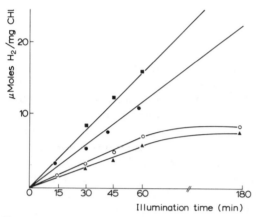

Fig. 10.3. Dependence of H_2 evolution on chlorophyll concentration. Two-ml reaction mixture contained components of the basic system, *C. pasteurianum* hydrogenase and varying amounts of *Ch. album* chloroplasts. ■————■, 50 μg chlorophyll; ●————●, 100 μg chlorophyll; ○————○, 200 μg chlorophyll; ▲————▲, 300 μg chlorophyll in 2 ml. (From Rao et al., 1978).

evolution reaction. This low coupling ratio is due to a combination of various factors. The electron donor to hydrogenase, viz. reduced ferredoxin, is capable of cycling electrons through various endogenous components of chloroplasts and reacts with oxygen (Hall et al., 1975); in some cases at a faster rate than the ferredoxins react with hydrogenase. King et al. (1977a) compared the K_m values of various ferredoxins in dark H_2 evolution from sodium dithionite using a hydrogenase prepared from *Chlamydomonas reinhardi* with the K_m values for NADP photoproduction catalysed by chloroplast membranes. Their data indicated a much more efficient transfer of electrons from ferredoxin to NADP than to protons for H_2 production. Reduced ferredoxin is also known to react with O_2 (evolved by PSII) forming hydrogen peroxide (Telfer et al., 1970). In a standard chloroplast— hydrogenase system containing added glucose, glucose oxidase, catalase and ethanol Packer and Cullingford (1978) measured the stoichiometry of O_2 disappearance by assaying the acetaldehyde produced. This provided a measure of the H_2O_2 production by either glucose oxidase or autoxidation of ferredoxin. Their data showed that the glucose/glucose oxidase trap only accounted for 60—80% of the O_2 produced by water photolysis linked to H_2 production, the remainder being linked to ferredoxin autoxidation. Earlier studies by Fry et al. (1977) showed that compared to the O_2 produced (measured as glucose consumed) about 70% of the theoretical H_2 production was observed during the first 2 h of illumination.

The amount of H_2 produced per hour from a particular quantity of chlorophyll is dependent on the chlorophyll concentration in the reaction vessel as shown in Fig. 10.3 (Rao et al., 1978). The reaction rate was propor-

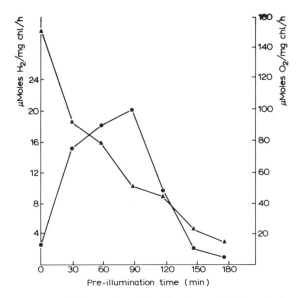

Fig. 10.4. Effect of pre-illumination of chloroplasts on $H_2 + O_2$ evolution. Basic reaction mixtures containing spinach chloroplasts were pre-illuminated, for the periods shown, in the Warburg bath (see Methods). Hydrogen evolution rates were measured after addition of *C. pasteurianum* hydrogenase and oxygen evolution rates were measured in the oxygen electrode after addition of potassium ferricyanide. (From Rao et al., 1978).

tional to the chlorophyll concentration up to a concentration of 25 μg chl per ml. Thereafter the amount of H_2 per mg of chlorophyll decreased with an increasing chlorophyll concentration in the reaction mixture. We normally use chloroplasts containing 100 μg chlorophyll in a 2-ml reaction mixture.

A question often asked is whether isolated chloroplasts can retain their electron transport activity under continuous illumination, for the lengthy periods of H_2 evolution that have been reported. Rao et al. (1976) found that in control experiments run parallel to the H_2 evolution assays, illuminated chloroplasts were able to reduce ferricyanide or NADP (both measured spectrophotometrically) as long as they were functional in H_2 evolution. The rates of reduction of these oxidants were similar to the rates of H_2 production measured under the same conditions. We have recently found that *Chenopodium* chloroplasts can retain more than 20% of their initial photosynthetic electron transport activity after 5 h in light at 20°C.

(iii) Effect of pre-illumination of chloroplasts. The effect of pre-illumination of chloroplasts on H_2 evolution is shown in Fig. 10.4. A series of vials containing the basic components for hydrogen evolution, except hydrogenase, were illuminated in a Warburg bath and the enzyme was injected into one set of vials at various intervals. Simultaneously the contents of a duplicate set were removed (at the time of enzyme injection) and assayed for electron transport to ferricyanide in an oxygen electrode (Rao et al., 1978). It is seen that while the oxygen reduction capacity of the chloroplasts decreased with time of illumination the initial rate of H_2 production increased with the period of illumination up to 90 min. This difference in the behaviour of electron transport can be explained by the differences in the nature of the two reactions and the conditions of measurement. The H_2 evolution reaction has an initial lag period and shows a steady rate only after 30 min of illumination. During this period there could be a steady accumulation of reduced ferredoxin and a reduction of other endogenous components (not participating in H_2 evolution) with which reduced ferredoxin reacts preferentially. After this the reduced ferredoxin can continuously donate its electrons to hydrogenase resulting in a steady rate of H_2 production. However, when the chloroplasts were pre-illuminated and then treated with hydrogenase there was already a pool of reduced ferredoxin for the enzyme to react with; thus the initial H_2 evolution rate was greater on injection of the hydrogenase. In the ferricyanide reduction assay by the oxygen electrode the electron flow is not catalyzed by the ferredoxin and the O_2 evolution can be measured soon after the injection of the oxidant. It should be mentioned that the total amount of H_2 produced in a vial was the same, whether the hydrogenase was added at the start or after a certain period of illumination. After two hours of exposure to light the rates of H_2 and O_2 evolution by the chloroplasts were parallel; probably the products of photolysis inhibited or inactivated the ferredoxin by this time.

   (iv) Protection of chloroplasts by BSA. There are numerous causes for the inactivation of isolated chloroplast membranes on exposure to light. Breakdown of the membranes by lipases and proteases releasing free fatty acids, hydrolysis of membrane phospholipids and glycolipids, generation of superoxide radicals, photoperoxidation of fatty acids and lipids generating inhibitory products, light-activated bleaching of photosynthetic pigments, etc. all contribute to the photo inactivation of chloroplasts (see review by Halliwell, 1978). Addition of BSA is known to confer stability and enhance the photosynthetic efficiency of chloroplasts (Friedlander and Neumann, 1968). Packer et al. (1976), during a study of the factors influencing the "in vitro" hydrogen evolution reaction, found that addition of BSA imparted stability to the system. Reeves et al. (1976) observed that addition of 5% BSA (Sigma Chemicals, Fraction V) to frozen and thawed chloroplasts increased the rates and duration of H_2 evolution by such chloroplasts. As shown in Fig. 10.5 the effect of BSA was less marked on fresh chloroplasts. Again the enhancement rate depends on the nature of the chloroplasts and the type of BSA used. We now add 0.1 to 1% of fat-free BSA (Sigma Chemicals) as a component of our reaction mixture for hydrogen evolution. The nature of action of BSA could be to bind and scavenge some of the fatty acids released thus creating a more favourable environment for electron transport.

309

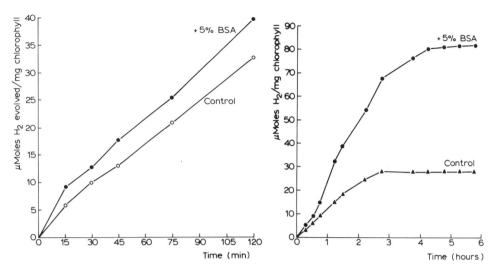

Fig. 5. Effect of BSA on H_2 evolution by fresh (left) and frozen and thawed chloroplasts (right).

The inhibition of hydrogen evolution by the oxygen released at PSII is due to a variety of causes; (a) photo oxidation of membrane components, (b) autoxidation of reduced ferredoxin, and (c) most importantly, oxygen-inactivation of hydrogenase. In the presence of a large excess of hydrogenase and ferredoxin (and no other added reagents) chloroplasts were able to evolve hydrogen, together with oxygen, for about 2 h. The rates of H_2 evolution from such a "simple" system were 10—50% of the rates obtained from a "complete" system containing oxygen and peroxide traps in addition to BSA: the wide variations in the rates being due to the variations in the chloroplast activity and in the concentration of the enzyme used in the assays. Addition of fresh chloroplasts and hydrogenase to the simple system, once H_2 evolution has ceased, did not restore the original rates of H_2 production. This suggests that some of the products formed under the aerobic conditions of the reaction completely inhibited the activity of the chloroplasts or hydrogenase, or both. Rates of electron transport measured in the oxygen electrode by an illuminated chloroplast—ferredoxin—hydroganse mixture were lower than the rates obtained from chloroplasts kept illuminated alone (without ferredoxin and hydrogenase) for the same period indicating that the products resulting from the addition of these two proteins to the chloroplasts do cause restrictions to the photosynthetic electron flow.

(v) Effect of oxygen scavengers. As shown in Fig. 10.6 in the presence of BSA, glucose, glucose oxidase, catalase and ethanol, the chloroplasts evolved H_2 at a much higher rate for two to three hours (Hall et al., 1977). When H_2

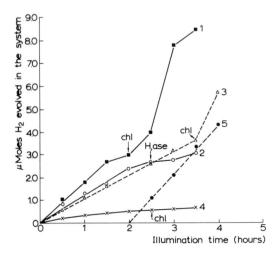

Fig. 10.6. Light-inactivation of the chloroplast-hydrogenase system. The reaction system contained, in a total volume of 2 ml, 100 μg spinach chloroplasts, 20 nmol spinach ferredoxin, 5 mg defatted BSA, 100 μmol glucose, 20 units glucose oxidase, 6000 units catalase and 50 μl ethanol in 50 mM HEPES, pH 7.5. Temp. 25°C, light intensity 11 000 lux. Reaction started by adding 0.01 mg of C. pasteurianum hydrogenase. At the times shown 100 μg fresh chloroplasts or 0.01 mg of fresh hydrogenase were injected into the system again. (1) Complete system, fresh chloroplasts added after 2 h. (2) Complete system, fresh hydrogenase added after 2.5 h. (3) Complete system—BSA, fresh chloroplasts added after 3.5 h. (4) System containing chloroplasts and ferredoxin only. (5) Complete system started at 2 h with spinach chloroplasts stored in the dark at 4°C. (From Hall et al., 1978).

evolution ceased, addition of fresh chloroplasts to such a "complete" system restored H$_2$ evolution to the initial rates; however, addition of fresh quantities of hydrogenase instead of fresh chloroplasts did not appreciably stimulate H$_2$ production. The data suggests that both ferredoxin and hydrogenase were still active when the H$_2$ evolution ceased but that the chloroplasts could not function under illumination for such long periods. The same chloroplasts, stored in ice in the dark for 2 h, did not lose any of their catalytic activity for H$_2$ evolution (Fig. 10.6).

A number of reagents known to protect biological materials against oxygen inactivation were tested in the H$_2$ evolution system. The results are shown in Table 10.1. The slight inhibitory effect observed in the presence of ascorbate may be due to a cycling of electrons between ascorbate, reduced ferredoxin and other components of PSI. Phenazine metho sulphate, a mediator of cyclic electron flow in PSI, has been shown to completely inhibit H$_2$ evolution presumably by diverting electrons from ferredoxin and cycling them around PSI (Reeves et al., 1976). The fact that superoxide

TABLE 10.1

EFFECT OF OXYGEN SCAVENGERS ON THE RATE OF HYDROGEN EVOLUTION
FROM A CHLOROPLAST—HYDROGENASE SYSTEM

Additions to the system	μmol H_2 evolved/mg chlorophyll/h at 25°
None	8.7
Dihydroxyfumarate (DHF)	12.3
β-carotene	10.0
Sodium ascorbate	2.2
Ascorbate + superoxide dismutase (SOD)	1.6
DHF + SOD	14.5
β-carotene + SOD	10.6
Glucose, glucose oxidase, catalase + ethanol	22.2
Glucose + glucose oxidase	11.4
Catalase + ethanol + SOD	23.0
Glucose, glucose oxidase, catalase, ethanol + SOD	22.5

Reaction mixture contained *Spinaceae oleracea* chloroplasts equivalent to 100 μg chlorophyll; *C. pasteurianum* hydrogenase, 25 μg; spinach ferredoxin, 50 μmol; bovine serum albumin, 5 mg; and where added DHF, 20 μmol; β-carotene, 10 mol; sodium ascorbate, 10 μmol; catalase, 6500 units; ethanol, 50 μl; glucose, 50 μmol; glucose oxidase, 20 units; bovine SOD; 3000 units; in 50 mM HEPES buffer, pH 7.5 to a final volume of 2 ml. (From Hall et al., 1978).

dismutase did not enhance the rate of H_2 evolution indicates that either there were not any free O_2^- radical in the system or this radical did not affect the electron transport activity. Both Rao et al. (1976) and Fry et al. (1977) have observed that addition of catalase and ethanol to the system considerably increased the H_2 evolution rate indicating that hydrogen peroxide (formed from O_2/glucose/glucose oxidase) is inhibitory to the H_2 evolution reaction. The latter authors also showed that neither glucose nor alcohol acted as PSI electron donors for H_2 evolution.

(vi) Effects of glutaraldehyde fixation of chloroplasts. One of the reagents that has often been tried to stabilise chloroplast membranes is glutaraldehyde (see Papageorgiou, chapter 7 of this volume). Rao et al. (1976) treated spinach chloroplasts isolated in a sorbitol medium with 0.05% glutaraldehyde. The glutaraldehyde-fixed chloroplasts were then stored under different conditions and their activities in the hydrogen evolution reaction were determined. The results are shown in Table 10.2. The glutaraldehyde fixed chloroplasts retained their hydrogen evolution capacity for weeks when stored at $-5°C$ in the dark. However, glutaraldehyde treatment did not improve the rates of H_2 evolution by the chloroplasts nor did they stabilize chloroplasts against light-inactivation.

TABLE 10.2

ACTIVITIES OF NORMAL AND FIXED CHLOROPLASTS AFTER 7 DAYS STORAGE

Type of chloroplasts and temperature of storage	μmol H_2/mg chlorophyll/h
Unfixed, 4° (cold room)	0
Fixed, 4°	5.2
Unfixed, −5° (freezer)	4.6
Fixed, −5°	10.0
Unfixed, −196° (liquid N_2)	8.4
Fixed, −196°	10.0

Spinach chloroplasts isolated in a sorbitol medium were fixed with 0.05% glutaraldehyde and stored under various conditions. Both freshly isolated and freshly fixed chloroplasts evolved 11.5 μmol H_2/mg chlorophyll/h. Assay conditions as described in Methods (From Rao et al., 1976).

10.4.3. Hydrogenases

Hydrogenases are enzymes which reversibly activate molecular hydrogen. $2H^+ + 2e^- \leftrightharpoons H_2$. They occur in many fermentative and photosynthetic bacteria, in sulphate reducing bacteria, hydrogen oxidising bacteria, in some aerobic bacteria, many blue-green algae and in rhizobia. They can also be induced in anaerobically adapted green alga (see Mortenson and Chen, 1974). All the hydrogenases purified so far are iron—sulphur proteins containing one or more ferredoxin-type [Fe-S] clusters in their molecules. Some of the properties of purified hydrogenases are listed in Table 10.3. As has been mentioned earlier many hydrogenases are inactivated on contact with oxygen. Recently hydrogenases stable in oxygen have been isolated from *Chromatium* (Gitlitz and Krasna, 1975), *Thiocapsa roseopersicina* (Gogotov et al., 1978), *Alcaligenes eutrophus* (Schneider and Schlegel, 1976), *Rhodospirillum rubrum* (Adams and Hall, 1977), *Desulfovibrio vulgaris* (van der Westen et al., 1978) and sulphate reducer strain 9974 * (Gogotov et al., unpublished). Though these hydrogenases retain their activity when stored in air their ability to evolve H_2 continuously in the presence of O_2 has not yet been demonstrated except in the case of *A. eutrophus*. We have also recently found that *A. eutrophus* hydrogenase (a gift of Schneider and Schlegel) can produce hydrogen from a chloroplast system, containing no oxygen scavengers, using NADH or reduced methyl viologen as electron donor.

---

* This strain was isolated by Biebl and Pfennig (1977) from the symbiont *Chloropseudomonas ethylica* N_2. The authors are indebted to J. Le Gall, CNRS, Marseille, for a gift of this bacterium.

TABLE 10.3

PROPERTIES OF BACTERIAL HYDROGENASES

	Photosynthetic			Non-photosynthetic			
	Rhodospirillum rubrum [a]	*Thiocapsa roseopersicina* [a,b]	*Chromatium* [a]	*Clostridium pasteurianum* [b]	*Alcaligenes eutrophus* [b]	*Desulphovibrio vulgaris* [a]	*Desulphovibrio vulgaris* [c]
Physiological role	H_2 uptake	H_2 uptake	H_2 uptake	H_2 evolution	H_2 uptake	H_2 uptake	H_2 uptake
Natural electron carrier	Unknown	Unknown	Unknown	Ferredoxin	NAD	Cytochrome c_3	Unknown
O_2 sensitivity	Relatively insensitive	Relatively insensitive	Relatively insensitive	Extremely sensitive	Insensitive [d]	Very sensitive	Relatively insensitive
Mol. wt	66 000	66 000	98 000	60 000	205 000	89 000	50 000
Fe/mol	4	4	4	12	12–16 [e]	7–9	12
S^{2-}/mol	4	4	4	12	12–16 [e]	7–8	12
Specific [g] Activity (electron carrier)	26 (MV_{red})	26 [f] (MV_{red})	35 (MV_{red})	500 (MV_{red})	54 (NAD)	610 (MV_{red} + cyto c_3)	3900 (MV_{red})
Reference	Adams and Hall (1978)	Gogotov et al. (1978a)	Gitlitz and Krasna (1975)	Chen and Mortenson (1974)	Schneider Schlegel (1976)	Yagi et al. (1976)	van der Westen et al. (1978)

[a] Particulate enzyme; [b] Soluble enzyme; [c] Located in preiplasmic space; [d] Contains a flavin component; [e] K. Schneider, unpublished work; [f] I.N. Gogotov, unpublished work; [g] Micromoles H_2 activated per mg per min.

TABLE 10.4

COMPARATIVE ACTIVITIES OF HYDROGENASE

Source of hydrogenase	μmol H_2 liberated/mg chlorophyll/h	
	(A) In the chloroplast system	(B) With 1.25 mM methyl viologen and 10 mM sodium dithionite
C. pasteurianum	38.5	80.2
T. roseopersicina	14.5	27.8
Chromatium	1.2	5.3
Sulphate reducer 9974	2.7	47.2
M. laminosus	0.7	4.2
S. maxima	negligible	13.1
E. coli	5.2	62.4
A. eutrophus	11.5	760

The assays were performed using the reagents mentioned under Methods at 25°C. Values of hydrogen liberated in column B are calculated for the amount of the respective hydrogenases used in the chloroplast system (A) and are not the maximum activities of these hydrogenases. Methyl viologen was used as electron mediator to E. coli and A. eutrophus hydrogenases in the chloroplast system; ferredoxin was the mediator to other hydrogenases.

Another problem in using these stable hydrogenases is that they do not readily react with reduced ferredoxins and hence the rates of H_2 evolution catalysed by these hydrogenases from an illuminated chloroplast system are low when compared to the respective rates obtained from a dithionite—methyl viologen system (see Table 10.4). It should be pointed out that the concentration of methyl viologen used in these assays was higher than the concentration of ferredoxin added to the chloroplast system.

10.4.4. Ferredoxins

Ferredoxins are low potential, low molecular weight, electron carrier proteins found in all bacteria, algae and plants so far examined (Hall et al., 1975). Ferredoxins from algae and plants ("chloroplast" ferredoxins) contain a single [2Fe—2S] active centre which participate in one electron transfers in many biological processes. Their midpoint redox potentials are close to that of the hydrogen electrode; usually in the range of —350 to —425 mV (Cammack et al., 1977). The bacterial ferredoxins contain one or two [4Fe—4S] clusters per molecule, each cluster mediating the transfer of a single electron. A few bacterial species have been shown to contain [2Fe—2S] ferredoxins. In fermentative bacteria such as the Clostridia, the ferredoxin in conjunction with the hydrogenase is involved in the hydrogen metabolism of these organisms. Both bacterial and chloroplast ferredoxins are involved in the energy metabolism and in a number of other catalytic reactions (see Rao and Hall, 1977). The interchangeability of plant and bacterial ferredoxins in the hydrogenase reaction was first demonstrated by Tagawa and Arnon (1962). These authors isolated ferredoxin from spinach and *C. pasteurianum* and showed that in the presence of added bacterial hydrogenase both proteins can (a) mediate the dark reduction of pyridine nucleotide by chloroplasts with H_2 as the electron donor, (b) mediate the light-induced production of H_2 gas by chloroplasts with ascorbate as electron donor and (c) mediate the dark production of H_2 gas with sodium dithionite as electron donor.

Ferredoxin is a normal constituent of chloroplasts in vivo. However, since it is water soluble most of the endogenous ferredoxin is washed out from the chloroplasts during their isolation; hence the need to add ferredoxin to the reaction mixture for H_2 evolution. As shown in Fig. 10.7 the rate of H_2 production is dependent on the concentration of ferredoxin present in the chloroplasts with optimum rates being obtained with a chlorophyll : ferredoxin ratio of 1 : 5 (w/w). Fry et al. (1977) have also found enhancement of H_2 production rates with increased ferredoxin concentration in both PSII and PSI catalysed electron transport. They suggested that low concentrations of ferredoxin may limit H_2 evolution (a) by partial loss of the ferredoxin due to its autoxidation by molecular oxygen and (b) by its slow diffusion from the low potential electron donor (chloroplast particle) to the low potential electron acceptor, viz. hydrogenase. Though higher concentrations of ferredoxin enhanced the rate of H_2 evolution they had no effect on the total lifetime of the system.

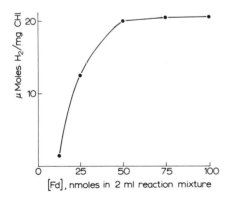

Fig. 10.7. Ferredoxin concentration and rate of H_2 evolution. 2 ml reaction mixture contained 100 μg chlorophyll from spinach chloroplasts and all the components of the basic system except varying quantities of spinach ferredoxin. (From Rao et al., 1978).

TABLE 10.5

HYDROGEN PRODUCTION FROM ILLUMINATED CHLOROPLAST SUSPENSIONS
IN THE PRESENCE OF DIFFERENT FERREDOXINS AND HYDROGENASES

Source of hydrogenase	Source of ferredoxin (μmol H_2/mg chlorophyll/h)			
	C. pasteurianum	R. rubrum	Spinach	Spirulina maxima
C. pasteurianum	11.9	15.0	10.6	8.7
T. rosepersicina	nil	2.7	14.5	10.1
R. rubrum	0.6	nil	0.3	0.6
Chromatium D	0.7	nil	0.3	1.2

Reaction mixture contained *Chenopodium album* chloroplasts equivalent to 200 μg
chlorophyll, 10 μmol ferredoxin, 1 mg bovine serum albumin, 100 nmol glucose, 20 units
glucose oxidase, 50 μl ethanol and 0.05 M HEPES buffer, pH 7.5 to make up to a final
volume of 2 ml. Hydrogenase 0.01 to 0.1 mg was added to the illuminated reaction mixture maintained at 25°. Experiments with *Spirullina maxima* ferredoxin were done using
Nicotiana sylvester chloroplasts in 20 mM phosphate buffer, pH 7 at 20°. (From Hall et
al., 1978).

Ferredoxins from all plants and algae have similar physicochemical properties and biological activities and so can be interchanged in the chloroplast—
hydrogenase system without causing appreciable differences in the H_2 evolution rates. However, although spinach and *C. pasteurianum* ferredoxins
mediated efficiently the electron transfer to *C. pasteurianum* hydrogenase,
the same was not true for reactions catalysed by other hydrogenases and
bacterial ferredoxins. The cross reactivity between some hydrogenases and
ferredoxins is shown in Table 10.5. It may be that the hydrogen-activation
reaction for *C. pasteurianum* hydrogenase is quite favourable at a redox
potential of about −400 mV, which is close to the midpoint potentials of
most chloroplast ferredoxins and of *C. pasteurianum* ferredoxin. Other
hydrogenases may have different midpoint potentials for hydrogen activation. Another possibility is that there is stronger binding affinity between
Clostridium hydrogenase and the *C. pasteurianum* ferredoxin and the
chloroplast-type ferredoxins. We still know very little about the mechanisms
involved in proton and electron transfer by hydrogenases.

10.4.5. Effect of pH and temperature on H_2 evolution

The effect of pH on H_2 evolution by the system is dependent on the pH
profile of electron transport to ferredoxin by the chloroplasts and on the pH
profile of hydrogenase activity. Mortenson and Chen (1976) have reported
that initial rates of H_2 evolution by *C. pasteurianum* hydrogenase from
reduced ferredoxin or methyl viologen were much faster at pH 6 than at
pH 8. However, the H_2 evolution rates declined with time more rapidly at

pH 6 than at pH 8 which was probably due to the slow rate of reduction of the carriers by sodium dithionite (electron donor). The pH optimum for electron transport from water to ferricyanide in our measurements was 7.6. The rates of H_2 evolution by the coupled chloroplast—hydrogenase system was maximum in the pH range 7 to 7.5 (Hall et al., 1978; Fry et al., 1977). Buffer systems made up of 20 mM phosphate, 50 mM Tris—HCl or 50 mM HEPES—NaOH were suitable for H_2 evolution. Many hydrogenases function efficiently in the temperature range 35—40°C. However, light-activated destruction of chloroplasts is quite marked above 30°C. A compromise between the activities of hydrogenase and chloroplasts can be obtained by performing the H_2 evolution reaction within the temperature range 20 to 25°C.

10.4.6. Effect of light intensity

One of the characteristic differences between H_2 evolution from illuminated chloroplasts and other lightdriven electron transport reactions of chloroplasts is the effect of incident light on the rate of the reaction. Though there is no H_2 evolution in the dark and lower light intensity reduces the total H_2 yield, the chloroplast—hydrogenase system saturates at light intensities that are about one-tenth of the intensity that saturates ferri-cyanide or $NADP^+$-supported O_2 evolution (Fry et al., 1977). The maximum rate of H_2 production and the maximum yield of H_2 appeared to be nearly saturated at light intensities as low as 8 Kergs/cm^2 · sec. According to Fry et al. (1977) the likely explanation for the low light intensity saturation threshold is a reaction bottleneck located beyond the step of the photo-production of low potential reducing equivalents, probably involving a diffu-sion limitation of the reduced components from the chloroplast particles to the ferredoxin and hydrogenase.

10.4.7. Effect of uncouplers

Addition of 10^{-6} M FCCP, an uncoupler of electron transport from photo-phosphorylation, to the standard reaction mixture had no effect on H_2 evolution (Reeves et al., 1976). This was due to the fact that the chloroplasts had already become uncoupled in the low osmotic and low ionic strength medium of the standard assay mixture. When the standard reaction medium was replaced by a more complex one suited to retain coupled chloroplast electron transport (330 mM sorbitol, 2 mM $MgCl_2$, 1 mM EDTA and 50 mM HEPES buffer, pH 7.6) the initial rate of H_2 evolution was increased three-fold by the addition of FCCP as shown in Fig. 10.8. Even in the complex medium the chloroplasts start to uncouple after 20 min illumination. Addi-tion of FCCP was found to stimulate H_2 evolution mediated by NADH from *Chlamydomonas reinhardii* chloroplasts indicating that the oxidation of

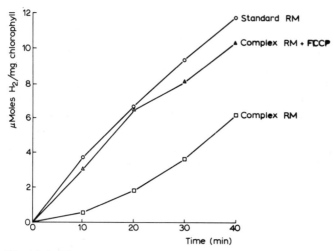

Fig. 10.8. The standard reaction mixture contained the components of the basic system (see Methods). The complex reaction medium contained in addition 330 mM sorbitol, 1 mM EDTA and 2 mM $MgCl_2$. FCCP concentration of 10^{-6} M. (From Hall et al., 1978).

NADH in this system was coupled through an energy conservation site (King et al., 1977a). King et al. (1977b) also studied photosynthetic electron transport by coupled and uncoupled chloroplast particles both in air and in an anaerobic nitrogen atmosphere. They found that the rate of ferricyanide reduction by spinach and pea chloroplast particles uncoupled with NH_4Cl or FCCP, and by *Chlamydomonas reinhardii* or *Euglena gracilis* particles uncoupled by sonication, was inhibited considerably by anaerobiocity. These authors suggest that extended anaerobiosis (as maintained in the H_2 evolution system) may account for the rapid loss of PSII activity by chloroplast particles which become uncoupled with time.

10.5. MODIFICATIONS AND IMPROVEMENTS OF THE COMPONENTS OF THE SYSTEM

We have so far discussed the requirements, the characteristics and the limitations of H_2 production by an in vitro chloroplast—hydrogenase system. Let us now turn our attention to some of the modifications and alternatives that have been tried and proposed to overcome the limitations and to increase the efficiency of the system.

10.5.1. Stabilization of chloroplast membranes

What are the possibilities of stabilizing chloroplast particles against photo-inactivation^ In chapter 7 of this book, Papageorgiou discusses in detail the

TABLE 10.6

ACTIVITY AND STABILITY OF TYPE C SPINACH CHLOROPLASTS TREATED WITH GLUTARALDEHYDE (MORRIS AND HALL UNPUBLISHED)

Treatment: conc of GA (μmole per mg Chl)	Osmotic response Δ A546 nm/mg Chl in 0 and 330 mM NaCl	Electron transport activity (μmol/mg Chl/h)		Stability in H_2O → ferricyanide + NH_4Cl (t½ in h)	
		H_2O → ferricyanide + NH_4Cl [PSII]	DCPI → MV + DCMU [PSI]	0°C, dark	20°C, 100 W m^{-2}
Freshly isolated (control)	5.4	326	400	4.8	
Washed (control)	5.2	197	300	2.1	—
5 GA, 20 min, 0°C, dark	3.6	222	378	2.5	—
100 GA, 20 min, 0°C, dark	0.1	44	126	0.5	—
Freshly isolated	4.3	210	228	—	—
Washed	4.3	123	252	12.0	—
50 GA + 100 μmole formaldehyde/ mg Chl 5 min, 0°C, dark	0.4	78	208	3.0	—
Freshly isolated	5.3	376	—	—	—
Washed	5.3	460	—	9.9	4.0
5 GA, 5 min, 0°C, dark	5.9	430	—	16.0	2.0
10 GA, 5 min, 0°C, dark	3.6	400	—	8.3	1.2
5 GA, 30 min, 0°C, dark	4.3	270	—	30.0	1.7
10 GA, 30 min, 0°C, dark	4.2	300	—	5.4	0.7
Washed	4.8	294	—	7.0	1.0
50 GA, 5 min, 0°C, dark	0.2	71	—	7.0	1.0

Chloroplasts were isolated in a pH 6.5 medium containing 0.33 M sorbitol, 0.01 M $Na_4P_2O_7$, 5 mM $MgCl_2$ 2.5 mM ascorbate. They were fixed and washed in a pH 7.6 medium with 0.33 M sorbitol, 50 mM HEPES, 2 mM EDTA, 5 mM $MgCl_2$ and 2 mM $MnCl_2$. Chl, chlorophyll; GA = glutaldehyde techniques employed for immobilization of chloroplasts and the properties of such fixed membranes. We will briefly mention some work done in our and other laboratories pertinent to the problem.

In the course of an extensive investigation of the factors contributing to the decay of chloroplast activity, in vitro, and methods to prevent this decay Morris and Hall (unpublished) treated spinach chloroplasts with glutaraldehyde under varying conditions. They determined the degree of fixation of the chloroplast membranes (by measuring the osmotic response) and their photosystem activities after storage. The results are shown in Table 10.6. Chloroplasts incubated with 5 μmol GA/mg chlorophyll for 30 min at 0°C, in the dark, had a half-life of 30 h estimated by measuring their electron transport capacity to ferricyanide. Hardt and Kok (1976) and Kulandaivelu and Hall (1976) have also reported that glutaraldehyde prolongs the biochemical lifetimes of chloroplasts stored in the dark. However, glutaraldehyde fixation does not improve the efficiency and longevity of the chloroplast membranes under illumination.

Kitajima and Butler (1976) encapsulated both chloroplast and PSI particles in small spheres of about 50 μm in diameter with an artificial membrane built up by cross-linking amino groups of protamine with toluene-diisocyanate. PSI activity was retained by the encapsulated particles. After storage for four weeks at 5°C the PSI activity of the encapsulated chloroplast fragments levelled off at about one-third of the initial rate, whereas the activity of the unencapsulated controls decayed to zero. PSI particles were encapsulated with *Chromatium* hydrogenase in a membrane made up of gelatin and toluenediisocyanate. The resultant capsules were able to evolve H_2 when supplied with ascorbate/DPIP as electron donor and methyl viologen as electron carrier to hydrogenase. The average activity of the encapsulated particles was 1 μmol H_2 per mg chlorophyll per h. This could be a very good method of producing hydrogen provided the chloroplast particles could be made to retain PSII activity during encapsulation and a more active hydrogenase could be incorporated into the capsules. However, it is quite encouraging to note that the artificial membranes of Kitajima and Butler were permeable to small substrate and product molecules.

Ochiai et al. (1977) treated spinach chloroplasts suspended in an isotonic medium with a mixture of 15% acrylamide, 1% BSA and some cross-linking and reducing agents at 10°C for 2 h. The chloroplasts thus immobilized under conditions of "redox polymerization" retained about 11% of their original PSII and PSI activity. The immobilized chloroplasts retained all their PSI activity even after storage for four weeks at 4°C though there was a gradual decrease of PSII activity with time.

10.5.2. Search for oxygen-insensitive hydrogenases

Research on ways to improve the performance of hydrogenase in the in vitro system is focused towards two main objectives; (a) isolation of oxygen-stable enzymes and (b) stabilization of highly active hydrogenases against oxygen inactivation by embedding the enzymes on solid supports. As mentioned earlier, hydrogenases stable in air have been purified from *Chromatium, T. roseopersicina, R. rubrum, A. eutrophus* and *D. vulgaris.* Though these hydrogenases are very active in H_2 evolution from a dithionite-methyl viologen mixture all, except *T. roseopersicina*, react very slowly with reduced ferredoxins from *C. pasteurianum, S. maxima* or *Spinacia* (Rao et al., 1978). This sluggishness in H_2 evolution from reduced ferredoxin was also observed with hydrogenases from many blue-green algae such as *Nostoc muscorum, Anabaena cylindrica* (Tel-Or et al., 1978) *Spirulina maxima* and *Mastigocladus laminosus* (Rao et al., 1978). It may be that all these hydrogenases will catalyse H_2 evolution in the chloroplast system provided we can find suitable carriers that would readily mediate electron flow from photosynthetically-reduced components of the chloroplasts to the respective hydrogenases. Thus, for example, the hydrogenase of *Desulfovibrio* strain 9974 liberated H_2 at a very low rate from a chloroplast—ferredoxin mixture. But when the system was supplemented with cytochrome "c_3" from the same organism the rate of H_2 evolution was considerably enhanced (Table 10.7). Chloroplasts reduce the cytochrome "c_3" on illumination and

TABLE 10.7

H_2 EVOLUTION FROM A CHLOROPLAST SYSTEM BY HYDROGENASES FROM *C. PASTEURIANUM* AND *DESULFOVIBRIO* (SULPHATE REDUCER) STRAIN 9974, MEDIATED BY VARIOUS ELECTRON CARRIERS

Hydrogenase source	Electron mediator/s	μmol H_2/mg chlorophyll/h
C. pasteurianum	*D.* 9974 Fd	2.5
C. pasteurianum	*D.* 9974 "c_3"	0.2
C. pasteurianum	*S. maxima* Fd	38.0
D. 9974	*D.* 9974 Fd	0.7
D. 9974	*D.* 9974 "c_3"	7.3
D. 9974	*D.* 9974 (Fd + "c_3")	19.0

Vessels contained components of the basic reaction mixture (Methods) except ferredoxin and where added *C. pasteurianum* hydrogenase 50 units, *D.* 9974 hydrogenase, 20 units, *S. maxima* ferredoxin (Fd) 50 nmol, *D.* 9974 Fd 5 nmol, *D.* 9974 cytochrome "c_3" ("c_3") 2.5 nmol. Temperature 24°C. Hydrogenase units represent μmoles of H_2 liberated per h from a dithionite—methyl viologen system at the same temperature. (Rao et al., 1978).

the electrons are readily taken up by the *Desulfovibrio* hydrogenase from reduced cytochrome. Similarly, *A. eutrophus* hydrogenase does not react with ferredoxin (Schneider and Schlegel, 1976) and did not evolve H_2 from a chloroplast—ferredoxin mixture. But *A. eutrophus* enzyme readily evolved H_2 when added to illuminated chloroplasts containing either NADH or methyl viologen (Rao and Schneider, unpublished). As already mentioned this hydrogenase is very stable in air and there was no need to add any oxygen-scavengers to the H_2-evolution system. A number of laboratories are involved in producing mutant strains of algae with oxygen-insensitive hydrogenase. McBride et al. (1977) have succeeded in isolating a mutant of *Chlamydomonas reinhardii* whose hydrogenase-dependent reaction was more oxygen resistant than the same reaction catalysed by wild type cells.

10.5.3. Immobilization of hydrogenases and ferredoxins

A wide variety of techniques are now available for attaching enzymes to solid supports (see Methods in Enzymology, Vol. XLIV). Enzymes can be immobilized by entrapment in gels or microcapsules, by adsorption in insoluble materials such as clay or ion exchange resins, by covalent attachment to specially treated glass or agarose, etc. Immobilization generally confers extra stability to the enzymes against pH and temperature changes though there is an appreciable loss in the specific activity of the enzyme as a result of immobilization. This loss of activity is partly due to the diffusion limitation of the substrates and products to and from the enzyme surface. Lappi et al. (1976) immobilized *C. pasteurianum* hydrogenase on glass beads using four types of linkages; diazo, glutaraldehyde, alkylaminecarbodiimide and succinyl carbodiimide. The immobilized enzymes had only less than 5%. of the activity of the native hydrogenase. All forms of binding stabilized the hydrogenase to oxygen relative to the native enzyme. Succinyl-bound hydrogenase was the most stable form increasing the time for the loss of 50% of the activity from 1—2 min for the native enzyme to several days for the immobilized enzyme. The glass-bound hydrogenases were able to produce H_2 from ferredoxin reduced by illuminated chloroplasts but unfortunately the actual rates of H_2 evolution were not reported. Berenson and Benemann (1977) extended this work and bound spinach and *C. pasteurianum* ferredoxins to alkylamine glass via glutaraldehyde. The immobilized ferredoxins were able to transfer electrons from ascorbate to soluble *C. pasteurianum* hydrogenase using illuminated chloroplasts. However, mixing of immobilized hydrogenase and immobilized ferredoxin or binding of the two proteins on the same glass beads resulted in the complete loss of H_2 evolution activity with dithionite or illuminated chloroplast as the electron source. The authors point out that this type of stearic blockage would pose a major problem in the utilization of immobilized enzymes for solar energy conversion. Rao et al. (1978) attached *S. maxima* ferredoxin to AH Sepharose (Pharmacia

Chemicals) and then bound *C. pasteurianum* hydrogenase on to the ferredoxin. The resultant immobilized ferredoxin-hydrogenase mixture was able to evolve H_2 from illuminated chloroplasts at a high rate but only when supplemented with soluble ferredoxin indicating the inactivation of the ferredoxin in the immobilized mixture. Yagi (1977) has trapped *D. vulgaris* hydrogenase in polyacrylamide gels in the presence of cross linking agents. The activity of the immobilized enzyme was more stable during storage at room temperature and during repeated cycles of use compared to the stability of the native enzyme.

10.5.4. Separation of O_2 and H_2 evolving systems

A completely different approach to solve the problem of in vitro inactivation of hydrogenase by the O_2 liberated from water photolysis is to construct a two-stage system as proposed by Krampitz (1973) where the light reactions of chloroplasts are separated from the dark evolution of H_2 by hydrogenase. In Kramptiz's experimental system, spinach chloroplasts, spinach ferredoxin and $NADP^+$ were illuminated in a glass chamber and the NADPH formed was removed by dialysis through a hollow fibre dialysis apparatus. The dialysate containing NADPH was pumped to a second chamber fitted with a dialysis unit containing *C. kluyveri* hydrogenase. NADPH reacted with the hydrogenase thus liberating hydrogen; the $NADP^+$ formed was pumped back to the first chamber. The process is cumbersome and the yield of H_2 very poor. However the idea is very sound and if the $NADP^+$ can be replaced by nonautoxidizable dyes that can be reduced by illuminated chloroplasts,

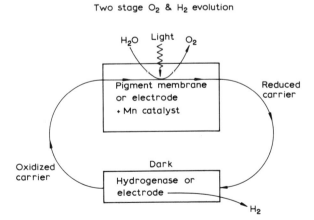

Fig. 10.9. A proposed model of a two stage system for the photolysis of water. The carriers may be of biological origin or their synthetic analogues or dyes.

324

the system could operate much more efficiently. A schematic representation of a proposed multistage biophotolysis reactor is shown in Fig. 10.9.

10.5.5. Preparation of synthetic analogues of the catalysts

We envisage that the ultimate goal in the construction of a solar reactor for the continuous production of H_2 from water is to replace all the biological components of the model system with synthetic substitutes. Many laboratories are actively searching for synthetic Mn-containing catalysts which could mimic the activity of the Mn enzyme involved in the water-splitting reaction in chloroplasts (see Harriman and Barber, chapter 8 of this volume). Considerable progress has been made in this field. Synthetic, active site analogues of many types of iron—sulphur proteins have been prepared by Holm and coworkers (Holm and Ibers, 1977). Many of the physical properties of these analogues are similar to those of their biological counterparts. However, these analogues are routinely prepared and tested in organic solvents such as DMSO and hexamethylphosphoramide. Their catalytic efficiency in biological electron transport could not be measured since they are unstable in a completely aqueous environment. Thus, if these analogues are to be used in photosynthetic reactions the reactions should occur in high concentrations of aprotic organic solvents such as DMSO. Reeves and Hall (1977) studied the effect of DMSO on electron transport in fresh and aged chloroplasts. Their results are shown in Fig. 10.10. Concentrations of DMSO

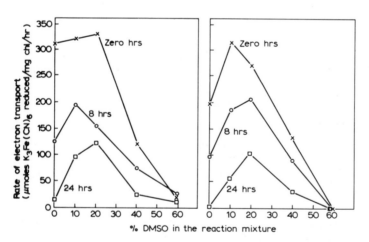

Fig. 10.10. The effect of DMSO on $K_3Fe(CN)_6$ reduction in fresh and "aged" chloroplasts. The chloroplasts were used soon after extraction, ×———×; after 8 h storage at 4°C, O———O; and after 24 h storage at 4°, □———□. The two experiments shown were carried out on two separate chloroplast preparations. (From Reeves and Hall, 1977).

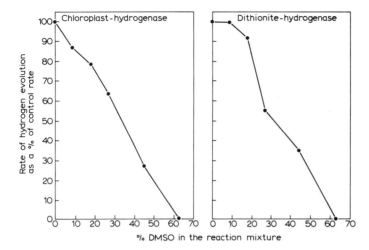

Fig. 10.11. The effect of DMSO on H_2 evolution. The chloroplast—hydrogenase system contained components of the basic system in varying concentrations of DMSO. Dithionite—hydrogenase system contained 1.25 mM methyl viologen and 200 mM sodium dithionite in 20 mM phosphate buffer, pH 7.0 and varying concentrations of DMSO. *C. pasteurianum* hydrogenase was used in both experiments. (From Reeves and Hall, 1977).

up to 20% (v/v) did not inhibit the electron transport capacity of freshly isolated chloroplasts. In chloroplasts that have been aged 8 to 24 h by storage at 4°C, the addition of DMSO at concentrations up to 20% caused a stimulation of electron transport, though the stimulated rates never reached those obtained with fresh chloroplasts. DMSO concentrations higher than 40% inhibited electron transport in both types of chloroplasts.

The effect of DMSO on H_2 evolution is shown in Fig. 10.11. There was a linear decrease in H_2 evolution rate with increasing concentrations of DMSO in the reaction mixture, both from a chloroplast/ferredoxin/hydrogenase system and from a dithionite/methyl viologen/hydrogenase system. Thus it seems that operation of a biophotolysis system in a DMSO/H₂O mixture using native hydrogenase, may not be a very efficient process if high concentrations of DMSO are necessary and if the hydrogenase reaction itself is the rate limiting step.

That synthetic analogues of ferredoxins can function in H_2 evolution when coupled to hydrogenase was demonstrated by Adams et al. (1977). The two water-soluble tetranuclear iron—sulphur clusters (shown below) undergo a reversible one electron reduction to the trianion in the presence of dithionite.

CLUSTER I

$$[Fe_4S_4(Ac \cdot Gly_2 \cdot Cys \cdot Gly_2 \cdot Cys \cdot Gly_2 \cdot NH_2)_2]^{2-}$$

CLUSTER II

$$[Fe_4S_4(S \cdot CH_2 \cdot CH_2OH)_4]^{2-}$$

The optical and e.p.r. spectra of the native and reduced clusters were similar to those of ferredoxins. These two clusters could replace ferredoxins in a H_2-evolving system using *C. pasteurianum* hydrogenase with dithionite as the electron donor (see Fig. 10.12). Though the clusters themselves could not be reduced in an illuminated chloroplast system, these results are significant in our attempts to use synthetic analogues in biological reactions.

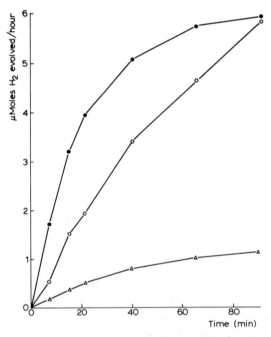

Fig. 10.12. Hydrogen evolution mediated by synthetic tetranuclear iron—sulphur analogues. The reaction mixture contained, in a volume of 2 ml, 50 mM Tris—Cl⁻ buffer pH 7.3, 10 μl of crude *C. pasteurianum* hydrogenase and 10 mM $Na_2S_2O_4$. the reaction was initiated by the addition of the electron mediator. △————△, Cluster I; ●————●, Cluster II (see text); ○————○, *Spirulina maxima* ferredoxin.

10.6. CONCLUSIONS

In the course of the past five years since Benemann et al. (1973) reported hydrogen production by a chloroplast—ferredoxin—hydrogenase system, very considerable progress has been made in improving the efficiency of the system in terms of rates and longevity of H_2 evolution. Many of the basic problems faced by Benemann et al. (1973) are still not satisfactorily solved. However, there has been a stimulus in the study of the characteristics of the various components of the system. Causes and prevention of the decay of isolated chloroplast membranes are being investigated by many groups. A number of stable hydrogenases have been isolated and the mechanism of action of these enzymes are studied in detail. Many hydrogenases have been successfully bound to solid supports. Synthesis of other active analogues of ferredoxins are being pursued as well as of nonautoxidizable viologen dyes which could mediate electron transfer between chloroplasts and hydrogenase. The collaborative effort of researchers in these different fields could result in a laboratory version of a biophotolytic reactor producing H_2 continuously in the not-too-distant future. Only then could we discuss the economics and photosynthetic efficiency of the whole operation.

The photolysis of water has three unique advantages not shown by any other energy system: (a) it uses an unlimited supply of substrate, water (b) an unlimited input of energy, sunlight, in order to (c) produce a storable and non-polluting source of energy, hydrogen gas (Hall, 1978; Bolton, 1978). Thus it seems very worthwhile to study the biophotolysis system seriously to see if we cannot overcome its stability problems in order to construct a synthetic system which mimics the biological components.

10.7. ACKNOWLEDGEMENTS

We thank the UK Science Research Council and the European Commission, Brussels, for support of our work reported here. We also thank Drs. I.N. Gogotov, S.G. Reeves, P. Morris and Mr. M.W.W. Adams for their help.

REFERENCES

Adams, M.W.W. and Hall, D.O. (1977) Biochem. Biophys. Res. Commun., 77, 730—737.
Adams, M.W.W., Reeves, S.G., Hall, D.O., Christou, G., Ridge, B. and Rydon, H.N. (1977) Biochem. Biophys. Res. Commun., 79, 1184—1191.
Adams, M.W.W. and Hall, D.O. (1978) Arch. Biochem. Biophys. (submitted).
Arnon, D., Mitsui, A. and Paneque, A. (1961) Science, 134, 1425.
Arnon, D.I. (1962) in Photosynthetic mechanisms in green plants (B. Kok and A.T. Jagendorf, eds.) pp. 196—212, Publ. No. 1145, NAS—NRC, Washington, DC.
Ben-Amotz, A. and Gibbs, M. (1975) Biochem. Biophys. Res. Commun., 64, 355—359.
Benemann, J.R., Berensen, J.A., Kaplan, N.O. and Kamen, M.D. (1973) Proc. Natl. Acad. Sci. U.S.A., 70, 2317—2320.
Berenson, J.A. and Benemann, J.R. (1977) FEBS Lett., 76, 105—107.
Biebl, H. and Pfennig, N. (1977) Arch. Microbiol., 112, 115—117.
Bolton, J.R. (1978) Sol. Energy, 20, 181—183.
Buchanan, B.B. and Arnon, D.I. (1970) Adv. Enzymol., 33, 119—176.
Cammack, R., Rao, K.K., Bargeron, C.P., Hutson, K.G., Andrew, P.W. and Rogers, L.J. (1977) Biochem. J., 168, 205—209.

328

Chen, J-S. and Mortenson, L.E. (1974) Biochim. Biophys. Acta, 371, 283—298.
Friedlander, M. and Neumann, J. (1968) Plant Physiol., 43, 1249—1254.
Fry, I., Papageorgiou, G., Tel-Or, E. and Packer, L. (1977) Z. Naturforsch., 32c, 110—117.
Gitlitz, P.H. and Krasna, A.I. (1975) Biochemistry, 14, 256—68.
Gogotov, I.N., Zorin, N.S., Serebriakova, L.T. and Kondratieva, E.N. (1978) Biochim. Biophys. Acta 523, 335—343.
Hall, D.O., Rao, K.K. and Cammack, R. (1972) Biochem. Biophys. Res. Commun., 47, 798—802.
Hall, D.O., Rao, K.K. and Cammack, R. (1975) Sci. Prog. (Oxford), 62, 285—327.
Hall, D.O. (1978) Fuel (in press).
Hall, D.O., Rao, K.K., Reeves, S.G. and Gogotov, I.N. (1977) in Alternative Energy Sources Symposium, Miami Beach (T.N. Veziroglu, ed.). McGraw-Hill, New York.
Halliwell, B. (1978) Prog. Biophys. Mol. Biol., 33, 1—54.
Hardt, H. and Kok, B. (1976) Biochim. Biophys. Acta, 449, 125—135.
Hoffmann, D., Thauer, R. and Trebst, A. (1977) Z. Naturforsch., 32c, 257—262.
Holm, R.H. and Ibers, J.A. (1977) in Iron-Sulphur Proteins (W. Lovenberg, ed.), Vol. III, pp. 206—282, Academic Press, New York.
Jones, W.L. and Bishop, N. (1976) Plant Physiol., 57, 659—665.
King, D., Erbes, D.L., Ben-Amotz, A. and Gibbs, M. (1977a) in Biological Solar Energy Conversion (A. Mitsui et al., eds.). Academic Press, New York.
King, D., Erbes, D.L. and Gibbs, M. (1977b) Biochem. Biophys. Res. Commun., 78, 734—738.
Kitajima, M. and Butler, W.L. (1976) Plant Physiol., 57, 746—750.
Krampitz, L.O. (1972) in An Inquiry into Biological Energy Conversion (A. Hollaender et al., eds.). The University of Tennessee, Knoxville, U.S.A.
Krampitz, L.O. (1973) in Biophotolysis of water NSF—RANN Report No. HA2, N-73-014 quoted by Benemann, J. (1977) in Living Systems as energy converters (Buvet et al., eds.), p. 288, North-Holland, Amsterdam.
Krasnovsky, A.A. (1976) in Research in Photobiology (A. Castellani, ed.), pp. 361—370. Plenum Press, New York.
Kulandaivelu, G. and Hall, D.O. (1976) Z. Naturforsch., 31C, 452—455.
Lappi, D.A., Stolzenbach, R.E., Kaplan, N.O. and Kamen, M.D. (1976) Biochem. Biophys. Res. Commun., 69, 878—884.
Lien, S. and San Pietro, A. (1975) An enquiry into biophotolysis of water to produce hydrogen. N.S.F., Indiana University, Indiana.
Mcbride, A.C., Lien, S., Togasaki, R.K. and San Pietro, A. (1977) in Biological Solar Energy Conversion (A. Mitsui et al., eds.) pp. 77—86, Academic Press, New York.
Mitsui, A. and Arnon, D.I. (1962) Plant Physiol., 37S, IV—V.
Mortenson, L.E. and Chen, J.S. (1974) in Microbial Iron Metabolism (J.B. Nielands, ed.), pp. 231—282, Academic Press, New York.
Mortenson, L.E. and Chen, K. (1976) in Microbial production and utilization of gases (G. Schlegel, G. Gottschalk, N. Pfennig, eds.), pp. 97—108, Goltze K.G., Gottingen.
Ochiai, H., Shibata, H., Matsuo, T., Hashinokuchi, K. and Yukawa, M. (1977) Agric. Biol. Chem., 41, 72—77.
Packer, L., Fry, I., Sarma, S. and Rao, K.K. (1976) Biophys. J., 16, 132a.
Packer, L. and Cullingford, W. (1978) Z. Naturforsch., 33C, 113—115.
Paneque, A. and Arnon, D.I. (1962) Plant Physiol., 37, S IV.
Rao, K.K., Cammack, R., Hall, D.O. and Johnson, C.E. (1971) Biochem. J. 122, 257—265.
Rao, K.K., Rosa, L. and Hall, D.O. (1976) Biochem. Biophys. Res. Commun., 68, 21—27.
Rao, K.K. and Hall, D.O. (1977) in The Evolution of Metalloenzymes (G.J. Leigh, ed.), pp. 39—66, Symposium Press, London.

Rao, K.K., Gogotov, I.N. and Hall, D.O. (1978) Biochimie, 60, 291—296.

Reeves, S.G. and Hall, D.O. (1973) Biochim. Biophys. Acta, 314, 66—78.

Reeves, S.G., Rao, K.K., Rosa, L. and Hall, D.O. (1976) in Microbial energy conversion (H.G. Schlegel, J. Barnea, eds.), pp. 235—243. Erich Goltze K.J., Gottingen.

Reeves, S.G. and Hall, D.O. (1977) Cell Biol. Int. Rep., 1, 353—361.

Schneider, K. and Schlegel, H.G. (1976) Biochim. Biophys. Acta, 452, 66—80.

Tagawa, K. and Arnon, D.I. (1962) Nature, 195, 537—543.

Telfer, A., Cammack, R. and Evans, M.C.W. (1970) FEBS Lett., 10, 21—24.

Tel-Or, E., Luijk, L.W. and Packer, L. (1978) Arch. Biochem. Biophys., 185, 185—194.

van der Westen, H.M., Mayhew, S.G. and Veeger, C. (1978) FEBS Lett., 86, 122—126.

Yagi, T. (1977) in Biological Solar Energy Conversion (A. Mitsui, et al., eds.), pp. 61—68, Academic Press, New York.

Yagi, T., Kimura, K., Daidoji, H., Sakai, F., Tamura, S. and Inokuchi, Hi. (1976) J. Biochem. (Tokyo), 79, 611—671.

Photosynthesis in relation to model systems, edited by J. Barber
© Elsevier/North-Holland Biomedical Press 1979

Chapter 11

Hydrogen from Algae

PATRICK C. HALLENBECK and JOHN R. BENEMANN

Sanitary Engineering Research Laboratory, University of California, Berkeley, 94720, U.S.A.

CONTENTS

Abbreviations

ATP, adenosine triphosphate; CCCP, carbonyl cyanide m-chlorophenyl hydrazone; DCMU, 3-(3,4-dichlorophenyl)-1,1-dimethylurea; DCPIP, 2,6-dichlorophenolindophenol; DMBIB, 2,5-dibromo-3-methyl-6-isopropyl-p-benzoquinone; DNP, 2,4-dinitrophenol; DSPD, disalicylidenepropanediamine; E'_0, oxidation—reduction midpoint potential at pH 7.0; FCCP, carbonyl cyanide p-trifluoromethoxyphenyl hydrazone; HOQNO, 2n-heptyl-4-hydroxyquinoline-N-oxide; MFA, monofluoroacetate; NAD, nicotinamide adenine dinucleotide; NADH, nicotinamide adenine dinucleotide, reduced form; NADP, nicotinamide adenine dinucleotide phosphate; NADPH, nicotinamide adenine dinucleotide phosphate, reduced form; pH$_2$, partial pressure of hydrogen; pO$_2$, partial pressure of oxygen; PMS, N-methylphenazonium methosulfate; Q$_{O_2}$, μl O$_2$ evolved per mg per hour.

11.1. INTRODUCTION

As already mentioned in previous chapters hydrogen evolution by algae was first studied by Gaffron and Rubin (1942). They found that after placing algae under dark anaerobic conditions, for periods of minutes to several hours, the algae evolved hydrogen. The reaction was stimulated by low light intensities and inhibited by high light and a number of metabolic inhibitors. The reader is referred to Spruit (1962), Bishop (1966) and Kessler (1974) for excellent reviews summarizing the early literature. Here we review only a few of the early studies that are necessary to introduce the more recent experimental work and to highlight present areas of knowledge.

Historically, algal hydrogen production has been studied mainly as an interesting metabolic reaction. The discovery of a hydrogen metabolism in organsims whose normal mode of metabolism involves oxygen production was certainly surprising as the hydrogenase enzyme and reaction are very sensitive to oxygen inhibition and inactivation. This pecularity lent the early impetus to the original work in this area. In addition to hydrogen evolution, anaerobically adapted algae were found capable of carrying out several metabolic activities in connection with the uptake of hydrogen, including the photoreduction of exogenous hydrogen acceptors such as nitrate, nitrite and carbon dioxide. Investigations of algal hydrogen metabolism, besides clarifying some of the metabolic processes involved, have contributed to our understanding of photosynthesis.

Currently, as concern over world energy supplies has led to a search for renewable fuel sources, the process of hydrogen evolution by algae has received considerable attention. Biophotolysis — the production of hydrogen and oxygen from water and sunlight by biological catalysts — is an attractive concept for the production of a renewable, non-polluting fuel. Hydrogen has been suggested as the fuel of the future, with several advantages over present, or other alternative energy supply systems (Brocksis, 1975). Electricity could be produced where needed with energy savings due to lower transmission losses. Hydrogen could be stored for use at peak periods of demand. Hydrogen itself would be directly useful in industrial applications such as iron ore reduction and ammonia (fertilizer) production.

Prerequisites for the development of any system for biophotolysis include (Weissman and Benemann, 1977): (i) sustained, catalytic H_2 and O_2 evolution for many weeks; (ii) a 2 : 1 ratio of hydrogen to oxygen; (iii) high specific rates of hydrogen production and photosynthetic efficiency; and (iv) no limitation to scaling up of the system. In the following pages, we will describe some of the physiological and biochemical aspects of hydrogen production by green algae and blue-green algae, and the biological systems that have been suggested for possible use in biophotolysis. Some of the characteristics, advantages and disadvantages of various potential biophotolysis approaches have been recently discussed (Benemann, 1977; Benemann and Weissman, 1977). After reviewing hydrogen production by green and blue-green algae, we return, in the conclusion, to a discussion of their possible use in a practical biophotolysis system.

As far as possible all data on hydrogen production reported below is expressed in terms of μl of H_2/mg dry wt/h to allow comparison of results obtained from different laboratories. This unit appears to be most satisfactory as data based on packed cell volumes is species (and even laboratory) dependent and chlorophyll based rates of hydrogen production seem inappropriate since cellular chlorophyll content is dependent upon the previous history of the cell and photomediated hydrogen reactions are often saturated at low light intensities. Thus the above unit should be adopted as the standard for biophotolysis research.

11.2. HYDROGEN PRODUCTION BY GREEN ALGAE

11.2.1. Physiology

Anaerobically adapted algae are capable of catalyzing a wide variety of metabolic reactions involving the gases H_2, O_2, and CO_2. These processes include both light and dark reactions (Stuart and Gaffron, 1972b);

Light reactions:
(1) H_2 photoevolution: $RH_2 + light \rightarrow R + H_2$
(2) photoreduction: $CO_2 + 2H_2 + \sim P \rightarrow (CH_2O) + H_2O$
(3) photosynthesis: $CO_2 + 2H_2O + \sim P \rightarrow (CH_2O) + O_2 + H_2O$

Dark reactions:
(4) H_2 production (or uptake) $RH_2 \leftrightharpoons R + H_2$
(5) oxy-hydrogen reaction: $O_2 + 2H_2 \rightarrow 2H_2O + \sim P$
(6) carbon fixation: $CO_2 + 2H_2 + \sim P \rightarrow (CH_2O) + H_2O$
(7) respiration: $\frac{1}{2} O_2 + RH_2 \rightarrow H_2O + R + \sim P$

The reactions 1, 2, 4, 5, 6, requiring the direct involvement of hydrogenase are discussed below.

11.2.1.1. Photoreduction

Photoreduction, first described by Gaffron (1940), is an alternative process for light-driven carbon fixation in which reductant is derived ultimately from H_2 instead of the water-splitting reaction. The carbon assimilation pathway in this case appears to be identical with the carbon reduction cycle operative in normal photosynthesis (Badin and Calvin, 1950; Gingras et al., 1963; Russell and Gibbs, 1968). Photoreduction is sensitive to light intensity, being abolished at moderate light intensities, presumably due to resumption of photosynthesis and oxygen inactivation of hydrogenase. Light inhibition of photoreduction commonly occurs near the

O_2 compensation point. Photosynthesis and photoreduction show different requirements for actively functioning photosystem components. The large body of available evidence indicates that, unlike photosynthesis, photoreduction is primarily, if not solely, dependent upon only photosystem I (Bishop and Gaffron, 1962). Mutants lacking photosystem II (Pratt and Bishop, 1968), manganese-deficient algae (Kessler, 1968, 1970), or cultures poisoned with various photosystem II inhibitors, all show normal photoreduction activity. The results of these, and similar studies, suggest that during the process of photoreduction carbon fixation proceeds along most of the pathways operative during photosynthesis with ATP being provided by cyclic photophosphorylation and with reductant provided at the level of ferredoxin or NADPH through the action of hydrogenase. However, both the low levels of oxygen production (Horwitz and Allen, 1957) observed during photoreduction by unpoisoned cultures, and results obtained with chloride deficient algae (Grimme, 1972) suggest that the situation is complex. In most cases, it is probable that with anaerobically adapted cultures of algae what is observed is really an apparent photoreduction, the sum of true photoreduction and low-level functioning of photosynthesis coupled with the oxy-hydrogen reaction.

11.2.1.2. Dark reactions

In the dark, anaerobically adapted algae can carry out a variety of metabolic processes mediated by hydrogenase. Under an atmosphere of hydrogen and in the absence of added exogenous hydrogen acceptors, a low level (0.35 μl/mg dry wt/h) of H_2 uptake has been observed (Ben-Amotz et al., 1975) which seems to be due to reduction of some cellular component. This activity is greatly stimulated by the addition of nitrite (Kessler, 1956; Stiller, 1966) or artificial acceptors such as methylene blue. Low levels of oxygen also stimulate this process, presumably by permitting a low level of respiration in which oxygen becomes the ultimate hydrogen acceptor for the reduced compounds generated by hydrogenase action. Additionally, if CO_2 is present, a chemosynthetic reduction of carbon can occur in which apparently hydrogen serves as the primary donor for the generation of the reductant for the carbon fixation process, and also for the necessary ATP synthesis (derived from the oxy-hydrogen reaction). In the absence of added electron donors, low rates (0.3—2.6 μl/mg dry wt/h) of hydrogen evolution by adapted algae in the dark have been observed (Ben-Amotz et al., 1975; Healey, 1970a), where it is likely that electrons generated through fermentative pathways are supplied to hydrogenase. Long periods of dark anaerobiosis (7 h) tend to depress rates of dark hydrogen evolution, presumably due to reductant depletion. Addition of suitable reductant-generating substrates, such as glucose, alleviates this effect (Kessler, 1974). The dark evolution of hydrogen is accompanied by CO_2 release and inhibited by both monofluoroacetate and uncouplers of phosphorylation (Healey, 1970,

Kaltwasser et al., 1969) indicating that this process is probably dependent upon the tricarboxylic acid cycle and phosphorylation (with an unknown acceptor).

11.2.1.3. Photoproduction of hydrogen

Photo-induced hydrogen production has been studied in greatest detail and is characterized by two phases (Stuart and Gaffron, 1971; Bishop et al., 1977); a fast initial phase consisting of the first few minutes after initiation of illumination, and a slow phase (hours) under prolonged illumination. Initial rates (during the first minute of illumination) as high as 208 μmoles/ ml cells/h (34 μl/mg dry wt/h) have been observed (Bishop et al., 1977), whereas rates determined after longer periods of illumination are typically around 5 μl/mg dry wt/h (Healey, 1970b). Thus there is an initial "burst" of hydrogen, with the rate of hydrogen evolution quickly declining after only one to two minutes of activity to a lower, more sustained rate. This rate decrease could be due to the rapid accumulation of oxygen, causing inactivation of hydrogenase, or to the depletion of the reductant pool available to hydrogenase, or both. Hydrogen evolution is saturated at relatively low-light intensities (100—200 μwatts/cm^2).

Moreover, high light intensities are inhibitory due to the resumption of normal photosynthesis and the accumulation of oxygen (Horwitz and Allen, 1957; Stuart and Gaffron, 1972b). Under low light intensities, and especially during the initial phase, any oxygen produced could be scavanged by the oxygen uptake processes of respiration.

11.2.2. Hydrogenase enzymes

It is generally believed that in green algae hydrogenase is a constitutive enzyme, becoming activated by some process during dark anaerobiosis. There is indirect evidence for residual hydrogenase activity under aerobic conditions (Kessler, 1970) and, in some cases, hydrogenase mediated reactions are observed within minutes of removing oxygen from the gas phase. On the other hand, in some cases periods of anaerobiosis as long as twenty four hours are required to obtain maximal expression of hydrogenase activity (Ward, 1970a). If the enzyme is indeed constitutive some activation process is necessary since low levels of oxygen cause inactivation. Hydrogenase is partially de-adapted at 0.8 μl O_2/ml and completely inactivated at 2.6 μl/ml (Stuart and Gaffron, 1972b). The adaptation process can be inhibited by dithionite, sulfite, iron chelating agents, and inhibitors of phosphorylation (Hartman and Krasna, 1963, 1964; Oesterheld, 1971). The adaptation process is temperature and pH dependent, with maximum adaptation in *Chlamydomonas* occurring at pH 7.5 (Ward, 1970a). In some cases, addition of sugars has been shown to decrease adaptation times (Stiller and Lee, 1964a). Thus, hydrogenase activation appears to be an energy-requiring process. However, the process may be more complex than just a simple activation. Changes in hydrogenase content with growth cycle periods have been noted (Yanagi and Sasa, 1966) and, in one case, addition of inhibitors of protein synthesis during adaptation inhibited subsequent hydrogenase activity by 75% (Oesterheld, 1971). Whether this is a primary effect, or due to the inhibition of synthesis of accessory enzymes is not clear. Other researchers have found that inhibitors of protein synthesis had no effect on development of hydrogenase activity when adaptation took place in the presence of glucose (Stiller and Lee, 1964b).

The algal hydrogenase itself is only poorly characterized biochemically. Adaptation of hydrogenase in cell-free extracts has been noted, and this adaptation process appears to be temperature dependent with complex inhibition below $2.5°C$ (Ward, 1970b). Early in vitro studies with iron chelating agents indicated that iron is probably part of the active site, and similar results have been obtained with sulfhydryl reagents (Hartman and Krasna, 1964). Thus, this hydrogenase, like the better characterized bacterial hydrogenase, is probably an iron—sulfur protein. Hydrogenase, unlike metal catalysts, has been shown to catalyze the heterolytic cleavage of hydrogen (Hartman and Krasna, 1963; Krasna, 1977). This suggests the formation of an enzyme hydride and a proton. Thus, in D_2O, the primary product is HD. However, with hydrogenase more D_2 is formed than is expected on the basis of the mechanism suggested above, and the ratio of HD : D_2 varies with the species tested (Krasna, 1977). This result suggests that the postulated enzyme hydride is exchangeable, to various extents, with water. Cell-free hydrogen uptake activities have been found (Lee and Stiller, 1967) to be up to 10-fold greater than whole-cell activities, indicating that added exogenous electron acceptors may have a hard time crossing intact cell membranes.

Three different hydrogenases are known: a reversible hydrogenase, a uni-directional "uptake" hydrogenase, and an ATP-dependent hydrogen evolution reaction of nitrogenase. Clearly most anaerobically adapted algae possess a reversible hydrogenase. Evidence has been presented (Ben-Amotz et al., 1975) that some anaerobically adapted algae, notably red algae, contain only an uptake hydrogenase. The unidirectional hydrogenase is important in studies of hydrogen evolution (and in potential biophotolysis systems) because it appears to catalyze only the uptake of hydrogen and thus acts to decrease the net hydrogen evolution of any system in which it is present. This hydrogenase may not be reversible due to its strong interaction with membranes (most reversible hydrogenases have been found to be soluble). Perhaps because it is a membrane bound enzyme, uptake hydrogenases have not been characterized in much detail. A particulate hydrogenase from *Chromatium* has been partially purified with sodium deoxycholate and characterized (Gitlitz and Krasna, 1975). Its reported properties were quite different than the properties of the more fully characterized, reversible hydrogenases; including a higher molecular weight (100 000 compared to

60 000) and a greater oxygen stability. Presumably, this enzyme functions only unidirectionally in vivo when bound to a membrane. When isolated, however, it is able to catalyze hydrogen evolution. Similarly, *Rhodospirillum rubrum* hydrogenase, after isolation with membrane solubilizing agents, yields a reversible hydrogenase activity which appears likely to only be present as an uptake hydrogenase in vivo (Adams et al., 1978). The uptake hydrogenase is half saturated at a relatively low pH_2 of 0.05 atm (Hyndman et al., 1953) and is incapable of reducing low potential electron acceptors. Whether or not hydrogen evolving green algae also usually contain an uptake hydrogenase is an open question. Different adaptation times for hydrogen evolution and the hydrogen uptake activities of photoreduction and the oxy-hydrogen reaction, and observed hydrogen uptake at very low pH_2's (<0.01 atm) indicate green algae may very well contain an uptake hydrogenase. The presence of two hydrogenases would also help explain apparently paradoxical data in which H_2 production appears to be balanced by simultaneous H_2 consumption (Stuart and Gaffron, 1972b). This enzyme appears to be present in some species of mosses and in red algae. These organisms, while apparently unable to evolve H_2, can catalyze the dark consumption of H_2 (Ben-Amotz et al., 1975).

A cell-free hydrogenase from *Chlamydomonas* has been described in some detail. Using cell-free extracts, light dependent hydrogen evolution could be demonstrated, with dithiothreitol and NADH being the most effective donors (Abeles, 1964; Ben-Amotz and Gibbs, 1975; King et al., 1977). Early reports indicated that the green algal hydrogenase would not react with ferredoxin (Ben-Amotz et al., 1975) which would be unusual, but not unprecedented. However, in later experiments, (King et al., 1977) in which light-driven hydrogen evolution was obtained with green algal chloroplasts prepared by mild disruption, a requirement for ferredoxin was demonstrated. DSPD (Disalicylidenepropanediamine) was markedly inhibitory. Apparently, ferredoxin is more tightly coupled with NADP reduction than with hydrogen evolution as the Km for ferredoxin in hydrogen evolution is seventy times higher than the Km for NADP reduction. Depending upon the method of preparation, hydrogenase is either sedimented by high speed centrifugation (Abeles, 1964) or remains soluble (Ben-Amotz et al., 1975). The cell-free enzyme is irreversibly inhibited by oxygen or ferricyanide, and will catalyze the reduction of methylene blue, benzyl viologen, methyl viologen, NAD, NADP, and triphenyl tetrazolium chloride (Abeles, 1964; Lee and Stiller, 1967).

11.2.3. Reductant pathways in photoproduction of hydrogen

The mechanism of photo evolution of hydrogen is still a matter of some uncertainty. Photosystem I is known to be directly involved, as photo hydrogen production shows an absolute dependence upon photosystem I functions, being similar in this respect to photoreduction. Thus mutants lacking photosystem I components show greatly impaired rates of hydrogen production (Stuart and Kaltwasser, 1970), and DSPD (Disalicylidene-propanediamine), a ferredoxin antagonist (Trebst and Burba, 1967) inhibits hydrogen evolution by normal algal cultures (Ben-Amotz and Gibbs, 1975). The action of photosystem I could be providing energy via cyclic photophosphorylation for a dark H_2-yielding reaction, or hydrogen evolution could be directly dependent on electron flow through photosystem I from reduced electron donors. Indirect evidence suggests that the latter possibility is correct. As has already been mentioned, light-induced electron flow from added electron donors to hydrogenase has been demonstrated in vitro. Addition of uncouplers of photophosphorylation, such as CCCP (carbonyl cyanide m-chlorphenylhydrazone) often stimulates hydrogen evolution one and a half to two-fold (Healey, 1970a, b; Kaltwasser et al., 1969). This effect would be expected if hydrogen evolution was dependent on electron flow through phosphorylating sites of the inter-photosystem electron transfer chain; uncoupling phosphorylation would tend to increase electron flow. However, it seems more likely that the action of CCCP is indirect through the inhibition of the competing reactions of photosynthesis and photoreduction. Thus, the inhibitory effects of gas phase concentrations of either 50% H_2 or 50% CO_2, thought to be due to stimulation of photoreduction, are eliminated by addition of CCCP (Healy, 1970a). This further supports the suggestion made earlier that green algae may contain two hydrogenases.

There is good evidence that many of the components of the interphotosystem electron trnasport chain are required for effective photoevolution of hydrogen. Genetic deletion of cytochrome f-553 inhibits photosynthesis, photoreduction and photohydrogen evolution (Bishop et al., 1977). Plastoquinone antagonist, DBMIB (dibromothymoquinone) causes almost complete inhibition of hydrogen evolution (Ben-Amotz and Gibbs, 1975; Bishop et al., 1977).

Whether or not water serves as the primary reductant in photohydrogen evolution has been a question of some theoretical importance, and is very relevant in considerations of this system for biophotolysis. The evidence for the involvement of photosystem II itself (not necessarily the water splitting reaction) is equivocal.

Numerous photosystem II mutants have been obtained which show normal rates of photoreduction. Photohydrogen production is as severely impaired as photosynthesis (Bishop et al., 1977). However, these mutations usually have pleiotropic effects; besides loss of the oxygen splitting reaction there is often loss of components of the inter-photosystem electron transport chain, which appear necessary for photoevolution of hydrogen (as discussed above). As explained in chapter 8, oxygen evolution during single-turnover flashes of light gives a characteristic pattern of zero yield on the first flash, maximum yield on the third flash, and yields on subsequent flashes whose amplitude is a damped oscillation of periodicity four about the steady state value (see Fig. 8.1). These observations have led to several hypotheses about the mechanism of photosynthetic oxygen evolution (Kok et al., 1970; Greenbaum, 1977a). However, single-turnover flashes give a different pattern with hydrogen evolution; the yield of hydrogen is fixed at the steady state value at the first flash and subsequent flashes give approximately the same yield for at least the next 26 flashes (Greenbaum, 1977b). The calculated Emerson and Arnold photosynthetic unit was chlorophyll : hydrogen ~ 1400 : 1, or 60% of the photosynthetic unit for oxygen evolution. Thus, unlike oxygen evolution, hydrogen evolution can be driven by two different photosynthetic units. This effect results from the accumulation of reducing equivalents in a common pool, perhaps reduced ferredoxin. The size of the determined photosynthetic unit for hydrogen evolution indicates a possible involvement of photosystem II in this process.

DCMU, a widely used inhibitor of photosystem II function, has variable effects dependent upon the species of algae tested. In general, hydrogenase containing algae can be divided into three groups on the basis of the inhibitory effects of DCMU (Healey, 1970b): those that show no inhibition (several species of *Chlamydomonas*), those that are inhibited by about 50% (several species of *Scenedesmus* and *Chlorella pyrenoidosa*) and some that are completely inhibited (*Chlorella vulgaris* and *Chlorella* sp.). The inhibitory effects of DCMU may be greater during the initial "fast" phase of hydrogen evolution (Bishop et al., 1977). Measurements of O_2 evolved during this process have given variable stoichiometries. Of course, the situation is complicated by the competing evolution and consumption reactions for both O_2 and H_2 (see section 11.2.7), and the O_2 inactivation of hydrogenase. Manipulation of experimental conditions, such as lengths of time between light flashes, can give a perhaps fortuitous stoichiometry that approaches the theoretical value of two (Bishop et al., 1977).

On the other hand, there is a great deal of evidence that the majority, if not all, the reductant necessary to drive photoevolution of hydrogen comes ultimately from organic donors. DCMU, which presumably blocks electron flow from Q, may be acting to block light induced electron flow from organic donors through photosystem II. Thus, in cultures of *Chlorella* and *Scenedesmus* in which O_2 evolution had been blocked by CCCP, addition of DCMU inhibited H_2 evolution up to 50% (Stuart and Gaffron, 1972a). This effect was more pronounced in autotrophically grown cells than in photoheterotrophically grown cells. Other evidence suggesting that the metabolism of reduced carbon compounds is involved is the concomitant release of CO_2 (Frenkel, 1952; Kaltwasser et al., 1969) and the inhibitory effect of high levels of CO_2 in the gas phase (Healey, 1970a). Prolonged periods of dark anaerobiosis, during which stored reductant may be depleted, decrease the yield of hydrogen during subsequent illumination. This effect can be overcome by adding glucose or acetate, or by a period of photosynthesis (Healey,

1970a, 1970b). Glucose appears to be degraded by the Embden-Meyerhof pathway as adapted cells release $^{14}CO_2$ from the 3 and/or 4 position of exogenously added, specifically labeled glucose (Kaltwasser et al., 1969). Of many possible metabolites, only glucose has been shown to be stimulatory. The rate of utilization of glucose is independent of its concentration, but the total amount of stimulation is directly proportional to the amount of added glucose. It is not clear what pathway leads from glucose catabolism to hydrogenase since in these experiments one μmole of glucose gave 0.5 μmol H_2 gas (Stuart and Gaffron, 1971). These results may not be totally unexpected as algae have a mixed fermentation. Monofluoroacetate inhibits hydrogen evolution in *Chlamydomonas*, indicating the involvement of the tricarboxylic acid cycle (Healey, 1970a).

Photosystem II function can be impaired in other ways without seriously affecting hydrogen evolution. Use of the heat treatment (55%°C for 1— 2 min) or salicylaldoxime has shown that hydrogen evolution can continue when photosynthesis, photoreduction, and cyclic photophosphorylation are completely abolished (Stuart, 1971). These treatments also have the advantage of simplifying the study of this system by abolishing many of the competing reactions. This can lead to a two-fold stimulation of hydrogen evolution.

In summary, the available evidence indicates that in most hydrogenase containing species of green algae the primary electron donor for photo-hydrogen evolution is reduced carbon compounds. These compounds, or some derived metabolite, donate to the inter photosystem electron transport chain. The action of light raises the donated electrons to a level where they are able to reduce hydrogenase, with probably reduced ferredoxin (or perhaps NADPH) as an intermediate electron carrier. In some cases it appears that the reduced organic compounds can donate electrons to photosystem II, accounting for at least some of the photosystem II contribution to hydrogen evolution. Whether or not there exist circumstances under which the splitting of water (a reaction which is basically incompatible with hydrogen evolution) can contribute to this process remains an open question. The fact that this process relies on previously stored reduced compounds formed during periods of photosynthesis means that the history of the culture is important, an aspect that has unfortunately been ignored too often in the past. Thus cultures grown with different degrees of light saturation would be expected to give different amounts of hydrogen produced after subsequent adaptation. A general scheme of electron flow in the hydrogen metabolism of green algae (based on Healey, 1970a) is shown in Fig. 11.1. (see also Fig. 9.2).

11.2.4. Possible physiological and ecological significance of H_2 production

Algae are usually thought of as living in an oxygen-rich environment because of their existence in habitats where gas exchange with the atmosphere

342

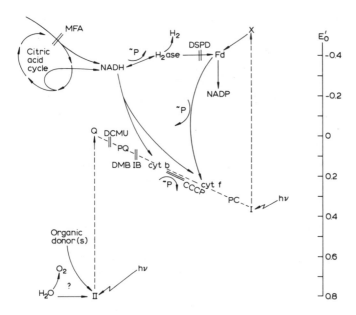

Fig. 11.1. Hydrogen metabolism of green algae.

takes place and their naturally high rate of oxygen production. However, during periods of darkness dissolved oxygen concentrations in some environments may actually reach zero due to high respiratory uptake by algae and associated microorganisms. Thus a capability for some type of anaerobic metabolism may, from time to time, confer a selective advantage to an organism whose main form of metabolism is aerobic. In the presence of hydrogen, carbon dioxide and a microaerobic environment, hydrogenase containing algae would be able to chemosynthetically fix carbon in the dark through hydrogen uptake and the oxy-hydrogen reaction. Additionally, in the light, further carbon fixation is possible. These may be the primary ecological advantages for possession of hydrogenase.

Photoproduction of hydrogen may not be important from an ecological point of view since it accomplishes little toward the maintenance or growth of algae in an anaerobic environment. However, some evidence has been presented that algae possessing hydrogenase retain most of their chlorophyll and continue to slowly grow under aerobic, Mn deficient conditions whereas algae without hydrogenase quickly become chlorotic (Kessler, 1968, 1970).

Light driven hydrogen evolution may serve an important physiological function. After a period of dark anaerobiosis, oxygen evolution is severely inhibited. This inhibition is due to the lowering of the redox level of a pool of electron carriers (i.e. plastoquinone) which prevents the effective oxidation of the primary acceptor of photosystem II (Diner and Mauzerall, 1973).

Under these conditions photolysis of water can occur only after substantial oxidation of this pool has occurred either through positive feedback by oxygen (Diner and Mauzerall, 1973) or addition of artificial oxidants (Greenbaum and Mauzerall, 1977). Photoproduction of hydrogen may help poise the photosynthetic apparatus for photosynthesis during illumination following a period of dark anaerobiosis by rapidly oxidizing this reduced pool. This is consistent with the short "burst" of hydrogen often observed upon initial illumination. At low light intensities after prolonged dark anaerobic incubation, algae without hydrogenase are not able to resume photosynthesis whereas closely related species of algae with hydrogenase are able to to evolve oxygen within minutes of illumination (Kessler, 1973). Thus photoproduction of hydrogen enables algae to recover quickly from periods of anaerobiosis and to resume photosynthesis and normal pathways of metabolism. It would be interesting to determine whether such bursts of hydrogen production occur in nature.

11.3. CYANOBACTERIA

11.3.1. Hydrogen evolution and uptake

The cyanobacteria (blue-green algae) are members of a great subclass of prokaryotes, gram-negative bacteria. Unlike the photosynthetic bacteria however, they are able to carry out oxygen evolving photosynthesis. Although some of the details of their photosynthetic process are unique (for example, accessory light harvesting pigments), in general the physiology and molecular mechanism of cyanobacterial photosynthesis is highly similar to the green algal process.

Most of the experiments on hydrogen metabolism in cyanobacteria have been carried out using heterocystous species. Heterocysts are morphologically and biochemically distinct cell types that are formed during a unique prokaryotic differentiation event. The correlation of heterocyst frequency with nitrogenase activity under different growth conditions has led to the hypothesis that nitrogen fixation is localized in the heterocyst. The heterocyst possesses several features that make it a likely candidate for this function. Some of the specialized features of heterocysts are summarized in Table 11.1. The most important feature is the presence of an anerobic environment in the heterocyst. Nitrogenase has been found to be a very oxygen labile enzyme, with an anaerobic environment necessary for active functioning, at least in vitro, and probably also in vivo, with the exception of *Azotobacter* (which is thought to have several specialized methods of oxygen protection). Because the heterocystous cyanobacteria contain two distinct cell types the interpretation of experimental results is often complicated by the very real possibility of compartmentalization of the involved enzymes

TABLE 11.1. Metabolic properties of heterocysts.

Heterocyst properties	Evidence	Reference
1. Low intracellular redox potential	— blackening of photographic emulsions	Stewart et al., 1969
	— differential straining by reduction of tetrazolium	Stewart et al., 1969; Fay and Kulasooriya, 1972
2. Gas diffusion barrier-due to thick cell wall of special biochemical composition	— indirect, e.g. O_2 resistance of nitrogen fixation	Weare and Benemann, 1972; Weare and Benemann, 1973
3. Lack of O_2 evolution and absence of photosystem II	— microphotospectroscopy	Thomas, 1970
	— pigment analysis	Wolk and Simon, 1969
	— low manganese content	Tel-Or and Stewart, 1977
4. Inability to fix CO_2	— absence of RUDP carboxylase in isolated heterocysts	Winkenbach and Wolk, 1973
	— failure of heterocysts to incorporate $^{14}CO_2$	Wolk, 1968
5. Functioning photosystem I	— heterocysts enriched in photosystem I components	Donze et al., 1972
	— light induced EPR signals of isolated heterocysts	Cammack et al., 1976
	— both cyclic photophosphorylation and photoreduction of acetylene	Tel-Or and Stewart, Wolk and Wojiuch, 1971
6. High levels of pentose phosphate	— enzymatic assays of isolated heterocysts	Winkenbach and Wolk, 1973
7. Glutamine synthetase	— incorporation of $^{13}NH_4$ by isolated heterocysts	Thomas et al., 1977
8. High levels of nitrogenase	— Isolated heterocysts exhibit high specific nitrogenase activity, up to 48% of intact filament	Peterson and Burris, 1976; Thomas et al., 1977

and accessory metabolic reactions. As a means of orientation during the following discussion, a model of cyanobacterial hydrogen metabolism is presented in Fig. 11.2.

Hydrogen production by cyanobacteria has been studied in detail only recently, although one report on hydrogen production by a natural sample of *Anabaena* sp. dates back to 1898 (Jackson and Ellms, 1898). Anaerobi-

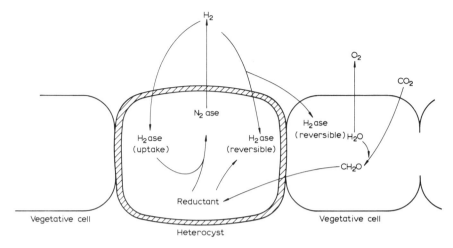

Fig. 11.2. Model of cyanobacterial hydrogen metabolism.

cally adapted cyanobacteria have been shown to be capable of carrying out the same metabolic reactions involving H_2, CO_2 and O_2 that are typical of anaerobically adapted green algae (see section 11.2.1). While green algal hydrogen metabolism is catalyzed solely by a reversible hydrogenase (with the possible contribution in some cases of an uptake hydrogenase), the situation in cyanobacteria is more complex owing to the presence of three different hydrogenases: a reversible hydrogenase, an "uptake" hydrogenase and nitrogenase (capable of acting as an ATP-dependent, unidirectional hydrogenase). While in some cases it is clearly possible to ascribe the observed effects to a single enzyme, in other cases a clear distinction has not been made, and this area requires further research. This section, while covering all known aspects of cyanobacterial hydrogen metabolism, emphasizes those reactions which differ markedly from the processes known from studies of green algae.

11.3.1.1. Photoreduction

When anaerobically adapted, some species of cyanobacteria carry out photoreduction (Frenkel, 1949; Frenkel and Rieger, 1951). Presumably this reaction may be catalyzed by a reversible hydrogenase since a period of anaerobic adaptation was necessary, but the uptake hydrogenase may also play a role in this process. This reaction, although not studied in much detail, appears to be identical to the reaction catalyzed by green algae. Photoreduction in a unicellular *Chroococcus* sp. has been shown to have the same dependence on light intensity as green algae with abolition of this reaction at moderate light intensities with the resumption of photosynthesis (Frenkel et al., 1949). Sodium sulfide inhibits the adaptation process; when added after adaptation it stabilizes photoreduction against reversal by high light intensities. Since pathways of carbon assimilation and photosynthesis are similar in cyanobacteria and green algae, it is likely that the same detailed mechanism is involved in each case. H_2 consumption in the light is partially inhibited by photosynthetic electron transport inhibitors such as DBMIB and DSPD and is stimulated about one and one-half fold by the presence of CO_2 (Tel-Or et al., 1977 — after correcting what is apparently a typographical error in their Table 1). Although hydrogen has some stimulating effect on CO_2 fixation by actively growing *Anabaena cylindrica* cultures, it does not significantly increase the levels of CO_2 fixation by DCMU treated cultures (Benemann and Weare, 1974a).

11.3.1.2. Dark hydrogen consumption

Evidence for a distinct uptake hydrogenase in cyanobacteria was first demonstrated by Benemann and Weare (1974a) who found that H_2 will supply reductant for nitrogenase in reductant depleted (DCMU treated) cultures. The reaction was characterized by a low $K_m^{H_2}$, about 2% H_2 in the gas phase (calculated from Fig. 2 in Benemann and Weare, 1974a). This enzyme system has been the subject of several recent investigations. Jones and Bishop (1976) found that uptake of hydrogen can proceed in the dark at hydrogen partial pressures as low as 0.05%. This activity is stimulated up to ten-fold by the presence of oxygen (Bothe et al., 1977b), and with optimal oxygen concentrations (about 2—10%) rates of H_2 consumption of 5.8 μmol H_2/mg chl/h have been observed. This oxy-hydrogen reaction is inhibited 50% by 7.3% CO (Bothe et al., 1977b, Fig. 2). The inhibitory effects of carbon monoxide may be due to direct inhibition of the algal hydrogenase, as is known to be the case in other organisms. However, indirect effects cannot be ruled out. Indeed, carbon monoxide inhibition of hydrogen uptake by isolated heterocysts is greatest when oxygen is supplied as an electron acceptor, suggesting that at least some of the inhibition is due to an inhibition of a cytochrome component of the respiratory chain (Peterson and Burris, 1978). Likewise, cyanide inhibited the oxy-hydrogen reaction by 90%, but methylene blue reduction was inhibited by only 30%.

Hydrogen consumption, although stimulated by oxygen, appears to be poorly coupled to respiration in cyanobacteria. Thus, O_2 consumption in the presence of hydrogen is two-fold greater than in its absence, and the presence of hydrogen gives a two-fold stimulation of radioactive phosphate incorporation (Peterson and Burris, 1978). However, uncouplers of phosphorylation, such as FCCP and DNP stimulate O_2 consumption up to 60% whereas gas consumption ($H_2 + O_2$) in the presence of hydrogen and the uncouplers is only 15% greater (Bothe et al., 1977b). These results support the earlier finding of a high degree of oxygen sensitivity of hydrogen-supported acetylene reduction activities (Benemann and Weare, 1974b). When the oxy-hydrogen reaction takes place under conditions in which the reaction is limited by oxygen, an apparent stoichiometry of $H_2 : O_2$ of 2.4 has been observed (Peterson and Burris, 1978).

11.3.1.3. Nitrogenase mediated photoproduction of hydrogen

Initial studies with cyanobacteria demonstrated that they were able to catalyze high rates (8.8 μl/mg dry wt/h) of hydrogen production in the light in a reaction which was linear for at least 3 hours (Benemann and Weare, 1974b). This process was not inhibited by DCMU. The properties of this reaction revealed that it was quite different from the process of photo-hydrogen production by anaerobically adapted green algae, and that the majority of the hydrogen evolved was produced by nitrogenase, which in this case acts as an ATP-dependent (unidirectional) hydrogenase. (The biochemistry of this enzyme is discussed in more detail in section 11.3.2). Hydrogen evolution was strongly inhibited by N_2 but only slightly inhibited by carbon monoxide or oxygen. Rates of hydrogen evolution were closely correlated with rates of acetylene reduction, with rates of hydrogen evolution being somewhat lower, presumably due to the action of an "uptake" hydrogenase. No adaptation process was necessary as rates under 18% O_2 were almost as high (80%) as rates under "anaerobic" conditions. (These cultures are not anaerobic in the strictest sense of the term since continued algal photosynthesis can lead to the build-up of appreciable oxygen tensions.) (Benemann and Weare, 1974b). The only requirement for expression of maximal hydrogen evolution activity is an atmosphere devoid of high levels of N_2 and O_2 (Benemann and Weare, 1974b; Jones and Bishop, 1976). Under N_2, H_2 evolution was inhibited by about 90%. The low rate of hydrogen evolution observed under these conditions is probably due to a side reaction of the nitrogenase-catalyzed reduction of atmospheric nitrogen to ammonia. Hydrogen evolution proceeds readily in atmospheres of N_2 or air if carbon monoxide, an inhibitor of the nitrogen fixation reaction of nitrogenase, is present (Benemann and Weare, 1974b; Daday et al., 1977). Hydrogen evolution in the air (plus carbon monoxide) is reported to be further stimulated by the presence of acetylene. Presumably this stimulation is due to acetylene inhibition of uptake hydrogenase activity (Smith et al., 1976). However, these results appear to be inconsistent with the observation that acetylene reduction assays conducted in the presence of hydrogen can restore acetylene reduction activities of severely reductant depleted *Anabaena cylindrica* cultures to near normal (non-starved) rates (Benemann and Weare, 1974a; see also Jones and Bishop, 1976). If the algal hydrogenase is inhibited by acetylene it must be much more refractory to acetylene inhibition than the hydrogenases of *Azotobacter* or *Klebsiella*. Gas phase concentrations of acetylene up to 17% are reported to have no effect on the oxy-hydrogen reaction catalyzed by isolated heterocysts (Peterson and Burris, 1978).

Although most of the studies of hydrogen evolution by cyanobacteria have been done with cultures of *Anabaena cylindrica*, it is reasonable to expect that any heterocystous (and nitrogenase containing) cyanobacteria would catalyze this reaction. Indeed, nitrogenase catalyzed hydrogen evolution has been described in the cyanobacteria-plant symbiosis *Azolla-Anabaena azollae* (Peters et al., 1976) and in several species of marine cyanobacteria (Mitsui and Kumazawa, 1977; Lambert and Smith, 1977). However, in some blue-green algae, such as *Chloroglea*, little hydrogen evolution is observed, apparently due to a high uptake hydrogenase activity (Benemann, unpublished).

Evolution of hydrogen attains a maximum rate within 15—30 s of the onset of illumination, and, in the absence of appreciable concentrations of oxygen, immediately ceases upon darkness (Jones and Bishop, 1976). Under the same conditions (i.e. a short period of anaerobic incubation in the dark), O_2 evolution shows a typical period of delay with the achievement of maximum and constant rates only after 4 to 5 min of illumination. In short-term assays, simultaneous polarographic measurement of both O_2 and H_2 has shown that there is no correlation of H_2 photoproduction with photosynthetic O_2 evolution (Jones and Bishop, 1976). Recently, sustained rates of hydrogen production (up to 19 days) have been demonstrated with nitrogen-starved cultures of *Anabaena cylindrica* (Weissman and Benemann, 1977). In contrast, photoproduction of hydrogen by green algae (see section 11.2.1.3) has a period of peak productivity which is only a few hours long and production ceases completely after one day. In nitrogen-starved cultures of *Anabaena cylindrica* maximum rates of hydrogen evolution ($32 \, \mu l$/mg dry wt/h) were obtained after one to two days of nitrogen starvation. After this period of peak productivity, rates slowly declined. Periodic ammonium additions stabilized this rate decrease, increased the rate of oxygen evolution and increased the total hydrogen production of the cultures. Rates averaged over the length of the experiment were about $14 \, \mu l$/mg dry wt/h. Under these conditions H_2—O_2 ratios vary with the degree of nitrogen starvation, being 4 : 1 with completely starved cultures and 1.7 : 1 after the addition of small amounts of ammonium. The photosynthetic efficiency of hydrogen production with this system is at least 0.4% (Weissman and Benemann, 1977).

11.3.1.4. Adaptation phenomena

As previously discussed above, there appears to be no adaptation requirement for hydrogen evolution by nitrogenase in cyanobacteria. Some cyanobacteria have been shown to have an adaptable, classical hydrogenase (Ward, 1970a; Fujita et al., 1964) but present evidence indicates that this enzyme does not make a significant contribution to the photoevolution of hydrogen in these organisms. The role of this enzyme in other reactions involving hydrogen is not well understood. In recent experiments a low rate of dark hydrogen evolution mediated by reversible hydrogenase has been demonstrated (Hallenbeck et al., manuscript in preparation). The hydrogenase involved in photoreduction is apparently inducible. After four to six days of growth under an atmosphere of H_2 and N_2 hydrogen consumption activities reach maximal values of 1.2 to 1.5 μl of H_2 consumed/mg chl/h (Tel-Or et al., 1977). This enzyme could be induced by the severe reductant limitation in these cultures. These cultures show appreciable hydrogen stimulation of acetylene reduction, typical of old, reductant depleted cultures (Benemann and Weare, 1974a).

11.3.2. The enzymology of hydrogen metabolism in cyanobacteria

11.3.2.1. "Uptake hydrogenase"

The enzyme that catalyzes hydrogen consumption in cyanobacteria appears to be an "uptake" (unidirectional) hydrogenase with properties similar to the hydrogenase of *Azotobacter*. Nitrogen fixation is a relatively energy intensive process and hydrogen is evolved as a side reaction to the nitrogen fixation reaction with almost one third of the energy as ATP and reductant utilized by nitrogenase being lost (Anderson and Shanmugam, 1977). It has been suggested that some nitrogen-fixing organisms contain an uptake hydrogenase that functions to increase the efficiency of nitrogen fixation by scavenging some of the energy lost through hydrogen evolution (Dixon, 1972). In the heterocystous, nitrogen-fixing cyanobacteria, possession of a unidirectional hydrogenase enables them to catalyze the hydrogen consumption reactions typical of anaerobically adapted algae, and at the same time, probably increases the efficiency of nitrogen fixation in these organisms. The nature of the integration of this hydrogenase with nitrogenase is largely unknown. The existence of an oxy-hydrogen reaction is fairly well established, and this reaction could serve a dual role in nitrogen fixation: supplying energy for the nitrogen fixation process as ATP generated through respiration, and indirectly as an oxygen scavenging process. Evidence has been presented that this hydrogenase is also capable of recycling some of the energy lost as hydrogen by supplying nitrogenase with some form of reductant (Wolk and Wojiuch, 1971; Benemann and Weare, 1974a), although this reaction has not been demonstrated with isolated heterocysts (Peterson and Burris, 1978).

Like other "unidirectional" hydrogenases, hydrogen consumption activity in cyanobacteria is membrane bound and has a high affinity for hydrogen, being half saturated at about 4% H_2 (Tel-Or et al., 1978). This enzyme activity may be localized preferentially in the heterocyst and preparations of isolated heterocysts have been used to study some of its characteristics. Hydrogen uptake is stimulated in the dark by addition of exogenous electron acceptors, and is enhanced in the light without added acceptors, indicating that the reduction of endogenous compounds is taking place in this case. Only acceptors with positive E_0''s appear to function efficiently in this system, with methylene blue, DCPIP and ferricyanide supporting high rates of hydrogen consumption (Peterson and Burris, 1978). The rate with ferricyanide is six times the rate with methylene blue and three times the rate with DCPIP. Apparently this enzyme does not couple with NADP, as addition of NADP did not stimulate hydrogen uptake.

The metabolic pathways that act as electron acceptors for the uptake hydrogenase are not known. Studies with isolated heterocysts and various metabolic inhibitors have shown that DSPD, a ferredoxin antagonist, does not inhibit hydrogen consumption (Tel-Or et al., 1978). HOQNO and DBMIB were both inhibitory, with HOQNO inhibiting consumption up to 95%.

11.3.2.2. Reversible hydrogenase

In addition to the membrane bound enzyme described above the heterocystous cyanobacteria are known to possess a soluble hydrogenase (Fujita et al., 1964; Ward, 1970a). Whereas hydrogen consumption activity appears to be localized primarily in the heterocyst, this soluble, reversible hydrogenase activity appears to be about equally concentrated in both cell types. Further evidence that two distinct hydrogenases exist is provided by the differential sedimentation of hydrogen consumption and hydrogen production activities in fragmented, isolated heterocyst preparations (Tel-Or et al., 1978). The reversible hydrogenase is easily solubilized, evolves hydrogen from reduced methyl viologen, and catalyzes the reduction of PMS, methylene blue, and DCPIP with PMS being the most effective electron acceptor (Fujita et al., 1964). Ferricyanide did not support hydrogen consumption by this enzyme, and their results suggest that the redox coupling between the hydrogenase and ferredoxin from this organism has a very low efficiency. Neither NAD or NADP were reduced by this preparation. When extracts were prepared in a different manner, subcellular fractions could be isolated that were effective in catalyzing a DCPIP Hill reaction, NADP photoreduction, and NADP reduction by the hydrogenase reaction (Fujita and Myers, 1965). NADP photoreduction was not affected by the addition of ferredoxin which, however, stimulated hydrogenase mediated NADP reduction. This hydrogenase has been partially purified (Hallenbeck et al., in preparation).

11.3.2.3. Nitrogenase

The biochemistry and molecular characteristics of nitrogenase from bacterial sources (principally *Clostridia* and *Azotobacter*) have been extensively studied. Excellent recent reviews are available on this subject (Burns and Hardy, 1975; Zumft, 1976); only a brief description will be given here. Nitrogenase catalyzes the reduction of H^+, N_2, C_2H_2 and numerous other triply bonded substrates in a reaction which requires both ATP and reductant. Nitrogenase has invariably been found to consist of two components, a molybdenum, iron and sulfur containing protein and an iron—

sulfur containing protein, both of which are required for activity. In the presence of both components, ATP, and reductant, nitrogenase turns over at a fixed rate; hydrogen is evolved in the absence of other reducible substrates, or concomitantly with the reduction of some substrates. The ratio of ATP utilized per two electrons transfered is somewhat variable, depending on pH, temperature, and component ratio, but approaches four under optimal conditions.

The presence of nitrogenase in cell-free extracts of cyanobacteria was described in some early publications (Schneider et al., 1960; Cox and Fay, 1967) but the enzyme activity was not characterized until Haystead et al. (1970) showed its dependence on ATP, Mg^{2+} and reductant as well as its inactivation by oxygen. The nitrogenase from *Anabaena cylindrica* is soluble and, similarly to that of *Azotobacter* (Benemann et al., 1971), has been shown to be capable of being driven by ferredoxin reduced by a NADPH generating system and ferredoxin-NADPH reductase (Bothe, 1970). The nitrogenase activity in cell-free extracts of the non-heterocystous cyanobacterium, *Gloeocapsa*, has been shown to be similar to that of other cyanobacteria and bacteria (Gallon et al., 1972). In general, the cyanobacterial nitrogenase appears to be similar to the better characterized bacterial enzyme. The nitrogenase from *Anabaena cylindrica* is composed of the normal two components, Fe protein and Fe—Mo protein (Smith et al., 1971), contains active center iron and sulfide groups, and ferredoxin reduced photochemically, or through the action of a bacterial hydrogenase, has been shown to be a fairly effective reductant (Bothe, 1970; Smith and Evans, 1971; Haystead and Stewart, 1972).

11.3.3. Pathways of reductant and ATP flow to nitrogenase

As has already been described, hydrogen evolution by cyanobacteria is catalyzed mostly, if not entirely, by nitrogenase. Like nitrogen fixation, hydrogen evolution is a highly energy consumptive process since both reductant (high energy electrons) and ATP are required. At least six electrons and twelve ATP's are utilized for the conversion of one molecule of N_2 to ammonia, or two electrons and four ATP's for the production of one molecule of hydrogen. Thus in any organism nitrogenase must be well integrated with the energy yielding processes of cell metabolism. In addition, when nitrogenase is acting to fix nitrogen, both reductant and ATP are required for the assimilation of the newly fixed nitrogen (as are carbon skeletons).

11.3.3.1. Dark reactions

Nitrogenase activity in the dark is dependent upon the presence of oxygen, indicating an absolute dependence upon respiration to supply the required energy. In addition, both the rate and duration of dark activity is dependent upon the rate of photosynthesis and carbon assimilation during the preceeding light period (Fay, 1976) (an absolute requirement for photo-autotrophic cyanobacteria; some heterotrophs are able to maintain a low level of nitrogen fixation in the dark when supplied with a suitable carbon source). The respiratory system seems to be unresponsive to exogenous substrates (other than oxygen) and respiratory rates are low with a Q_{O_2} of about 5—10 (Webster and Frenkel, 1953). Oxidative phosphorylation has been found to be primarily associated with NADPH oxidation; rates with NADPH being at least two-fold higher than rates with NADH (Leach and Carr, 1970). Several reasons probably exist for the lack of response of the respiratory chain to exogenous substrates (some of which are known to be assimilated under some conditions). Effective utilization of the tricarboxylic acid cycle is blocked by a biochemical lesion (Fogg et al., 1973), and cyanobacteria in general show a lack of adaptability since many of their enzymes appear to be regulated by feedback inhibition and not by transcriptional control (Hood and Carr, 1970).

The cyanobacteria possess the enzymes of the reductive pentose phosphate pathway of sugar catabolism, with the majority of the activity being associated with the heterocyst (Winkenbach and Wolk, 1973). Although no connection of this pathway with nitrogenase activity has been demonstrated yet, the utilization of glucose-6-phosphate to produce large quantities of NADPH (12 NADPH/G-6-P) is possible, and NADPH has been found to be the primary source of reductant for nitrogenase in some bacteria (Benemann et al., 1971).

Pyruvate has been reported to stimulate nitrogen fixation in cell-free extracts of Anabaena cylindrica (Cox and Fay, 1967). This conclusion was based on an observed correlation of CO_2 release and $^{15}N_2$ incorporation upon the addition of pyruvate. However, this correlation was variable, as only a small amount of decarboxylation occurred under an argon atmosphere; conditions under which nitrogenase should continue to function and produce hydrogen. Several other researchers have been able to substantiate this claim (Smith and Evans, 1970; Haystead and Stewart, 1972). Evidence for the presence of pyruvate: ferredoxin oxidoreductase in Anabaena variabilis, has been presented (Leach and Carr, 1971). Whole cells decarboxylated pyruvate (from the 1st and 2nd carbon) and cell-free extracts formed acetyl-CoA and reduced ferredoxin from pyruvate; ATP was found to be required as an activator. The presence of this enzyme has also been described in nitrogen-fixing cultures of Anabaena cylindrica (Bothe and Falkenburg, 1973) and was reported to stimulate nitrogenase activity in low activity extracts prepared in the absence of dithionite (Codd et al., 1974). Ferredoxin, coenzyme A and ATP were all required for maximum pyruvate dependent activity.

Although similar in some respects to the pyruvate oxidase activity found in some anaerobic nitrogen-fixing bacteria, the enzyme system for the cyanobacteria appears to differ in some important aspects relevant to its ability to support nitrogenase activity in this organism. In organisms with a *Clostridial* type metabolism, utilization of pyruvate is coupled to ATP synthesis, a composite reaction termed phosphoroclastic cleavage of pyruvate. This reaction efficiently converts pyruvate into reductant and ATP, both required for nitrogenase activity. There is no evidence that pyruvate decarboxylation is coupled to ATP synthesis in the cyanobacteria. Instead, in vivo studies (Bennett et al., 1975) have shown that exogenous pyruvate is probably metabolized through the interrupted tricarboxylic acid cycle. It was found that pyruvate stimulated nitrogenase activity only in the dark under low pO_2. The requirement for the presence of oxygen for maximal stimulation suggests that in vivo pyruvate acts by supplying ATP through stimulated oxidative phosphorylation rather than by supplying reductant. This is further substantiated by the failure, in their experiments, of pyruvate to alleviate the effects of DCMU inhibition on nitrogenase activity in the light.

11.3.3.2. Light reactions

Light is required for maximal nitrogenase activity, indicating at least an indirect dependence on photosynthesis. Depending upon the amount of previously fixed carbon available ("reductant pools") nitrogenase activity is either inhibited or unaffected by DCMU at concentrations that completely inhibit carbon fixation (Lex and Stewart, 1973). A primary involvement of photosystem I in light driven acetylene reduction has been demonstrated using mono-chromatic light (Fay, 1970). This is consistent with the hypothesis that nitrogenase (at least in aerobically grown cultures) is localized in the heterocysts, specialized differentiated cells that appear unable to catalyze oxygen evolving water photolysis. However, these cells appear to have a fully active photosystem I as isolated heterocysts have been found to reduce acetylene in the light (Wolk and Wojiuch, 1971) and have been found to contain the necessary photosystem I components as well as most of the known components of the photosynthetic electron transport chain linking photosystem II and I (Tel-Or and Stewart, 1977).

Isolated heterocysts are capable of carrying out cyclic photophosphorylation (Tel-Or and Stewart, 1976), and it is widely assumed that the primary support of nitrogenase activity bia photoreactions in the heterocyst is the supply of ATP through the cyclic photophosphorylation process. However, since heterocysts contain active components of the interphotosystem electron transport chain (Tel-Or and Stewart, 1977), photosystem I in the heterocysts may also function to transfer electrons from organic substrates to nitrogenase with the concomitant synthesis of ATP non-cyclically. Evidence for this possibility is two-fold. It has been found that in cell-free extracts of a blue-green alga several organic acids (glycolate, malate, succinate and isocitrate) can serve as electron donors to photosystem I (Murai and Katoh, 1975). More important are the observations that mono-fluoroacetate, an inhibitor of the tricarboxylic acid cycle, inhibits acetylene reduction in the dark, and eliminates the light stimulation of nitrogenase activity in the presence of DCMU (Lex and Stewart, 1973).

As mentioned above, in some cases pyruvate has been said to reduce nitrogenase in a presumably ferredoxin linked reaction. Other studies have also implicated ferredoxin as the physiological electron donor to nitrogenase in these organisms. Ferredoxin reduced photochemically by chloroplasts or photosynthetically active *Anabaena* particles, or by NADPH has been found to drive acetylene reduction (Smith et al., 1971; Smith and Evans, 1971). *Anabaena* ferredoxin can also mediate electron transfer from hydrogen to nitrogenase via *Clostridial* hydrogenase (Haystead and Stewart, 1972). Phytoflavin, a flavodoxin from iron deficient cyanobacteria, has also been found to be an effective mediator of photochemical reduction of *Anabaena* nitrogenase. The primary involvement of a ferredoxin molecule in the supply of reductant and/or ATP to nitrogenase is substantiated by the inhibitory effects of DSPD (a ferredoxin antagonist) in vivo (Bothe and Loos, 1972; Tel-Or and Stewart, 1977).

Many nitrogen-fixing organisms have been found to contain more than one electron carrier capable of reducing nitrogenase. These organisms include *Azotobacter* which has a flavodoxin (azotoflavin) and several ferredoxins (Benemann et al., 1971; VanLin and Bothe, 1972), *Mycobacterium flavum* (Bothe and Yates, 1976), and the photosynthetic bacterium *Rhodospirilum rubrum* (Yoch and Arnon, 1975). In some cases one carrier has been found to be up to five times as effective in promoting acetylene reduction as the other carrier (see Yoch and Arnon, 1975). The cyanobacterium *Nostoc* has been reported to contain two plant-type ferredoxins (Hutson and Rojers, 1975). Thus, the possibility exists that nitrogenase containing cyanobacteria contain more than one ferredoxin, one of which may be more effective at transferring electrons to nitrogenase. This might help explain the finding that sometimes in vitro ferredoxin is not nearly as effective a reducing agent as dithionite.

Throughout this discussion we have alluded to the various mechanisms by which the cyanobacteria can provide nitrogenase with ATP. However, only a few studies have been done on this problem per se. With the development of a simple, sensitive and quantitative assay for extractable ATP in filamentous cyanobacteria, ATP pools have been measured under a variety of physiological conditions (Bottomley and Stewart, 1976a). Under most conditions ATP is maintained at a constant level by a dynamic balance between the rate of ATP utilization and ATP synthesis. Four general mechanisms of generating ATP have been described in cyanobacteria; total photophosphorylation, cyclic phosphorylation, oxidative phosphorylation and substrate level phosphorylation, with any of the first three mechanisms alone being able to maintain the ATP pool in the short-term (Bottomley and Stewart, 1976). With a change in environmental conditions, the cyanobacterium maintains the ATP pool by rapidly switching from one form of

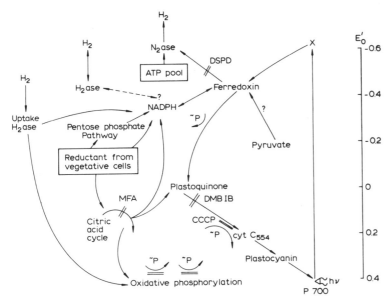

Fig. 11.3. Electron transport pathways in heterocysts of cyanobacteria. For simplicity, the ATP generated by the various reactions is shown as entering a common pool which supplies nitrogenase. However, evidence has been presented that the ATP pool is compartmentalized inside the cell, and some of the ATP may not be available to nitrogenase.

ATP synthesis to another. This switch-over is accompanied by a transient (several seconds) drop in the level of ATP. However, even though the total extractable ATP pool is maintained constant under different conditions, evidence has been presented which suggests that there is a compartmentalization of the ATP in the cell (Stewart and Bottomley, 1976). ATP generated by different phosphorylating mechanisms does not seem to be equally available to nitrogenase. In particular, as has often been noted, nitrogenase activity is lower when the ATP is provided by oxidative phosphorylation.

The possible mechanisms of ATP and reductant supply to nitrogenase are summarized in Fig. 11.3. Evidence for the presence in heterocysts of the components of the inter-photosystem electron transport chain, of ferredoxin and ferredoxin; NADPH oxidoreductase, and of cyclic photphosphorylation has been presented recently (Tel-Or and Stewart, 1977). Of course, all these postulated mechanisms presented in the figure are probably not operative under a single physiological state, and it is quite possible that different mechanisms are utilized under different metabolic conditions. Under conditions of nitrogen starvation, which have been shown to give the maximal transient and sustained rates of hydrogen evolution, stores of reduced carbon compounds are initially relatively high, sufficient to sustain hydrogen production for one to two days. After prolonged periods of nitrogen starvation and hydrogen production (two weeks), reductant stores are depleted as addition of DCMU abolishes hydrogen production within hours (Weissman and Benemann, 1977). This effect can also be seen in the increased oxygen sensitivity of older cultures. Photorespiration in the heterocysts may be considered as an oxygen protective mechanism, and increased oxygen sensitivity in these cultures can perhaps be attributed to competition between nitrogenase and photorespiration for a depleted reductant. Light levels sufficient to support the exponential growth of aerobic cultures does not saturate the increased capacity for nitrogen fixation of these cultures. Increasing the light intensity increases hydrogen production by stimulating photosystem I activity in the heterocysts, demonstrating that the nitrogenase activity of the cultures was probably initially APT limited (Weissman and Benemann, 1977).

11.4. ENGINEERING AND ECONOMIC FEASIBILITY OF BIOPHOTOLYSIS

11.4.1. Concepts of biophotolysis

The proposal to produce hydrogen from water using solar energy and a biological catalyst has been the major impetus behind most recent research in the hydrogen metabolism of algae, including that in the Soviet Union (Kondratieva, 1977; Evstigneev, 1977; Krasnovsky, 1976). On the basis of present knowledge of algal physiology, two basic concepts may be considered: two-stage systems and single-stage systems. In a two-stage system photosynthetic CO_2 fixation would lead to O_2 evolution and bring about the accumulation of stable reduced carbon compounds which would then be transfered by pumping to an anaerobic chamber where a hydrogen fermentation would take place. In single-stage systems hydrogen and oxygen would be released under a transparent covered collector either simultaneously or alternatively.

The key advantage of a two-stage system is that it does not require a H_2 collecting cover over the photosynthetic solar conversion area in which O_2 is released. However, this concept suffers from two basic drawbacks: the need to pump between the stages and the thermodynamic inefficiency of most hydrogen fermentations (Thauer, 1977). The first might be overcome by optimizing the engineering of a two-stage system. The drawbacks of the fermentation stage could be avoided by the development of either a specific intermediate that can be efficiently converted into hydrogen in the dark, or, preferably, a photofermentation stage. Photofermentations are capable of complete dissimilation of organic compounds and could be accomplished either through the light-driven evolution of hydrogen by anaerobically adapted algae, or, as has been suggested previously, by photosynthetic bacteria (Benemann, 1977). If suitable organic substrates were produced by an algal photosynthetic stage, they would be converted to hydrogen by the action of nitrogen starved photosynthetic bacteria in a second state that would require only about 1/10th the area of the first stage. Of course, a method must be devised to prevent the intermixing of the two stages; either filteration or immobilization of the microorganisms could be considered.

An alternative two-stage system is the cycling of an algal culture between the photosynthetic and photohydrogen-fermentative stages. Such a system would be based on the spatial separation of photosynthetic carbon fixation (and oxygen evolution) in the photosynthetic stage, and hydrogen evolution in a fermentation stage. The relative areas, depth, cell and hydraulic detention times, and gas phases would have to be adjusted such that in the second stage the algae can anaerobically adapt to hydrogen evolution and completely dissimilate the organic substrates accumulated intracellular in the first, photosynthetic stage. Of course, a great deal of operational control over cell detention times and relative light intensities would be necessary to achieve maximum production of hydrogen.

This concept uses presently known properties of anaerobically adapted algae; genetic engineering, which in the foreseeable future is of doubtful applicability to this area, is not required. Despite drawbacks because of operational costs of culture or reductant transfer, the concept of a two-stage biophotolysis system is attractive since hydrogen and oxygen are produced separately. This eliminates the requirement of O_2 and H_2 separation and allows storage of the H_2 gas. A two-stage system employing a single species of algae could readily be converted to a single-stage system by separating the photosynthetic and fermentative stages temporally rather than spatially. Cycling between photosynthesis and hydrogen production would coincide with the natural diurnal solar cycle. While possible in theory, this system has not yet been demonstrated in practice.

In the simplest conceivable biophotolysis concept system, the process of hydrogen and oxygen production proceed simultaneously in a single stage without temporal or spatial separation, thus avoiding the operational difficulties and costs involved in liquid transfers or timing of the adaptation cycle. However, with this system in order for the gas output to be useful, it must either be used immediately or the gases separated. More importantly, this approach necessitates the use of an oxygen stable biophotolysis catalyst. The oxygen stability and resistance of the hydrogen evolving reactions of a biophotolysis system is a key problem. The green algae cannot maintain a sustained simultaneous production of both oxygen and hydrogen. At present one oxygen stable biophotolysis catalyst has been demonstrated — the heterocystous nitrogen-fixing cyanobacteria — in which oxygen and hydrogen production are separated spatially on the microscopic level. Use of these organisms has already demonstrated a catalytic, sustained, stoichiometric production of O_2 and H_2 using sunlight in an outdoor test (Benemann et al., 1978; Hallenbeck et al., manuscript in preparation), extending

358

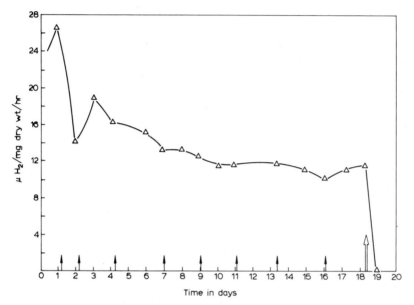

Fig. 11.4. Hydrogen production by 2-l cultures of *Anabaena cylindrica* maintained with ammonia addition. Culture density was initially 60 Klett units and ranged down to 50 during experiment. Chlorophyll *a* content: 5.6 μg/mg dry wt on day 0 and 6.7 μg/mg dry wt on day 1. Small arrows indicate addition of 1.0×10^{-4} M NH_4Cl. The large arrow (day 19) indicates addition of 2.1×10^{-5} M DCMU.

previous laboratory experiments (Benemann and Weare, 1974; Weissman and Benemann, 1977). The experiments involved sparging the cultures with a mixture of CO_2 (0.3%), N_2 (1%) and Argon (balance) in vertical glass tubes containing 1-liter of culture. The low N_2 concentrations maintained the algae in a healthy condition, substituting for the periodic additions of ammonia used in the laboratory experiments (Fig. 11.4). Hydrogen production rates followed the sunlight intensities, decreasing to zero after sunset.

11.4.2. General economic considerations

The economic constraints of biophotolysis can be readily calculated from incident sunlight intensities and assumed photosynthetic conversion effiencies from the equation:

H_2(l m^{-2} day^{-1}) = fraction of sunlight converted to H_2 × insolation

(K cal m^{-2} day^{-1}) × 0.47 K cal l^{-1} (energy content of H_2 at 290°K)

Considering the practical limitations of photosynthesis, a reasonable goal for a biophotolysis converter is operation at a 3% solar conversion efficiency. Also, in a fairly sunny location (e.g. a desert near the equator) a value of 5500 Kcal m^{-2} day^{-1} may be expected (yearly average) resulting in a production rate of 28 000 l H_2 m^{-2} y^{-1}. In an energy content basis of \$ 20/10^6 Kcal (approximately twice current oil price) this H_2 would be worth about \$ 1.20 m^{-2} y^{-1}. Therefore, very low costs collectors will be required. Flat plate solar hot water heaters, for example, achieve a 30 to 60% conversion efficiency, allowing a proportionally much higher capital investment, particularly since they require only negligible maintenance or operation cost. Assuming that a biophotolysis system will require about half of allowable costs for operations and maintenance, and making favorable assumptions about capital charges, a total capital investment of about \$ 5/m^2 would be allowable on the basis of the above calculations and assumptions. This is the key economic constraint on biophotolysis systems.

11.4.2.1. Bioengineering aspects of biophotolysis

As single-stage systems based on heterocystous blue-green algae are conceptually the simplest and experimentally the best developed they will be discussed first. The vertical glass tubes used in current experiments will have to be replaced by horizontal glass tubes, where gas flows over the surface of the algae culture. This minimizes the energy required to pump the gas phase through the system. Thin tubular glass is presently manufactured on a large scale for the fluorescent lamp industry and is available in various sizes, diameters and glass thickness for costs of about \$ 1.50/m^2. These tubes could be assembled in a close-packed array with terminal plastic connecting pipes for a cost as low as \$ 2.50/m^2. It appears feasible that with installation (on, for example, the roofs of light industrial buildings) and accessory blowers, pumps, holding tanks, etc., a total cost near \$ 5/m^2 may be possible. A conceptual representation of such a glass tube array is shown in Fig. 11.5.

One unknown in the operation of such a system is the required gas transfer rates of H_2 and O_2 out of, and CO_2 into, solution. If CO_2 is enriched tenfold more above atmospheric it appears that its passive diffusion rate will be high enough not to limit the rate of photosynthesis. The oxygen build-up in solution may present a problem, however, this would need to be determined experimentally. The required mixing-pumping energy to maintain the algae in suspension and healthy, as well as that required in two-stage systems, has not yet been analyzed.

Biophotolysis systems could provide hydrogen for storage and off-site use or fuel for immediate consumption. In the latter case it might not be necessary to separate hydrogen and oxygen as they may be recombined directly in an internal combustion system. For the first types of uses, and for more efficient use of hydrogen, a more or less pure hydrogen product is desireable. The key requirement is separating the H_2 from the O_2; some con-

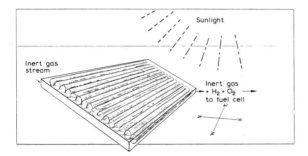

Fig. 11.5. Conceptual design of a glass tube biophotolysis converter.

tamination by CO_2 or N_2 may be tolerable. (CO_2 could be easily scrubbed where necessary). Such separations of H_2 and O_2 can be expensive, particularly considering the economic limitations under which a biophotolysis system must operate. Even relatively inexpensive methods of gas evaporation through molecular sieves or membranes are too high in cost for biophotolysis. Therefore, two-stage systems or alternating biophotolysis systems that produce H_2 and O_2 separately would be favored, although at present the single-stage heterocystous blue-green algae system producing a mixture of H_2 and O_2 appears to be the simplest to operate. The bioengineering aspects of biophotolysis obviously require considerable attention in the near future.

11.5. CONCLUSIONS

In conclusion, this review has demonstrated the strong scientific basis from which the development of applied algal hydrogen producing biophotolysis systems could be pursued. A number of different approaches were conceptualized and others may still be proposed. The bioengineering aspects of such solar converters have not yet been subjected to even a preliminary analysis; this should be an immediate aim of biophotolysis research. Without such an effort the real potential, and limitations, and thus research aims cannot be decided. The technical and economic constraints that must be met in the development of practical systems suggests the need for a relatively long-term high-risk research and development program which combines both basic and applied research.

11.6. ACKNOWLEDGEMENTS

This research was supported in part by the United States Department of Energy, Contract Ey-76-S-03-0034.

REFERENCES

Abeles, F. (1964) Plant Physiol., 39, 169—176.
Adams, M.W. and Hall, P.O. (1977) Biochem. Biophys. Res. Commun., 77, 730—737.
Andersen, K. and Shanmugam, K.T. (1977) J. Gen. Microbiol., 103, 107—122.
Badin, E.J. and Calvin, M. (1950) J. Am. Chem. Soc., 72, 5266—5270.
Ben-Amotz, A., Erbes, D.L., Riederer-Henderson, M.A., Peavey, D.G. and Gibbs, M. (1975) Plant Physiol., 56, 72—77.
Ben-Amotz, A. and Gibbs, M. (1975) Biochem. Biophys. Res. Commun., 64, 355—359.
Benemann, J.R., Yoch, D.C., Valentine, R.C. and Arnon, D.I. (1971) Biochim. Biophys. Acta, 226, 205—212.
Benemann, J.R., Berenson, J.A., Kaplan, N.D. and Kamen, M.D. (1973) Proc. Natl. Acad. Sci. U.S.A., 70, 2317—2320.
Benemann, J.R. and Weare, N.M. (1974) Arch. Microbiol., 101, 401—408.
Benemann, J.R. and Weare, N.M. (1974) Science, 184, 174—175.
Benemann, J.R. and Weissman, J.C. (1976) in Microbial Energy Conversion (H.G. Schlegel, J. Barnea, eds.), pp. 413—426, Erich Goltz K.G., Gottingen.
Benemann, J.R. (1977) in Living Systems as Energy Converters (R. Buvet et al., eds.), pp. 285—297, Elsevier North-Holland Biomedical Press, Amsterdam.
Benemann, J.R., Weissman, J.C., Koopman, B.L. and Oswald, W.J. (1977) Nature, 268, 19—23.
Benemann, J.R., Weissman, J.C., Koopman, B.L., Eisenberg, D.M., Murray, M. and Oswald, W.J. (1977) Species Control in Large-Scale Algal Biomass Production, SERL Report No. 77-5. University of California, Berkeley.
Benemann, J.R., Hallenbeck, P.C., Weissman, J.C., Kochian, L.V., Kostel, P.J. and Oswald, W.J. (1978) Solar Energy Conversion with Hydrogen Producing Algae, SERL Report No. 78-2. University of California, Berkeley.
Benemann, J.R., Hallenbeck, P.C., Miyamoto, K., Kochian, L. and Kostel, P. (1978) Solar Energy Conversion Through Biophotolysis, SERL Report No. 78-6. University of California, Berkeley.
Benemann, J.R. (1978) Biofuels: A Survey. Electric Power Research Institute, Palo Alto.
Bennett, K.J., Silvester, W.B. and Brown, J.M.A. (1975) Arch. Microbiol., 105, 61—66.
Bishop, N. and Gaffron, H. (1962) Biochem. Biophys. Res. Commun., 8, 471—476.
Bishop, N.I. (1966) Annu. Rev. Plant Physiol., 17, 185—208.
Bishop, N.I., Frick, M. and Jones, L. (1977) in Biological Solar Energy Conversion (A. Mitsui et al., eds.), pp. 3—22, Academic Press, New York.
Bothe, H. (1970) Ber. Dtsch. Bot. Ges., 83, 421—432.
Bothe, H. and Loos, E. (1972) Arch. Microbiol., 86, 241—254.
Bothe, H. and Falkenberg, B. (1973) Plant Sci. Lett., 1, 151—156.
Bothe, H. and Yates, M.G. (1976) Arch. Microbiol., 107, 25—31.
Bothe, H., Tennigkeit, J. and Eisbrenner, G. (1977) Arch. Microbiol., 114, 43—49.
Bottomley, P.J. and Stewart, W.D.P. (1976a) Br. Phycol. J., 11, 69—82.
Bottomley, P.G. and Stewart, W.D.P. (1976b) Arch. Microbiol., 108, 249—258.
Brockris, T.O.M. (1975) The Solar Hydrogen Alternative. Academic Press, New York.
Burns, R.C. and Hardy, R.W.F. (1975) Nitrogen Fixation in Bacteria and Higher Plants. Springer Verlag, New York.
Cammack, R., Tel-Or, E. and Stewart, W.D.P. (1976) FEBS Lett., 70, 241—244.
Codd, G.A., Rowell, P. and Stewart, W.D.P. (1974) Biochem. Biophys. Res. Commun., 61, 424—431.
Cox, G.R. and Fay, P. (1967) Arch. Microbiol., 58, 357—365.
Daday, A., Platz, R.A. and Smith, G.D. (1977) Appl. Environ. Microbiol., 34, 478—483.
Diner, B. and Mauzerall, D. (1973) Biochim. Biophys. Acta, 305, 329—352.
Dixon, R.O.D. (1972) Arch. Microbiol., 85, 193—201.

362

Donze, M., Haveman, J. and Schiereck, P. (1972) Biochim. Biophys. Acta, 256, 157—161.
Evstigneev, V.B. (1977) in Living Systems as Energy Converters (R. Buvet, M.J. Allen, J.P. Massué, eds.), pp. 275—284, North Holland Publishing Co., Amsterdam.
Fay, P. (1970) Biochim. Biophys. Acta, 216, 353—356.
Fay, P. and Kulasooriya, S.A. (1972) Arch. Microbiol., 87, 341—352.
Fay, P. (1976) Appl. Environ. Microbiol., 31, 376—379.
Fogg, G.E., Stewart, W.D.P., Fay, P. and Walsby, A.E. (1973) The Blue-Green Algae, pp. 171—174. Academic Press, New York.
Frenkel, A. (1949) Biol. Bull., 97, 261—262.
Frenkel, A., Gaffron, H. and Battley, E.H. (1949) Biol. Bull., 97, 269.
Frenkel, A. and Rieger, L. (1951) Nature, 167, 1030.
Frenkel, A. (1952) Arch. Biochem. Biophys., 38, 219—230.
Fujita, Y., Ohama, H. and Hattori, A. (1964) Plant Cell Physiol., 5, 305—314.
Fujita, Y. and Myers, J. (1965) Arch. Biochem. Biophys., 111, 619—625.
Gaffron, H. (1940) Am. J. Bot., 27, 273—283.
Gaffron, H. and Rubin, J. (1942) J. Gen. Physiol., 26, 219—240.
Gallon, J.R., La Rue, J.A. and Kurz, W.G.W. (1972) Can. J. Microbiol., 18, 327—332.
Gingras, G., Goldsby, R. and Calvin, M. (1963) Arch. Biochem. Biophys., 100, 178—184.
Gitlitz, P.H. and Krasna, A.I. (1975) Biochemistry, 14, 2561—2568.
Greenbaum, E. and Mauzerall, D.C. (1976) Photochem. Photobiol., 23, 369—372.
Greenbaum, E. (1977a) Photochem. Photobiol., 25, 213.
Greenbaum, E. (1977b) Science, 196, 879—878.
Gingras, G., Goldsby, R. and Calvin, M. (1963) Arch. Biochem. Biophys., 100, 178—184.
Grimme, L.H. (1972) Proc. 2nd Int. Congr. Photosyn. Res., pp. 2011—2019. Stresa.
Hartman, H. and Krasna, A. (1963) J. Biol. Chem., 238, 749—757.
Hartman, H. and Krasna, A.I. (1964) Biochim. Biophys. Acta, 92, 52—58.
Haystead, A., Robinson, R. and Stewart, W.D.P. (1970) Arch. Microbiol., 74, 235—243.
Haystead, A. and Stewart, W.D.P. (1972) Arch. Microbiol., 82, 325—336.
Healey, F.P. (1970a) Plant Physiol., 45, 153—159.
Healey, F.P. (1970b) Planta, 91, 220—226.
Hood, W. and Carr, N.G. (1970) J. Gen. Microbiol., 73, 417—426.
Horwitz, L. and Allen, F.L. (1957) Arch. Biochem. Biophys., 66, 45—63.
Hutson, K. and Rojers, L.J. (1975) Biochem. Soc. Trans., 3, 377—379.
Hyndman, L.A., Burris, R.H. and Wilson, P.W. (1953) J.Bacteriol., 65, 522—531.
Jackson, D.P. and Ellms, J.W. (1896) On Odors and Tastes of Surface Waters. Rep. Mass. State Board of Health.
Jones, L. and Bishop, N. (1976) Plant Physiol., 57, 659—665.
Kaltwasser, H., Stuart, J.S. and Gaffron, H. (1969) Planta, 89, 309—322.
Kessler, E. (1956) Arch. Biochem. Biophys., 62, 241—242.
Kessler, E. (1968) Arch. Mikrobiol., 63, 7—10.
Kessler, E. (1970) Planta, 92, 222—234.
Kessler, E. (1973) Arch. Mikrobiol., 93, 91—100.
Kessler, E. (1974) in Algal Physiology and Biochemistry (W.D.P. Steward, ed.), p. 456. Blackwell Scientific Publ. Ltd., London.
King, D., Erbes, D., Ben-Amotz, A. and Gibbs, M. (1977) in Biological Solar Energy Conversion (A. Mitsui et al., eds.), pp. 69—75. Academic Press, New York.
Kok, B., Forbush, B. and McGloin, M.P. (1970) Photochem. Photobiol., 11, 457.
Kondratieva, E.N. (1976) in Microbial Energy Conversion (H.G. Schlegel, J. Barnea, eds.), pp. 205—216. Erich Goltz K.G., Göttingen.
Krasna, A.I. (1977) in Biological Solar Energy Conversion (A. Mitsui et al., eds.), pp. 53—60. Academic Press, New York.
Krasnovsky, A.A. (1976) in Research in Photobiology (A. Castellani, ed.), Plenum Press, New York and London.

Lambert, F.R. and Smith, D.G. (1977) FEBS Lett., 83, 159—162.

Leach, C.K. (1970) J. Gen. Microbiol., 64, 55—70.

Leach, C.K. (1971) Biochim. Biophys. Acta, 245, 165—174.

Lee, J. and Stiller, M. (1967) Biochim. Biophys. Acta, 132, 503—505.

Lex, M. and Stewart, W.D.P. (1973) Biochim. Biophys. Acta, 292, 436—443.

Mitsui, A. and Kumazawa, S. (1977) in Biological Solar Energy Conversion (A. Mitsui et al., eds.), pp. 23—51. Academic Press, New York.

Murai, T. and Katoh, T. (1975) Plant Cell Physiol., 16, 789—797.

Oesterheld, H. (1971) Arch. Mikrobiol., 79, 25—43.

Oswald, W.J. and Golueke, C.G. (1960) Adv. Appl. Microbiol., 2, 223—262.

Peters, G.A., Evans, W.R. and Toia, R.E. (1976) Plant Physiol., 58, 119—126.

Peterson, R.B. and Burris, R.H. (1976) Arch. Microbiol., 108, 35—40.

Peterson, R.B. and Burris, R.H. (1978) Arch. Microbiol., 116, 125—132.

Pratt, L.H. and Bishop, N.I. (1968) Biochim. Biophys. Acta, 153, 664—674.

Russell, G.K. and Gibbs, M. (1968) Plant Physiol., 43, 649—652.

Schneider, K.C., Bradbeer, C., Singh, R.N., Wang, C.C., Wilson, B.W. and Burris, R.H. (1960) Proc. Natl. Acad. Sci. U.S.A., 46, 726—733.

Smith, L.A., Hill, S. and Yates, M.G. (1976) Nature, 262, 209—210.

Smith, R.V. and Evans, M.C.W. (1970) Nature, 225, 1253—1254.

Smith, R.V. and Evans, M.C.W. (1971) J. Bacteriol., 105, 913—917.

Smith, R.V., Telfer, A. and Evans, M.C.W. (1971) J. Bacteriol., 107, 574—575.

Spruit, C.J.P. (1962) in Physiology and Biochemistry of Algae (R.A. Lewin, ed.), pp. 47—60. Academic Press, New York and London.

Stewart, W.D.P., Haystead, A. and Pearson, H.W. (1969) Nature, 224, 226—228.

Stewart, W.D.P. and Bottomley, P.J. (1976) in International Symposium on Nitrogen Fixation (W.E. Newton, C.J. Nyman, eds.), Vol. 1, pp. 257—273, Washington State Univ. Press.

Stiller, M. and Lee, J.K.-H. (1964a) Biochim. Biophys. Acta, 93, 174—176.

Stiller, M. and Lee, J.K.-H. (1964b) Plant Physiol., 39, xv.

Stiller, M. (1966) Plant Physiol., 41, 348—352.

Stuart, T.S. and Kaltwasser, H. (1970) Planta, 91, 302—313.

Stuart, T.S. (1971) Planta, 96, 81—92.

Stuart, T.S. and Gaffron, H. (1971) Planta, 100, 228—243.

Stuart, T.S. and Gaffron, H. (1972a) Planta, 106, 101—112.

Stuart, T.S. and Gaffron, H. (1972b) Plant Physiol., 50, 136—140.

Tel-Or, E. and Stewart, W.D.P. (1976) Biochim. Biophys. Acta, 423, 189—195.

Tel-Or, E. and Stewart, W.D.P. (1977) Proc. R. Soc. London, Ser. B, 198, 61—83.

Tel-Or, E., Luijk, L.W. and Packer, L. (1977) FEBS Lett., 78, 49—52.

Tel-Or, E., Luijk, L.W. and Packer, L. (1978) Arch. Biochem. Biophys., 185, 185—194.

Thauer, R. (1976) in Microbial Energy Conversion (H.G. Schlegel, J. Barnea, eds.), pp. 201—204. Erich Goltz K.G., Göttingen.

Thomas, J., Meeks, J.C., Wolk, C.P., Shaffer, P.W., Austin, S.M. and Chien, W.S. (1977) J. Bacteriol., 129, 1545—1555.

Thomas, J. (1970) Nature, 228, 181—183.

Trebst, A. and Burba, M. (1967) Z. Pflanzenphysiol., 57, 419—433.

Van Lin, B. and Bothe, H. (1972) Arch. Microbiol., 82, 155—172.

Ward, M.A. (1970) Phytochemistry, 9, 259—266.

Weare, N.M. and Benemann, J.R. (1972) Arch. Microbiol., 90, 323—332.

Weare, N.M. and Benemann, J.R. (1973) Arch. Microbiol., 93, 101—112.

Weare, N.M. and Benemann, J.R. (1974) J. Bacteriol., 119, 258—265.

Webster, G.C. and Frenkel, A.W. (1953) Plant Physiol., 28, 63—69.

Weissman, J.C. and Benemann, J.R. (1977) Appl. Environ. Microbiol., 33, 123—131.

Winkenbach, F. and Wolk, C.P. (1973) Plant Physiol., 52, 480—483.

Wolk, C.P. (1968) J. Bacteriol., 96, 2138—2143.
Wolk, C.P. and Simon, R. (1969) Planta, 86, 92—97.
Wolk, C.P. and Wojiuch, E. (1971) Planta, 97, 126—134.
Yanagi, S. and Sasa, T. (1966) Plant Cell Physiol., 7, 593—598.
Yoch, D. and Arnon, D. (1975) J. Bacteriol., 121, 743—745.
Zumft, W.G. (1976) Struct. Bonding (Berlin), 29, 1—65.

Photosynthesis in relation to model systems, edited by J. Barber
© Elsevier/North-Holland Biomedical Press 1979

Chapter 12

The Photosynthetic Reduction of Nitrate and its Regulation

MANUEL LOSADA and MIGUEL G. GUERRERO

Departamento de Bioquímica, Facultad de Ciencias y C.S.I.C., Universidad de Sevilla, Spain

CONTENTS

12.1. INTRODUCTION

Photosynthesis consists essentially (Whatley and Losada, 1964; Losada and Arnon, 1964; Losada, 1976a, 1977, 1978b,c; Arnon, 1977) in the transduction of radiant energy at the level of visible light into electronic energy and subsequently into redox energy. Alternatively photosynthesis can be expressed in other terms, that is, it is the endergonic synthesis, at the expense of sunlight energy, of unstable products, namely cell material and molecular oxygen, from fully oxidized substrates with no useful chemical potential, namely water, carbon dioxide, nitrate, sulfate and phosphate. In the process, carbon is reduced from carbon dioxide (oxidation state +4) to carbohydrate (oxidation state 0), nitrogen from nitrate (+5) to ammonia (−3), and sulfur from sulfate (+6) to sulfide (−2), whereas oxygen is stoichiometrically oxidized from water (−2) to molecular oxygen (0). By contrast, phosphorus does not change its valence, but is only converted from inorganic phosphate to high-energy phosphate. The overall reaction taking place can accordingly be represented by the following equation:

$$H_2O + aCO_2 + bNO_3^- + cSO_4^{2-} + dPO_4^{3-} \xrightarrow{\text{light}} \text{cell material} + \tfrac{1}{2} O_2$$

The redox energy stored in cell material and molecular oxygen can be further transformed during respiration into phosphate-bond energy, the biogenic elements becoming thus again completely oxidized.

Before going into details regarding the photosynthetic reduction of nitrate, let us first analyze, in its more fundamental aspects, the overall cycle closed by these two energy-transducing processes from a potentiometric point of view (Losada, 1977, 1978b,c; Relimpio et al., 1977).

Only green plants among living organisms are capable of splitting water into hydrogen and oxygen during the light phase of photosynthesis (Van Niel, 1941; Hill, 1965; Rao et al., 1976). Since life on our planet depends energetically on this photochemical reaction, the photolysis of water can be considered as the simplest and primordial endergonic reaction of bioenergetics. Equally simple and essential for life is the opposite exergonic reaction that aerobic organisms carry out in the dark during respiration in order to transform the energy stored in hydrogen and oxygen into energy-rich phosphate bonds. As shown in Fig. 12.1, water is thus the primary substrate of green plant photosynthesis for the conversion of light energy into redox energy, and the final product of the aerobic respiration of hydrogen for the conversion of redox energy into phosphate-bond energy. Photoergonic and chemoergonic organisms carry out in this way what can be regarded as the basic cycle of bioenergetics (Losada, 1976a, 1978a,b,c) spanning back and forth over 1.24 V between the potential of the hydrogen electrode (E_0', pH 7, −0.42 V) and the potential of the pair H_2O/O_2 (E_0', pH 7, +0.82 V).

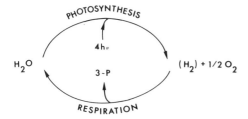

Fig. 12.1. The basic cycle of bioenergetics. First, light energy is converted into redox energy by the photosynthetic splitting of water into hydrogen and oxygen, and second, redox energy is converted into pyrophosphate-bond energy by the aerobic respiration of hydrogen.

Since four einsteins of red light of 680 nm — a wavelength that can still promote efficiently photosynthesis — are required ($\Delta G_0' = -4 \times 42 = -168$ kcal) to split one mol of water into hydrogen and oxygen ($\Delta G_0' = 57$ kcal), and the pertinent redox reaction is also accompanied by the formation of at least one mol of high-energy phosphate bond ($\Delta G_0' = 7.5$ kcal), the energy-yield of this primary reaction of photosynthesis is 64.5/168, i.e. about 40%. On the other hand, the oxidation of one mol of hydrogen by oxygen during respiration ($\Delta G_0' = -57$ kcal) is coupled with the production of three mol of high-energy phosphate bonds ($\Delta G_0' = 3 \times 7.5 = 22.5$ kcal), which means that the energy-yield of this primary reaction of respiration is 22.5/57, i.e. about 40%. As a consequence, chemoergonic organisms benefit indirectly, through photoergonic organisms, from the radiant energy of the sun with a yield that cannot exceed the value of 16% (Losada, 1977; Relimpio et al., 1977).

Nevertheless, neither does photosynthesis come to an end with the photolysis of water into electrons at the potential level of the hydrogen electrode and molecular oxygen, nor does respiration always start with the oxidation of electrons at such a level. Actually, the reducing power supplied by the photo-decomposition of water is not accumulated by the green cell as molecular hydrogen, but is transferred through suitable electron carriers (ferredoxin and pyridine nucleotides) and incorporated into cell material (carbohydrates, hydrocarbons, lipids, proteins, nucleic acids, etc.) by reduction of the oxidized bioelements during the dark phase of photosynthesis (Arnon, 1977; Losada, 1978a,b,c).

Only in the case of nitrate-nitrogen, is the photosynthetically generated reducing power at the level of ferredoxin or pyridine nucleotide energetic enough to promote reduction. In all the other cases, the reduction of the oxidized bioelements cannot proceed unless the reducing power is fortified with adenosine triphosphate (simultaneously formed by transduction of part of the photosynthetically generated redox energy into phosphate-bond energy (Arnon, 1977) during the processes of cyclic and non-cyclic phosphorylation), as summarized by the following overall equations:

$$HNO_3 + 4 (H_2) \rightarrow NH_3 + 3 H_2O$$

$$CO_2 + 2 (H_2) \xrightarrow{3 \sim P} (CH_2O) + H_2O$$

$$H_2SO_4 + 4 (H_2) \xrightarrow{3 \sim P} H_2S + 4 H_2O$$

These equations show also that most of the energy stored in the reduced bioelements is supplied via redox energy (100% in the case of ammonia and about 84% in the case of sugar) and only a very small fraction via phosphate-bond energy (Losada, 1978a,b,c).

As pointed out recently by Losada (1978a,b,c), ferredoxin or nicotinamide adenine dinucleotides do not constitute the most negative final potential stages for the electrons raised up from water during the light phase of photosynthesis, since these electrons are pushed further up to more negative potentials at the expense of phosphate-bond energy during the assimilation of carbon and sulfur reaching potential as low as —0.7 V (Fig. 12.2).

During the catabolic reactions of the carbon-dissimilatory pathways (glycolysis, pyruvate dehydrogenation and Krebs cycle) and respiration, electrons drop first to redox coenzymes, especially pyridine nucleotide, and then stepwise down to oxygen. This sequence of reactions close the cycle and allow the transformation of redox energy into phosphate-bond energy through phosphorylation at the substrate and membrane levels, respectively. Since the dehydrogenation of certain reduced carbon compounds by oxidized pyridine nucleotide during the preparatory phase of respiration is exergonic enough, it can be coupled (when the potential span amounts up to the value of 0.2 to 0.4 V) to phosphorylation. Although the dehydrogenation of reduced inorganic sulfur compounds can also conserve some of the free energy as ATP, via substrate level phosphorylation, the dehydrogenation of reduced inorganic nitrogen compounds supplies electrons at higher potential levels and can exclusively yield phosphorylation at the membrane level (Losada, 1978, in press).

It should be stressed against the proposal of several distinguished authorities in bioenergetics (Wald, 1966; Racker, 1976) that water is only a source of reducing power in the light phase of photosynthesis but not in respiration (Losada, 1978a,b,c).

Overall equations for these processes are paradoxically misleading and should be systematically avoided. Let us consider, for example, the simplified equations for the photosynthetic reduction of carbon, nitrogen and sulfur:

$$CO_2 + H_2O \xrightarrow{light} (CH_2O) + O_2$$

$$HNO_3 + H_2O \xrightarrow{\text{light}} NH_3 + 2\,O_2$$

$$H_2SO_4 \xrightarrow{\text{light}} H_2S + 2\,O_2$$

Even if the number of electrons transferred in each case is taken into consideration, the overall resulting equations are not yet complete and correct

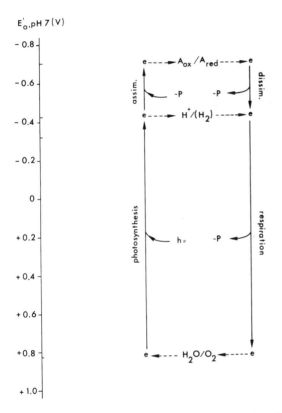

Fig. 12.2 .The electron cycle of bioenergetics. First, electrons are moved uphill from water to the potential level of the hydrogen electrode during the light phase of photosynthesis at the expense of light energy, and further up to the level of intermediate redox metabolites during the assimilatory reduction of the biogenic elements (specially carbon) at the expense of chemical-bond energy. Second, electrons fall downhill from the level of intermediate redox metabolites during the dissimilatory anaerobic oxidation of the reduced bioelements to the level of redox coenzymes (specially pyridine nucleotides), and subsequently further down in cascade to the level of oxygen during the aerobic phase of respiration. Redox energy is transduced into chemical-bond energy through coupled phosphorylation at the substrate and membrane level, respectively, during these exergonic processes.

enough:

$$CO_2 + 2\,H_2O \xrightarrow{\text{light}} (CH_2O) + H_2O + O_2$$

$$HNO_3 + 4\,H_2O \xrightarrow{\text{light}} NH_3 + 3\,H_2O + 2\,O_2$$

$$H_2SO_4 + 4\,H_2O \xrightarrow{\text{light}} SH_2 + 4\,H_2O + 2\,O_2$$

Actually, since the partial redox reactions (photooxidation of water and reduction of the bioelements) take place on each side of the thylakoid membrane, the generation of a significant proton gradient is always concomitant with the photosynthetic reduction of a natural Hill reagent (Fig. 12.3). To say it in another way, the photosynthetic reduction of carbon in chloroplasts takes place according to the following partial and total equations, which have been simplified (as noted by the omission of cyclic and noncyclic photophosphorylation) in order to underline the compulsory liberation and fixation of protons that occur in addition to and as a consequence of the transduction of light into redox energy across the membrane:

intrathylakoid space (water oxidation):

$$12\,H_2O \xrightarrow{48\ h\nu} 24\,H^+ + 6\,O_2 + 24\,e$$

extrathylakoid space (carbon dioxide reduction):

$$6\,CO_2 + 24\,H_2O + 24\,e \xrightarrow{18\ \sim\ P} (C_6H_{12}O_6) + 6\,H_2O + 24\,HO^-$$

_____

overall redox reaction (carbon dioxide reduction by water):

$$6\,CO_2 + 36\,H_2O \xrightarrow{\text{light}} (C_6H_{12}O_6) + 6\,H_2O + 6\,O_2 + 24\,H^+ + 24\,HO^-$$

and, likewise, the photosynthetic reduction of nitrogen:

$$4\,H_2O \xrightarrow{16\ h\nu} 8\,H^+ + 2\,O_2 + 8\,e$$

$$NO_3^- + 10\,H_2O + 8\,e \rightarrow NH_4^+ + 3\,H_2O + 10\,HO^-$$

_____

$$NO_3^- + 14\,H_2O \xrightarrow{16\ h\nu} NH_4^+ + 3\,H_2O + 2\,O_2 + 8\,H^+ + 10\,HO^-$$

and the photosynthetic reduction of sulfur:

$$4\,H_2O \xrightarrow{16\ h\nu} 8\,H^+ + 2\,O_2 + 8\,e$$

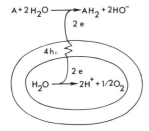

Overall:

$$A + 3H_2O \xrightarrow{4h\nu} AH_2 + 1/2\,O_2 + 2H^+ + 2HO^-$$

Fig. 12.3. Schematic representation of the photooxidation of water (intrathylakoid space) coupled to the reduction of a bioelement (extrathylakoid space), taking also into account the simultaneous liberation of 1 proton and 1 hydroxyl ion, respectively, on each side of the membrane per electron transferred.

$$SO_4^{2-} + 10\,H_2O + 8\,e \xrightarrow{3 \sim P} H_2S + 4\,H_2O + 10\,HO^-$$

$$SO_4^{2-} + 14\,H_2O \xrightarrow{light} H_2S + 4\,H_2O + 2\,O_2 + 8\,H^+ + 10\,HO^-$$

As a consequence, for each molecule of water that is photolyzed for the reduction of the biogenic elements, at least two are dissociated into hydrogen and hydroxyl ions (Losada, 1978a,b,c).

12.1.1. Historical background

In 1920, Warburg and Negelein made the fundamental discovery that living *Chlorella* cells suspended in a nitrate-containing solution, to which no CO_2 was added, evolved O_2 in the light with the concomitant reduction of nitrate to ammonia, according to the equation:

$$HNO_3 + H_2O \xrightarrow{light} NH_3 + 2\,O_2$$

Nevertheless, since the alga could also reduce nitrate in the dark with the simultaneous oxidation of some organic substrate to CO_2, they interpreted the photoreduction of nitrate on the basis of a dark reaction, which yielded CO_2, and a photolysis of CO_2, which yielded O_2. In other words, the photochemical reduction of nitrate coupled to the evolution of O_2 was visualized to result from the two following partial reactions: (1) a dark reaction in which the carbohydrates of the cells were oxidized to CO_2 by nitrate, and (2) a light reaction in which CO_2 was subsequently converted to carbohydrate and O_2:

$$HNO_3 + 2\,C + H_2O \xrightarrow{dark} NH_3 + 2\,CO_2$$

$$2\,CO_2 \xrightarrow{light} 2\,C + 2\,O_2$$

$$HNO_3 + H_2O \xrightarrow{light} NH_3 + 2\,O_2$$

Warburg's conclusion (1964) was that the balanced Hill reaction written above was deceptive, in the sense that what appears to be a photolysis of H_2O is in fact due to the photolysis of CO_2. Warburg pointed out that nitrate was the first Hill reagent discovered and extended his reasoning to all the Hill reactions found afterwards, even to the photoreduction of nicotinamide adenine dinucleotide phosphate.

In discussing the light-dependent production of O_2 by living *Chlorella* in the presence of nitrate, Van Niel (1941) and Rabinowitch (1945) offered an alternate explanation to the one advanced by Warburg and interpreted the reduction of nitrate in the light as a reaction essentially analogous to the photochemical reduction of CO_2, with nitrate replacing CO_2 as the final hydrogen acceptor. Thus, according to Van Niel (1941) light energy would be used in photosynthesis to decompose water, and the hydrogen released would subsequently reduce either CO_2 or nitrate by dark enzymic reactions. Although Warburg himself (1948) accepted earlier this interpretation as a possible alternative to his original hypothesis, he repeated in 1965 his classical experiment with more refined techniques, but, having reached identical results as previously, became even more committed to his early interpretation (Warburg et al., 1965).

With respect to the metabolic pathway leading from nitrate to ammonia, biochemists traditionally accepted (mainly on the basis of chemical considerations (Meyer and Schulze, 1894) but with insufficient evidence and without much criticism) that the biological reduction of nitrate to ammonia should proceed in four consecutive steps, each one involving two electrons, according to the following sequence of reactions:

$$HNO_3 \xrightarrow{2e} HNO_2 \xrightarrow{2e} (HNO) \xrightarrow{2e} NH_2OH \xrightarrow{2e} NH_3$$

$$(+5) \qquad (+3) \qquad (+1) \qquad (-1) \qquad (-3)$$

Thus, nitrite, hyponitrite and hydroxylamine, and their corresponding reductases, were (Nason, 1962; Hewitt and Nicholas, 1964; Kessler, 1964) and are still cited by some authors as intermediates in the reductive pathway from nitrate to ammonia.

In the last few years, however, the assimilatory nitrate-reducing system of a great variety of organisms has been more critically and intensively investigated at the cellular, subcellular and molecular levels in different laboratories. The results have firmly established (Losada, 1976b; Hewitt et al., 1976; Vennesland and Guerrero, in press) that this enzyme system consists only of two metalloproteins, nitrate reductase and nitrite reductase, which use ferredoxin or pyridine nucleotide as electron donor to catalyze the stepwise reduction of nitrate to nitrite and to ammonia, through a sequence of reactions unexpectedly much shorter than the one first postulated:

$$NO_3^- \overset{2e}{\Rightarrow} NO_2^- \overset{6e}{\Rightarrow} NH_4^+$$

In spite of these findings, scientists, predisposed by historical and quantitative reasons and, paradoxically, by Warburg's own experiments, have habitually accepted that the reduction of nitrate to ammonia in green cells is a process far removed from the light reactions of photosynthesis and dependent on carbohydrates as the source of reducing power. Now it seems, however, that, not only in blue-green algae but also in eukaryotic algae and higher plants, the photosynthetic reduction of nitrate is even more direct and simpler than that of carbon dioxide, since it depends partially or totally on ferredoxin as the electron donor and does not require adenosine triphosphate. Furthermore, the two enzymes involved in the reduction of nitrate are, at least in the prokaryotic algae, tightly bound to the pigment-containing particles (Losada, 1976b; Candau et al., 1976; Ortega et al., 1976).

The ferredoxin-dependent photosynthetic reduction of nitrite to ammonia coupled to oxygen evolution by the chloroplast enzyme nitrite reductase was first reported by Losada et al. in 1963, and shown afterwards to be of universal occurrence both in eukaryotic and prokaryotic organisms (Losada, 1976b; Hewitt et al., 1976; Vennesland and Guerrero, in press).

The ferredoxin-dependent photosynthetic reduction of nitrate to nitrite coupled likewise to oxygen evolution has only recently been achieved with pigment-containing particles of the blue-green algae *Anacystis nidulans* and *Nostoc muscorum* (Losada, 1976b; Candau et al., 1976; Ortega et al., 1976). Since nitrate reductase from eukaryotes is a NAD(P)H-dependent enzyme and its location in the cell (i.e. as a chloroplast or cytoplasmic enzyme) is yet a matter of controversy (see below), the first step in the photosynthetic reduction of nitrate in these organisms seems to be more complex and indirect than in prokaryotes (Losada, 1976b; Hewitt et al., 1976; Vennesland and Guerrero, in press). In any case, as diagrammatically represented in Fig. 12.4, reduced ferredoxin, acting by itself or indirectly through pyridine nucleotide, is always the primary electron donor for the photosynthetic reduction of nitrate to ammonia with water as the ultimate source of electrons.

According to Losada's recent proposal (Losada, 1978a,b,c), this scheme takes into account the repeated transduction of light energy into redox energy by reaction center chlorophyll *a* of photosystems I and II,

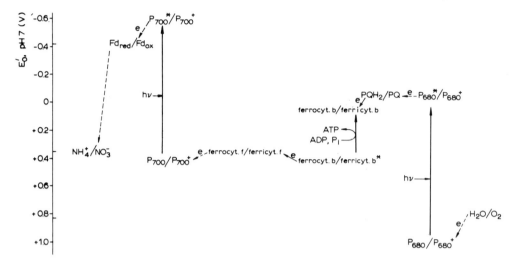

Fig. 12.4. Diagrammatic representation of non-cyclic photophosphorylation with water as the initial electron donor and nitrate as the terminal electron acceptor.

during the electron flow from water to ferredoxin and nitrate. The overall span includes also a short downhill stretch between the two photosystems, which allows the additional transduction of redox energy into phosphate-bond energy by cytochrome b-559. Chlorophyll a and cytochrome b-559 correspond, respectively, to the energy-transducing redox systems of the energized reduced-form type (which decreases its potential upon activation) and of the energized oxidized-form type (which increases its potential upon activation) (Losada, 1978a).

It should finally be considered (Cramer and Myers, 1949; Losada, 1977) that in photosynthetically growing microorganisms which contain about 10% nitrogen versus 50% carbon per dry weight, the proportion of electrons going to nitrate is only 3 times less than that going to carbon dioxide. In effect, the photosynthetic reduction of nitrate to ammonia and its incorporation into an amino acid by reductive amination of a α-keto acid (see Leech and Murphy, 1976, in Vol. I of this series) requires 10 electrons, whereas the reduction of carbon dioxide to carbohydrate requires only 4.

12.2. THE ASSIMILATORY NITRATE-REDUCING SYSTEM

As mentioned above, recent studies carried out in different laboratories have given a clear insight into the nature of the enzymatic reduction of nitrate to ammonia in bacteria, fungi, algae and higher plants (Beevers and Hageman, 1969, 1972; Kessler, 1971, 1974, 1976; Payne, 1973; Morris,

1974; Hewitt, 1975; Hewitt et al., 1976; Losada, 1976b; Zumft, 1976; Vennesland and Guerrero, in press). In this section we shall discuss in detail the two enzymes of the assimilatory nitrate-reducing system, namely nitrate reductase and nitrite reductase, from both prokaryotic and eukaryotic organisms.

12.2.1. Reduction of nitrate to nitrite

The two-electron reduction of nitrate to nitrite is catalyzed by the molybdoprotein nitrate reductase. Two main types of assimilatory nitrate-reducing enzymes can be distinguished: (a) pyridine nucleotide-dependent nitrate reductase, which is present in eukaryotic cells, either photosynthetic or non-photosynthetic, and (b) ferredoxin-dependent nitrate reductase, which is found in blue-green algae and chemoergonic and photosynthetic bacteria (Hewitt, 1975; Losada, 1976b; Vennesland and Guerrero, in press).

12.2.1.1. NAD(P)H : nitrate reductase

12.2.1.1.1. Pyridine nucleotide specificity. The nitrate reductase of fungi, green algae and higher plants catalyzes the reduction of nitrate to nitrite by reduced pyridine nucleotides (Hewitt, 1975; Losada, 1976b; Vennesland and Guerrero, in press), according to the equation:

$$NO_3^- + NAD(P)H + H^+ \rightarrow NO_2^- + NAD(P)^+ + H_2O$$

$$\Delta G_0', pH\ 7 = -34\ kcal/mol$$

Since most of the pyridine nucleotide-dependent nitrate reductases present in eukaryotes operate with both NADH and NADPH, the enzyme has been classified as NAD(P)H : nitrate oxidoreductase (EC 1.6.6.2). Notwithstanding, the enzyme in molds and yeasts has been sometimes designated NADPH : nitrate oxidoreductase (EC 1.6.6.3), because it usually exhibits a preference for NADPH over NADH; this denomination should, however, be reserved for those enzymes exhibiting an absolute specificity for NADPH. On the other hand, the enzyme of photosynthetic eukaryotes is commonly named NADH : nitrate oxidoreductase (EC 1.6.6.1), but it reacts also to some extent with NADPH, and only in a few cases is NADH-specific. In some higher plants, two different nitrate-reducing enzymes, NADH- and NAD(P)H-dependent, are simultaneously present in the same tissues (Campbell, 1976; Jolly et al., 1976; Shen et al., 1976).

The stereospecificity of nitrate reductases for hydrogen removal from reduced pyridine nucleotides has been determined to be of the A-type regardless of the source of the enzyme (Guerrero and Vennesland, 1975; Guerrero et al., 1977). B-stereospecificity was reported by Davies et al.

(1972) for the nitrate reductase of *Candida utilis* in experiments carried out with a crude particulate enzyme preparation, but Guerrero et al. (1977) have shown that extensive enzyme purification is an important requirement to avoid interferences in stereospecificity determinations with nitrate reductases.

12.2.1.1.2. Activities of the enzyme complex. In addition to catalyzing the normal reaction, i.e. reduction of nitrate by reduced pyridine nucleotides, NAD(P)H : nitrate reductase exhibits two other activities (Hewitt, 1975; Losada, 1976b; Vennesland and Guerrero, in press). These activities can be assayed separately and involve only part of the overall electron-transport chain of the enzyme complex. The action of the diaphorase (or NAD(P)H dehydrogenase), which constitutes the first moiety of the complex, results in the reduction by NAD(P)H of a variety of one- and two-electron acceptors, such as cytochrome *c*, ferricyanide, several redox dyes and other oxidants. The so-called terminal nitrate reductase or nitrate reductase proper constitutes the second moiety of the complex and is expressed as a pyridine nucleotide-independent activity which results in the reduction of nitrate by reduced flavin nucleotides or viologens. Both moieties participate jointly in the sequential transfer of electrons from NAD(P)H to nitrate as shown in Fig. 12.5.

Although physical separation of the two functional moieties of the enzyme has not yet been achieved, there is a strong evidence in favor of the above model, that may be summarized as follows. (1) Both partial activities of the NAD(P)H : nitrate reductase complex respond very differently to selective inhibitors and treatments, all of which obviously hinder the overall activity of nitrate reduction by reduced pyridine nucleotide. The pyridine nucleotide-activating (diaphorase) moiety can be completely inhibited by treatment with sulfhydryl-binding reagents or heating at mild temperatures, without affecting the activity of the terminal nitrate-activating moiety.

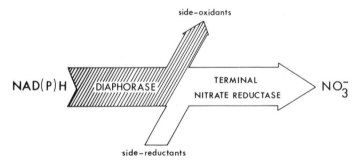

Fig. 12.5. Schematic diagram of the NAD(P)H : nitrate reductase complex showing the two moieties which sequentially participate in the transfer of electrons from NAD(P)H to nitrate.

Conversely, the activity of the second moiety can be totally and specifically inhibited competitively with nitrate by azide, cyanate, cyanide and other inhibitors which do not affect the diaphorase activity (Losada, 1976b; Vennesland and Guerrero, in press). (2) Mutants which possess inactive NAD(P)H : nitrate reductase have been isolated from fungi and green algae; some of these mutants are devoid of diaphorase activity whereas others are lacking in terminal nitrate reductase activity. In vitro complementation of extracts of both types of mutants results in a wild-type nitrate reductase having all the activities (see Coddington, 1976; Hewitt et al., 1976). (3) In vivo inactivated nitrate reductase exhibits a fully active diaphorase but is inactive in the reduction of nitrate. The lost activities are recovered after subjecting the inactive enzyme to the action of oxidizing conditions (Losada, 1976b; Vennesland and Guerrero, in press).

Chlorate, an inhibitor competitive with nitrate of nitrate reductase, can replace nitrate as a substrate for the purified enzyme (Solomonson and Vennesland, 1972a; Vega et al., 1972). The reaction product, chlorite, is a toxic substance, which is thought to be responsible for the toxicity of chlorate in cells with active nitrate reductase (Åberg, 1947; Solomonson and Vennesland, 1972; Tromballa and Broda, 1971; Cove, 1976).

Nitrite, the product of the reduction of nitrate, is not a substrate of the enzyme but has been reported to act as an inhibitor of NAD(P)H : nitrate reductase. The inhibition is of the competitive type with respect to nitrate in the enzymes of *Chlorella vulgaris* (Solomonson and Vennesland, 1972b), *Aspergillus* (McDonald and Coddington, 1974) and *Rhodotorula* (Guerrero and Gutierrez, 1977). In the two latter cases, the K_i value for nitrite is similar to the K_m value for nitrate. Product inhibition has also been reported for the spinach enzyme (Eaglesham and Hewitt, 1975) but the affinity of the enzyme for nitrite is far lower than for nitrate, the inhibition being of the uncompetitive type.

12.2.1.1.3. Purification, molecular properties and reaction mechanism. In spite of many attempts there has been relatively little success in the purification of assimilatory pyridine nucleotide-dependent nitrate reductases, though many partial purifications have been reported (Hewitt, 1975). The enzyme of *C. vulgaris* has been the first to be purified to homogeneity (Solomonson et al., 1975). The development by Solomonson (1975) of a simple procedure for purifying *C. vulgaris* nitrate reductase by affinity chromatography on Blue-Dextran Sepharose has provided a new approach which has been successfully applied to the enzyme from other sources (Campbell, 1976; Guerrero and Gutierrez, 1977; Guerrero et al., 1977; Notton et al., 1977).

The molecular weight of purified NAD(P)H : nitrate reductases estimated from values of Stokes radii and sedimentation coefficients, range from 220 000 to 360 000 (see Vennesland and Guerrero, in press). These proteins behave anomalously on gel filtration, suggesting assymmetry in the enzyme molecule. Exceptionally high values of molecular weight reported for some nitrate reductases were determined by molecular sieving procedures and should be regarded as inaccurate.

The enzyme of *C. vulgaris* (MW 356 000) is composed of at least three subunits of 100 000 daltons (Solomonson et al., 1975). The presence of sub-units of 60 000 in a native enzyme of 475 000 has been reported for *Ankistrodesmus braunii* nitrate reductase (Ahmed and Spiller, 1976). In fungi, two subunits of 118 000 seem to constitute the nitrate reductase of *R. glutinis* (Guerrero and Gutierrez, 1977), whereas different subunits of 115 000 and 130 000 seem to be present in the enzyme of *N. crassa* (Pan and Nason, 1976); both enzymes have a molecular weight of 230 000 daltons.

Before an homogeneous preparation of NAD(P)H : nitrate reductase was available, indirect evidence spoke for the enzyme as a molybdo-flavo-heme-containing protein, as it was later confirmed by direct analysis (Solomonson et al., 1975; Pan and Nason, 1976).

The specific requirement of added FAD for full diaphorase and NAD(P)H : nitrate reductase activities in both crude and purified enzyme preparations of a variety of sources (Hewitt, 1975) early suggested this nucleotide to be a functional component of the enzyme. However, added FAD does not always stimulate the activity of the enzyme, as is the case with partially purified nitrate reductase of *Chlorella fusca* and leaves of higher plants which do not exhibit the requirement of FAD for catalytic activity of the diaphorase moiety of the complex. Nevertheless, added FAD has a stimulatory effect after these enzymes are subjected to gel filtration (Zumft et al., 1970; Relimpio et al., 1971; Castillo et al., 1976). In the case of *C. vulgaris* nitrate reductase, the flavin prosthetic group remains bound to the enzyme even after a thorough purification procedure, and the addition of FAD has no effect on enzymatic activity (Vennesland and Solomonson, 1972; Solomonson et al., 1975). This variability seems to reflect a difference in the strength of binding of the flavin component to the protein. The nitrate reductase of *C. vulgaris* contains two mol of FAD per mol of enzyme (Solomonson et al., 1975), whereas only one mol of the flavin seems to be present in fungal nitrate reductase (Downey, 1973).

A *b*-type cytochrome with absorption maxima at 557 nm (α), 527 nm (β) and 423 nm (Soret) for the reduced form, and at 412 nm for the oxidized (Fig. 12.6), appears to be an ubiquitous component in NAD(P)H : nitrate reductase. Since Garrett and Nason (1967) suggested the association of cytochrome *b*-557 with the enzyme of *Neurospora*, this prosthetic group has been reported to be present in all nitrate reductases which have been adequately purified (Vennesland and Guerrero, in press). Two mol of heme per mol of enzyme were reported by Solomonson et al. (1975) for the nitrate reductase of *C. vulgaris*, but again the enzyme from fungi contains only one mol of cytochrome per mol of enzyme (Guerrero and Gutierrez, 1977).

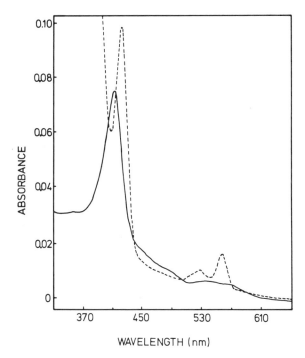

Fig. 12.6. Typical absorption spectra of the NAD(P)H : nitrate reductase complex in its reduced (- - - - - -) and oxidized (———) forms (Guerrero and Gutierrez, 1977).

The essentiality of molybdenum for activity of nitrate reductase was early recognized, and early experiments at both the cellular and the enzymatic level showed this metal to be a functional component of the enzyme. The use of radioactive isotopes of molybdate and tungstate — the latter an antagonist of molybdate in nitrate assimilation — has confirmed the key role of molybdenum in nitrate reductase and located the participation of this metal in the activity of the nitrate-reducing moiety of the enzyme (Hewitt, 1975; Losada, 1976b). Quantitative analysis indicate the presence of two molybdenum atoms per molecule of *C. vulgaris* nitrate reductase (Solomonson et al., 1975) and about one atom in fungal nitrate reductase (Downey, 1973; Pan and Nason, 1976). It is believed that molybdenum is bound to a small polypeptide, thus forming a low molecular weight (ca. 1000 daltons) cofactor which appears to be common to many molybdo-enzymes (see Hewitt, 1975; Hewitt et al., 1976, 1977; Losada, 1976b; 1977; Zumft, 1976).

From present evidence, the electron-transport chain of NAD(P)H : nitrate reductase can be pictured as:

$$NAD(P)H \rightarrow \{FAD \rightarrow cyt\ b\text{-}557 \rightarrow Mo\} \rightarrow NO_3^-$$

Whereas it is clearly established that FAD is an essential prosthetic group of the first moiety of the enzyme and that molybdenum participates in the catalytic activity of the terminal moiety, the localization of the heme group in the enzyme molecule and the functional role of this component have not yet been clearly determined. From the reported requirement for added FAD in order to get full reduction of the cytochrome of some nitrate reductases by NAD(P)H (Garrett and Nason, 1969; Guerrero and Gutierrez, 1977), it has been inferred that FAD preceeds the heme in the electron-transport chain of the enzyme. In supporting the idea of an active participation of the heme group in the electron-transport chain of the enzyme, it has been reported that the cytochrome of different nitrate reductases is reduced by added NAD(P)H and reoxidized on addition of nitrate (see Guerrero and Gutierrez, 1977) or acceptors of the diaphorase activity (Notton et al., 1977). Cytochrome b-557 has also been considered as the link between the two moieties of the enzyme, i.e. the carrier which mediates the electron transfer from the diaphorase to the terminal nitrate reductase (Hewitt et al., 1976).

The activity of the diaphorase moiety of the NAD(P)H : nitrate reductase complex is severely inhibited by sulfhydryl-binding reagents such as p-hydroxymercuribenzoate, and both NAD(P)H and FAD protect against this inhibition (Relimpio et al., 1971; Losada, 1976b). It has been suggested from these data that essential sulfhydryl groups of the enzyme participate in the binding of NAD(P)H to the enzyme and are also related to the flavin prosthetic group (see Barea et al., 1976). Recent work with the nitrate reductase of the mold *Neurospora* reinforces this interpretation and suggests the active participation of sulfhydryl groups in the electron flow from NAD(P)H to FAD (Amy et al., 1977). Absence of labile sulfide has been reported in the enzyme of *C. vulgaris* (Solomonson et al., 1975).

No general agreement has been reached with regard to the oxidation state change of the molybdenum present in nitrate reductase during the enzymatic reduction of nitrate to nitrite. Mainly on the basis of chemical models, it has been proposed that the oxidation state of molybdenum changes from Mo (IV) to Mo (VI) (Stiefel, 1973), from Mo (III) to Mo (V) (Ketchum et al., 1976), or even from Mo (V) to Mo (VI) (Nicholas and Stevens, 1955; Garner et al., 1974). Stiefel (1973) has postulated that Mo in the enzyme mediates a coupled electron—proton transfer and that Mo (V) could be an intermediate in the reduction of Mo (VI) to Mo (IV) by the electron donor.

Cyanide is a potent inhibitor of nitrate reductase. When NAD(P)H : nitrate reductase is treated in the absence of reductant with cyanide, this compound behaves as a competitive inhibitor with nitrate. The same behavior is exhibited by other inhibitors of the terminal nitrate reductase moiety. However, when the enzyme is kept in the reduced state by natural (NAD(P)H) or artificial (dithionite) electron donors, cyanide combines tightly with the enzyme and the inhibition becomes of the non-competitive type. The stable reduced enzyme—HCN complex is inactive for nitrate reduction but retains full diaphorase activity. Removal of excess cyanide and reductant by gel filtration does not lead to activation, but the enzyme is instantaneously re-activated by oxidation with ferricyanide, and more slowly by other oxidants. Nitrate and the competitive inhibitors cyanate, azide and carbamylphosphate protect the enzyme against the cyanide binding under reducing conditions, presumably by keeping the enzyme in the oxidized form (Relimpio et al., 1971; Vega et al., 1972; Lorimer et al., 1974; Losada, 1976b; Vennesland and Guerrero, in press). The site of action of cyanide seems to be the molybdenum of the terminal nitrate reductase moiety (Relimpio et al., 1971; Vega et al., 1972; Lorimer et al., 1974; Solomonson, 1974), but the nature of the chemical binding is not known. Solomonson (1974) has suggested that binding of cyanide to the enzyme in the presence of reductant would result in the formation of a cyanide complex with molybdenum in the stable Mo (III) state, which would not be able to reduce nitrate.

The kinetic behavior of nitrate reduction by reduced pyridine nucleotide catalyzed by the spinach enzyme seems to correspond to a mechanism of the ordered sequential ping-pong (bi-bi) type (De la Rosa, 1975; Eaglesham and Hewitt, 1975), the reduced form of the nucleotide being attached, and its oxidized form being released, before nitrate binds to the enzyme. For the nitrate reductase of *Aspergillus*, however, a random order rapid-equilibrium mechanism has been determined (McDonald and Coddington, 1974).

12.2.1.2. Ferredoxin : nitrate reductase

With the exception of the early reports by Katoh (1963a) about the presence of a flavo-heme-containing NADH-dependent nitrate reductase in the photosynthetic bacteria *Rhodospirillum rubrum*, where the function-assimilatory or respiratory- of the enzyme was not clearly determined (Katoh, 1963a, b) and by Hattori and Myers (1967) in *Anabaena cylindrica* (see below), it is well established that the assimilatory nitrate-reducing enzyme from prokaryotic organisms cannot accept electrons from NAD(P)H or reduced flavins and seems to be rather dependent on ferredoxin as the physiological reductant (Losada, 1976b; Vennesland and Guerrero, in press).

The nitrate reductase of prokaryotes catalyzes the reduction of nitrate to nitrite by reduced ferredoxin, according to the equation:

$$NO_3^- + 2\ Fd_{red} + 2\ H^+ \rightarrow NO_2^- + 2\ Fd_{ox} + H_2O$$

$$\Delta G_0',\ pH\ 7 = -38{,}6\ kcal/mol$$

The aerobic chemiotrophic bacteria *Azotobacter* (Guerrero et al., 1973; Kadam and Naik, 1975, 1976; Tortolero et al., 1975), and *Acinetobacter* (Villalobo et al., 1977) contain a soluble nitrate reductase (MW ca. 100 000) which is inactive with reduced pyridine nucleotide but can accept electrons from reduced ferredoxin and viologens. Cyanate has a striking activating effect on the reduction of nitrate by reduced methyl viologen catalyzed by this molybdenum-containing nitrate reductase (Guerrero et al., 1973; Villalobo et al., 1977).

The presence of a ferredoxin-linked nitrate reductase has been reported in the phototrophic bacteria *Ectothiorhodospira shaposhnikovii* (Malofeeva et al., 1975). Most of the total enzyme activity was associated with a particulate fraction containing chromatophores. It was found that either dithionite-reduced methyl viologen or *Ectothiorhodospira* ferredoxin reduced with illuminated pea chloroplasts could act as electron donors for the reduction of nitrate to nitrite in the presence of this particulate nitrate reductase. A soluble nitrate reductase has been partly purified from *Rhodopseudomonas capsulata* cells grown on nitrate (Alef and Klemme, 1977). The enzyme is active with reduced viologens but it cannot use reduced pyridine nucleotide as reductant.

In the blue-green algae, nitrate reductase has been shown to be present in a particle-bound form. Particulate preparations of *A. cylindrica* obtained by sonication or acetone treatment could catalyze the reduction of nitrate with reduced ferredoxin or NADH, respectively (Hattori and Myers, 1967). However when the enzyme was solubilized by Triton X-100 treatment, both reductants were without effect and only reduced flavins or viologens served as electron donors (Hattori, 1970). In *Anacystis nidulans* and *Nostoc muscorum*, reduced ferredoxin, but not NAD(P)H, could donate electrons to nitrate reductase, whether solubilized (Manzano et al., 1976) or bound to particles (Manzano et al., 1976; Candau et al., 1976; Ortega et al., 1976). Also flavodoxin, which physiologically substitutes for ferredoxin under conditions of iron starvation (Bothe, 1977), can act as an electron source for the nitrate reduction by *Anacystis* nitrate reductase (Manzano, 1977).

By using affinity chromatography on ferredoxin—Sepharose gel as the main step, the ferredoxin : nitrate reductase of *A. nidulans* has been recently purified to homogeneity and partly characterized. The enzyme is constituted by only one polypeptide chain with a molecular weight of 75 000 daltons and does not seem to contain flavin or cytochrome as prosthetic groups (P. Candau and C. Manzano, unpublished).

Cyanide and *p*-hydroxymercuribenzoate are also powerful inhibitors of the nitrate reductase from prokaryotic organisms (Guerrero et al., 1973; Manzano et al., 1976; Villalobo et al., 1977).

The ferredoxin : nitrate reductase of prokaryotes may be envisaged as a precursor of the nitrate reductase of eukaryotic organisms, which had further evolved and acquired the ability to use NAD(P)H by incorporation of an initial FAD-containing diaphorase moiety.

12.2.2. Reduction of nitrite to ammonia

Nitrite reductase is the enzyme responsible for the second step in the assimilatory reduction of nitrate, i.e. the six-electron reduction of nitrite to ammonia. Regarding the nature of the physiological electron donor, there is a clear difference between the enzyme found in photosynthetic cells and the one present in non-photosynthetic tissues and chemoergonic organisms. The nitrite reductase of bacteria and fungi has been characterized as NAD(P)H : nitrite reductase (EC 1.6.6.4), whereas the enzyme of algae and leaves of higher plants is dependent on reduced ferredoxin as reductant (Hewitt, 1975; Losada, 1976b; Vennesland and Guerrero, in press).

12.2.2.1. Ferredoxin : nitrite reductase
12.2.2.1.1. Electron donors and acceptors. The photosynthetic enzyme of both eukaryotic and prokaryotic cells has been classified as ferredoxin : nitrite oxidoreductase (1.7.7.1) because its marked specificity for reduced ferredoxin as the electron donor. It catalyzes the following exergonic redox reaction according to the equation:

$$NO_2^- + 6\ Fd_{red} + 8\ H^+ \rightarrow NH_4^+ + 6\ Fd_{ox} + 2\ H_2O$$

$$\Delta G_0',\ pH\ 7 = -103,5\ kcal/mol$$

Disregarding early reports which have not been further confirmed (Roussos and Nason, 1960; Czygan, 1963), photosynthetic nitrite reductase has been shown to be completely inactive with reduced pyridine or flavin nucleotides as immediate reductants. On the other hand, ferredoxin reduced either chemically with dithionite, enzymatically with H_2 plus hydrogenase, or photosynthetically with illuminated chlorophyll-containing particles, supports very good rates of nitrite reduction (Losada and Paneque, 1971; Vega et al., in press). Photosynthetically reduced flavodoxin can also act as electron source for the nitrite reduction catalyzed by nitrite reductase (Bothe, 1969; Zumft, 1972; Manzano, 1977). Artificial one-electron reductants such as reduced viologens are also effective electron donors and are extensively used for the routine assays of the enzyme activity (Losada and Paneque, 1971; Vega et al., in press).

In addition to nitrite, hydroxylamine can also be reduced by nitrite reductase, although with very low rates as compared with those for nitrite reduction (cf. Vega et al., in press). Sulfite can not be utilized as a substrate by nitrite reductase in spite of the fact that nitrite and hydroxylamine are both good electron acceptors for the enzyme sulfite reductase. As advanced by Zumft (1972), striking similarities exist between nitrite reductase and sulfite reductase. Both enzymes catalyze six-electron reductions directly coupled to the electron transport system of chloroplast via the electron carrrier ferredoxin, and both have similar prosthetic groups (see Schwenn and Trebst, 1976, in Vol. I of this series; Vega et al., in press).

12.2.2.1.2. Purification, molecular properties and reaction mechanism.
Purification to homogeneity of ferredoxin : nitrite reductase from a variety of sources, which include green algae and leaves of higher plants, has been reported (see Vega et al., in press; Vennesland and Guerrero, in press). Only partial purification has been achieved for the enzyme of blue-green algae (Hattori and Uesugi, 1968; Manzano, 1977).

The enzymes from organisms belonging to diverse groups have values of molecular weight ranging from 60 000 to 70 000 daltons, and appear to be constituted by a single polypeptide chain with about 600 amino acid residues (Cárdenas et al., 1972b; Zumft, 1972; Ho et al., 1976; Vega and Kamin, 1977; Vega et al., in press).

Before direct determination of metals in homogenous nitrite reductase preparations could be verified, iron was identified as an essential constituent of the enzyme by the in vivo incorporation of ^{59}Fe into the nitrate reductase

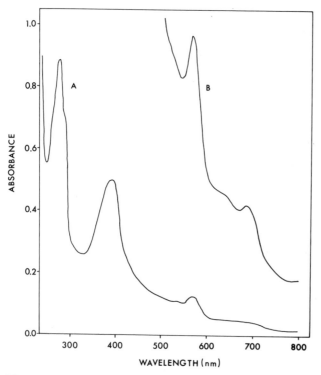

Fig. 12.7. Absorption spectrum at two different concentrations (A and B) of spinach nitrite reductase in its oxidized form (Vega and Kamin, 1977).

molecule (Aparicio et al., 1971). Purified ferredoxin : nitrite reductases are red brown in color and show similar absorption spectra with peaks in the regions 380—390 nm and 570—580 nm, characteristic of a heme-containing protein (Fig. 12.7). Murphy et al. (1974) have identified the heme-prosthetic group in spinach nitrite reductase as "siroheme", an iron tetrahydro-porphyrin of the isobacteriochlorin type with eight carboxylic acid-containing side chains, previously shown to be a component of the enzyme sulfite reductase (Murphy et al., 1973; Murphy and Siegel, 1973). EPR spectroscopy has confirmed the presence of siroheme and provided evidence for the additional presence of an iron-sulfur center in the spinach enzyme (Aparicio et al., 1975; Stoller et al., 1977; Vega and Kamin, 1977; Vega et al., in press). The value of three iron atoms per molecule of spinach nitrite reductase has been recently reported, and it has been proposed (Vega and Kamin, 1977) that one atom is in siroheme and the other two, together with two labile sulfides, in the iron—sulfur cluster ($Fe_2—S_2^*$).

In contrast to the NAD(P)H-dependent enzyme of non-photosynthetic organisms (Losada, 1976b; Vennesland and Guerrero, in press), ferredoxin : nitrite reductase is not a flavoprotein.

Present evidence supports an interaction between siroheme and nitrite, and because complete reduction of the iron—sulfur center can be only obtained in the presence of inhibitors (cyanide and CO) which bind to siroheme, it is believed that the electron donor first reduces the iron—sulfur center, which in turn transfers the electrons to the siroheme prosthetic group, where nitrite is directly reduced to ammonia (Aparicio et al., 1975; Losada, 1976b; Vega and Kamin, 1977; Vega et al., in press; Vennesland and Guerrero, in press). The electron-transport chain from reduced ferredoxin to nitrite can be thus depicted as:

$$Fd_{red} \rightarrow \{(Fe_2—S_2^*) \rightarrow siroheme\} \rightarrow NO_2^-$$

The recently reported values (Stoller et al., 1977) for the midpoint potentials of the iron—sulfur center and siroheme (—550 mV and —50 mV, respectively) support the view that the iron—sulfur center preceeds the siroheme in the electron-transport chain. The problem of how chloroplast ferredoxin (E_0', pH 7, —420 mV) can reduce the iron—sulfur center has been discussed by Stoller et al. (1977). They have proposed that perhaps ferredoxin controls the potential of the iron—sulfur center through the formation of a reduced ferredoxin—nitrite reductase complex, or that the midpoint of the iron—sulfur cluster shifts to a more electropositive value on interaction with the nitrite—siroheme complex.

Cyanide, carbon monoxide and sulfhydryl reagents such as p-chloromercuribenzoate and mersalyl inhibit ferredoxin : nitrite reductase from different sources (Vega et al., in press; Vennesland and Guerrero, in press). The inhibition by cyanide is of the competitive type with respect to nitrite

(Cárdenas et al., 1972a), and the inhibitor seems to bind to the siroheme prosthetic group (Vega and Kamin, 1977). Carbon monoxide also inhibits nitrite reductase activity and forms a complex with the reduced enzyme by binding to the siroheme under reducing conditions (Murphy et al., 1974; Vega and Kamin, 1977). The inhibition by p-chloromercuribenzoate and mersalyl is correlated with an irreversible alteration of the absorption bands corresponding to siroheme, which could suggest degradation of this prosthetic group by such sulfhydryl reagents (Hucklesby et al., 1976; Vega and Kamin, 1977).

12.3. LOCALIZATION OF THE NITRATE-REDUCING SYSTEM

Although most of the studies concerning the enzymes of the nitrate-reducing system have been carried out using photosynthetic cells and tissues, namely algae and leaves, as the starting material, both nitrate and nitrite reductases are also normally present in nonchlorophyllous tissues such as the roots of many plants (Hewitt et al., 1976).

The leaves of the C-4 plants possess a Kranz-type anatomy, distinct biochemical functions being associated with each of the photosynthetic mesophyll and bundle sheath cell layers (see Coombs, 1976, in Vol. I of this series). Studies of the distribution of nitrate and nitrite reductases carried out with leaves of different C-4 species have shown that both enzymes are predominantly localized in the mesophyll cells, which thus represent the major site of nitrate assimilation in C-4 plants (Mellor and Tregunna, 1971; Rathnam and Das, 1974; Rathnam and Edwards, 1976; Harel et al., 1977).

The intracellular localization of nitrate reductase and nitrite reductase in green cells is a matter of controversy that has not yet been clearly resolved. Regarding nitrite reductase, the bulk of the evidence indicates that it is a chloroplastic enzyme (see Miflin, 1974) and that chloroplasts can reduce nitrite in the light (Paneque et al., 1963; Losada et al., 1963; Ramirez et al., 1966; Swader and Stocking, 1971; Magalhaes et al., 1974; Neyra and Hageman, 1974; Plaut et al., 1977). Additional support for the conclusion of nitrite reductase being associated with the chloroplast is provided by the well-known association of ferredoxin, the natural electron donor for the enzyme, and by the quenching of chlorophyll fluorescence caused by nitrite addition to cells of green algae (Kessler and Zumft, 1973; Kulandaivelu et al., 1976).

Some authors have concluded that nitrate reductase is also localized in the chloroplast (Del Campo et al., 1963; Coupé et al., 1967; Plaut and Littan, 1974) and several others have reported that intact chloroplasts are able to reduce nitrate (Heber and French, 1968; Grant and Canvin, 1970; Swader and Stocking, 1971). Alternatively, the cytoplasm and the microbodies have also been considered likely places for the localization of this enzyme within the cell (Ritenour et al., 1967; Dalling et al., 1972; Lips and Avissar, 1972; Lips, 1975). An interesting and relatively well-supported proposal states that nitrate reductase is localized in the outer membrane of the chloroplast, in both C-3 and C-4 plants (Ritenour et al., 1967; Grant et al., 1970; Eaglesham and Hewitt, 1971; Rathnam and Das, 1974).

In the prokaryotic blue-green algae, both enzymes of the ferredoxin : nitrate-reducing system have been found to be tightly associated with chlorophyll-containing particles (Guerrero et al., 1974; Candau et al., 1976; Manzano et al., 1976; Ortega et al., 1976). This proves that in these organisms both enzymes are associated with the photosynthetic apparatus, and considering the proposed evolution of chloroplasts from blue-green algae which had developed some form of symbiosis (Lipmann, 1976), the possibility of the existence in higher plant chloroplasts of a ferredoxin : nitrate reductase of the blue-green algal type can not be disregarded. Whereas, to the best of our knowledge, many experiments aimed to detect NAD(P)H : nitrate reductase activity in chloroplasts have been carried out, no attempts have yet been conducted to the search for a ferredoxin-dependent enzyme.

In supporting the view of a close relationship between higher plant chloroplasts and nitrate reductase are the results of Sawhney et al. (1972). By using chlorophyll mutants of barley, these authors have shown that the level of nitrate reductase activity in leaves is closely related to the presence of chloroplasts. The enzyme was absent from white albino leaves whereas it was present in pale green chlorina leaves, which in turn exhibited lower nitrate reductase levels than those found in the normal green leaves.

12.4. PHOTOSYNTHETIC REDUCTION OF NITRATE AND NITRITE

Since Warburg and Negelein (1920) reported a stimulation by light of nitrate utilization by *Chlorella* cells, the close connection between photosynthesis and nitrate assimilation in a great variety of photosynthetic organisms and tissues (see e.g. Hattori, 1962; Grant and Turner, 1969; Thacker and Syrett, 1972; Stevens and Van Baalen, 1973; Canvin and Atkins, 1974; Losada, 1976a, b) has become a well substantiated fact.

Intact chloroplasts isolated by the method of Jensen and Bassham (1966) or by slightly modified procedures (Robinson and Stocking, 1968; Grant et al., 1970) contain nitrate and nitrite reductase and, when illuminated, are able to reduce nitrate and nitrite with concomitant oxygen evolution (Heber and French, 1968; Grant and Canvin, 1970; Swader and Stocking, 1971). When these chloroplasts are subjected to further washing, although still intact they lose the ability to reduce nitrate, but retain the ability to reduce nitrite (Swader and Stocking, 1971). Nitrite reduction by intact chloroplasts has also been reported by different laboratories (Magalhaes et al., 1974; Neyra and Hageman, 1974; Plaut et al., 1977).

Reconstituted chloroplast systems can also carry out the photoreduction of nitrate to nitrite, nitrite to ammonia, and the overall reduction of nitrate to ammonia, with water as the ultimate electron donor (Losada et al., 1963, 1965; Losada and Paneque, 1966; Paneque et al., 1969). Spinach grana supplemented with NAD$^+$, ferredoxin, NADP reductase and the corresponding nitrate-reducing enzymes can reduce nitrate to ammonia with the stoichiometric evolution of four oxygen atoms per molecule of nitrate reduced. In the presence of orthophosphate and ADP, the photoreduction of nitrate and the coupled photooxidation of water are accompanied by the formation of ATP (Paneque et al., 1969). With nitrite as electron acceptor, three mol of ATP are produced per mol of ammonia formed (Paneque et al., 1964).

Even if the consensus has been reached that in higher plants nitrite reduction is intimately linked with the light reactions of photosynthesis, such an agreement does not exist regarding the reduction of nitrate to nitrite. Consequently, it has been proposed that the effect of light in the generation of the reductant (NAD(P)H) for the reduction of nitrate is indirect, through the formation of reduced carbon compounds (e.g., triose phosphates) which would be oxidized in the cytoplasm by NAD(P)$^+$ (Klepper et al., 1971; Beevers and Hageman, 1972). This theory assumes a cytoplasmic localization for NAD(P)H : nitrate reductase.

The clearest evidence so far reported concerning the photosynthetic nature of the process of nitrate reduction comes from recent experiments carried out with blue-green algal systems. Nitrate reductase and nitrite reductase appear to be tightly bound to the lamellar system of these photosynthetic prokaryotes, thus making possible the preparation of subcellular chlorophyll-containing particles which are able to photoreduce nitrate to nitrite and ammonia with water as electron donor (Candau et al., 1976; Ortega et al., 1976).

Hattori and Myers (1967) and Hattori and Uesugi (1968) had previously shown that particles isolated from *Anabaena* together with ferredoxin and the corresponding reductase, could carry out the ferredoxin-dependent photoreduction of nitrate or nitrite. In addition Manzano et al. (1976) have prepared particulate fractions from *Anacystis* which had tightly associated with them both enzymes of the nitrate-reducing system and could photoreduce nitrate and nitrite with the sole requirement of added ferredoxin. Both types of particles, either from *Anabaena* or from *Anacystis*, did not, however, exhibit activities of photosystem II and were therefore unable to use water as electron donor. In consequence, electrons had to be supplied by donors for photosystem I, such as the couple ascorbate—indophenol dyes.

The coupling of nitrate reduction with the photolysis of water was recently achieved by using pigment-containing particles from *A. nidulans* cells disrupted under mild conditions in a French press (Candau et al., 1976). The cell-free preparations so obtained are active in the photoreduction of several artificial electron acceptors of photosystems I and II and of the

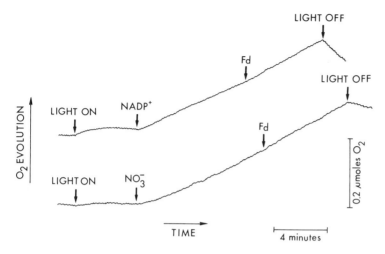

Fig. 12.8. Time course of the light-dependent evolution of oxygen by *Anacystis nidulans* particles with either NADP$^+$ or NO$_3^-$ as "natural" Hill reagents (Candau et al., 1976).

natural Hill reagents NADP$^+$, nitrate and nitrite, with the concomitant evolution of oxygen (Fig. 12.8). One oxygen atom is evolved per molecule of nitrate reduced to nitrite, and four atoms are evolved when complete reduction to ammonia takes place. Similar results have been achieved by Manzano et al. (1977) using a reconstituted system consisting of spinach grana and ferredoxin-nitrate and nitrite reductases from *Anacystis*. The latter system was able to reduce nitrate to ammonia at a rate of 40 μmol per mg chlorophyll per hour (C. Manzano and P. Candau, unpublished), a value which matches with those reported for the photoreduction of NADP$^+$ and other Hill reagents. Ortega et al. (1976) have also reported the photosynthetic reduction of nitrate to nitrite by photosynthetic subcellular particles of another blue-green alga, *N. muscorum*.

The photoreduction of nitrate to ammonia with water can thus be considered one of the simplest and most relevant examples of photosynthesis (see Fig. 12.4).

It has to be pointed out that not only the reduction of nitrate to ammonia, but also the incorporation of the latter compound into cell material by reductive amination of α-keto acids is dependent on photosynthetically reduced ferredoxin, both in algae and in higher plants (see Leech and Murphy, 1976, in Vol I of this series). In this context, it has been shown that intact chloroplasts carry out oxygen evolution dependent on ammonia and α-ketoglutarate (Anderson and Done, 1977).

12.4.1. *Photoproduction of ammonia and hydrogen peroxide by photosynthetic systems*

Conversion of sunlight energy into suitable redox energy by biological or physico-chemical systems capable of using water as the electron donor is at the present time of great significance for trying to solve the world energy crisis (Buvet et al., 1977). Fig. 12.9 diagrammatically represents different redox systems (real or possible) whose reduction may be photosynthetically coupled with the photooxidation of water to molecular oxygen (Losada, 1977, 1978a,b,c). All of those which have potential levels similar to the hydrogen electrode are ferredoxin-dependent, thus requiring the sequential cooperation of photosystems II and I. Several of them have, in addition,

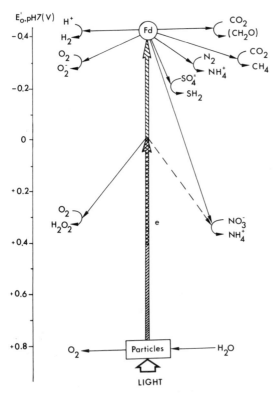

Fig. 12.9. Different redox systems for the bioconversion of light energy. Electrons from water are moved in two stages by the sequential cooperation of the two photosystems up to the redox level of reduced ferredoxin and eventually to a variety of electron acceptors. In principle, O_2 could accept electrons at the reducing side of photosystem II at a redox potential of about 0 V, being directly reduced to H_2O_2. Due to its high potential, the redox pair NH_4^+/NO_3^- might also behave like the pair H_2O_2/O_2, by passing photosystem I (dashed line).

ATP and/or anaerobic requirements, which makes them particularly complicated. By contrast, the redox system ammonia/nitrate involves two pairs of positive redox potential (NO_2^-/NO_3^-, E_0', pH 7, +0.42 V; NH_4^+/NO_2^-, E_0', pH 7, +0.33 V), does not have ATP requirement, and can steadily operate under aerobic conditions, both in vivo and in vitro. Similar peculiarities are exhibited by the system H_2O_2/O_2 (E_0', pH 7, +0.30 V) and will therefore also be briefly considered.

As discussed above, the photosynthetic reduction of nitrate in its more simple version occurs as follows (see Fig. 12.4). Firstly, the electrons derived from water are raised by light at photosystem II to the level of plastoquinone, transported downhill to photosystem I, and raised again by light to the level of ferredoxin (see chapter 2 of this book). Then, the stepwise reduction of nitrate to ammonia, catalyzed by the ferredoxin-dependent nitrate-reducing system, takes place in the dark. As a result of the overall redox reaction, oxygen is simultaneously evolved in stoichiometric amounts, according to the following simplified equations:

$$NO_3^- + H_2O \xrightarrow{2e} NO_2^- + \tfrac{1}{2} O_2 + H_2O$$

$$(\Delta E_0' = -0.40 \text{ V}; \Delta G_0' = +18.4 \text{ kcal})$$

$$NO_2^- + 3 H_2O + 2 H^+ \xrightarrow{6e} NH_4^+ + \tfrac{3}{2} O_2 + 2 H_2O$$

$$(\Delta E_0' = -0.49 \text{ V}; \Delta G_0' = +67.6 \text{ kcal})$$

$$NO_3^- + 4 H_2O + 2 H^+ \xrightarrow{8e} NH_4^+ + 2 O_2 + 3 H_2O$$

$$(\Delta E_0' = -0.47 \text{ V}; \Delta G_0' = +86 \text{ kcal})$$

Since the uphill transfer of 1 electron from the potential level of water to that of ferredoxin requires 2 quanta of light, it can be calculated that the photosynthetic reduction of 1 mol of nitrate to ammonia by water requires 16 einsteins of red light (672 kcal) and occurs with a yield of 13% (disregarding coupled photophosphorylation and the generation of any transmembrane electrical and proton gradient). It should be considered that the potential difference between ferredoxin (E_0', pH 7, −0.42 V) and the pair NH_4^+/NO_3^- (E_0', pH 7, +0.35 V) is unnecessarily high from a thermodynamic point of view ($\Delta E_0'$, pH 7, +0.77 V) and that the redox potential at the reducing site of photosystem II (about zero volts) is sufficiently low to allow the complete reduction of the pair NH_4^+/NO_3^-. In consequence, it should be possible by using, instead of ferredoxin, appropriate artificial or natural electron carriers of adequate potential, capable of interacting at this site with the nitrate-reducing system (see dashed line in Fig. 12.9), to bypass photosystem I and to achieve the photoproduction of ammonia with only 8 einsteins per mol, which means doubling the energy yield to 26%. It can also be imagined that the natural photosystems and the enzymes and carriers involved in the photosynthesis of ammonia may eventually be, at least partly, substituted by their synthetic chemical counterparts.

Besides its biochemical and chemical interest, ammonia is an excellent and powerful fuel and has been prominently mentioned as a constituent of various types of mixtures both for internal-combustion engines and for jet propulsion (Jones, 1975).

It can react with molecular oxygen and be oxidized either to molecular nitrogen or to nitrate, according to the following highly exergonic equations:

$$NH_3 + \tfrac{3}{4} O_2 \xrightarrow{3e} \tfrac{1}{2} N_2 + \tfrac{3}{2} H_2O$$

$(\Delta E_0' = +1{,}10 \text{ V}; \Delta G_0' = -76 \text{ kcal})$

$$NH_3 + 2 O_2 \xrightarrow{8e} HNO_3 + H_2O$$

$(\Delta E_0' = +0.47 \text{ V}; \Delta G_0' = -86 \text{ kcal})$

These reactions demonstrate in practical terms that 1 atom of nitrogen reduced to the state of ammonia can supply either 3 electrons at the potential level of -0.28 V and be oxidized to molecular nitrogen, or 8 electrons at the potential level of $+0.35$ V and be oxidized to nitrate.

In addition, ammonia can undergo thermal dissociation in the presence of certain catalysts to give molecular nitrogen and molecular hydrogen (E_0', pH 7, -0.42 V). Decomposition into the elements can also be effected by photochemical means and by passing a silent electrical discharge through the gas. The reaction is weakly endergonic, as shown by the following equation:

$$2 NH_3 \xrightarrow{6e} N_2 + 3 H_2$$

$(\Delta E_0' = -0.14 \text{ V}; \Delta G_0' = +19 \text{ kcal})$

In this reaction 2 mol of ammonia yield 3 mol of hydrogen, this is to say, liquid ammonia supplies more hydrogen than could be compressed in the same volume. The hydrogen resulting from the lysis of ammonia can be utilized in the oxy-hydrogen blowpipe for producing intensely hot flame for welding metals and special steels, taking advantage of the strong exergonicity of its reaction with oxygen:

$$H_2 + \tfrac{1}{2} O_2 \xrightarrow{2e} H_2O$$

$(\Delta E_0' = +1{,}24 \text{ V}; \Delta G_0' = -57 \text{ kcal})$

For the above reasons, ammonia can also be used for storing and transporting hydrogen in a convenient and compact way.

A photosynthetic system that has strong similarities with the one reducing
nitrate to ammonia and that for its simplicity and requirements may likewise
be suitable for the bioconversion and storage of light energy, is that which
reduces oxygen to hydrogen peroxide (Losada, 1977, 1978b,c).

Since the classical experiments of Mehler (1951) demonstrated the photoproduction of H_2O_2 by isolated chloroplasts, the role of O_2 as a Hill reagent
has been well established both in subcellular preparations (Arnon et al.,
1961; Elstner and Frommayer, 1978; Allen, 1977) and in algal cells (Patterson and Myers, 1973; Radmer and Kok, 1976).

O_2 can be monovalently reduced to the unstable superoxide free radical
anion O_2^- (E_0', pH 7, -0.33 V) by natural or artificial low potential electron
carriers accepting electrons at the reducing side of photosystem I (see
Fig. 12.9). Dismutation of the superoxide anion by superoxide dismutase
(Allen, 1977) eventually results in the formation of O_2 and H_2O_2 with a great
loss of free energy, according to the equation:

$$2 \ O_2^- + 2 \ H^+ \xrightarrow{1e} O_2 + H_2O_2$$

The photosynthesis of H_2O_2 would, however, be more favorable from an
energetic point of view, doubling its yield, if O_2 could be directly reduced
to H_2O_2 (E_0', pH 7, $+0.30$ V) by electron carriers operating at the reducing
side of photosystem II (Losada, 1977, 1978), a suggestion that has recently
been experimentally realized (Elstner and Frommayer, 1978). The direct
photoreduction of 1 molecule of O_2 to H_2O_2 by electrons derived from water
and supplied by photosystem II (see Fig. 9) requires only 2 quanta and converts light energy into redox energy with a yield of 28%.

$$H_2O + O_2 \xrightarrow{2e} H_2O_2 + \tfrac{1}{2} O_2$$

$(\Delta E_0' = -0.52 \text{ V}; \Delta G_0' = +24 \text{ kcal})$

H_2O_2 tends to decompose into H_2O and O_2, particularly in the presence
of catalysts or enzymes (catalase). The reaction involves the transfer of
2 electrons and is strongly exergonic:

$$H_2O_2 + H_2O_2 \xrightarrow{2e} 2 \ H_2O + O_2$$

$(\Delta E_0' = +1.05 \text{ V}; \Delta G_0' = -48 \text{ kcal})$

For its high energy content, hydrogen peroxide, like ammonia, is a good
fuel, and has been used as a propellant for the launching of V-1 missiles and
in submarines. From a biological point of view it is remarkable that certain

representatives of the genus *Brachinus* (known to naturalists as "bombardier beetles") have developed two-compartmented reactor glands that open at the tip of the abdomen and contain an aqueous solution of H_2O_2 (25%) and hydroquinones in the inner compartment (reservoir) and a mixture of catalase and peroxidase in the outer one. In order to effect a discharge, the beetle squeezes some reservoir fluid into the vestibule, triggering what is essentially an instantaneous and explosive set of events and ejecting under the pressure of the free oxygen liberated a defensive spray at $100°C$ (Eisner, 1970).

12.5. REGULATION OF NITRATE REDUCTION

Regulation of nitrate reduction is achieved by several control mechanisms acting at different levels. Since nitrate uptake must occur before the nitrate-reducing process itself, control at this step is of outstanding importance. Additional relevant means for controlling the reduction of nitrate include regulation of enzyme synthesis and degradation and of enzyme activation and inactivation. The rate of in vivo nitrate and nitrite reduction will presumably be limited by the availability of reductant and by the competence of other metabolic processes that simultaneously make use of the specific reductants. Intermediate metabolites or final products of different pathways may equally have a regulatory role in nitrate reduction (see e.g., Plaut et al., 1977; Vennesland and Guerrero, in press), and feedback inhibition of nitrate reductase by its reaction product is a situation which might also take place in vivo, as nitrite behaves as an active inhibitor of the purified enzyme. Beevers and Hageman (1969) have regarded nitrate reductase as the rate-limiting enzyme in the reduction of nitrate, and considered this enzyme as the logical point of control of the input of reduced nitrogen for the plant. In fact, nitrate reductase appears to be the target for most of the control mechanisms of nitrate utilization.

In algae and higher plants, the capacity for nitrate utilization has been shown to vary in response to changes in environmental conditions such as light intensity, CO_2 and oxygen levels, temperature, nitrogen source, and other factors (Beevers and Hageman, 1969, 1972; Losada, 1974, 1976a, b; Hewitt, 1975; Hewitt et al., 1976; Vennesland and Guerrero, in press). Many of the reports regarding these effects do not allow, however, a direct conclusion about the nature of the control mechanisms in each case, for it has not been always possible to determine whether they were brought about by changes in the net amount of the enzyme proteins or by changes in enzyme activity.

Ammonia, the end product of nitrate reduction, has no inhibitory effect in vitro on nitrate reductase and nitrite reductase activities. Nevertheless, it behaves in vivo as a very active antagonist of nitrate metabolism in practically all types of organisms. In the last few years considerable attention has been paid to the study of this so-called "ammonia effect", and a good amount of information is available at present (Beevers and Hageman, 1969; Losada, 1974, 1976a, b; Hewitt, 1975; Hewitt et al., 1976; Syrett and Leftley, 1976; Vennesland and Guerrero, in press). Interpretation of the data has led, in some cases to controversy regarding the mechanism of ammonia action on the regulation of nitrate assimilation. This has been due, at least partly, to the fact that ammonia seems to affect practically all the different control mechanisms implied in the regulation of the in vivo reduction of nitrate (see below).

12.5.1. Uptake of nitrate

The absorption of nitrate by algal cells and higher plants appears to be an active process which requires metabolic energy (Eisele and Ullrich, 1975, 1977; Rao and Rains, 1976a, b). For a variety of plant species, nitrate has been reported to act as an inducer of the nitrate uptake system (Heimer and Filner, 1971; Schloemer and Garrett, 1974; Chantarotwong et al., 1976; Rao and Rains, 1976a, b). The light induced enhancement of nitrate absorption in plants is also well documented (Beevers and Hageman, 1969, 1972; Ullrich-Eberius, 1973; Hewitt et al., 1976; Rao and Rains, 1976a, b), and the existence of a diurnal cycle of nitrate uptake with a maximum in the middle of the photoperiod and a minimum in the dark period has been recently reported (Pearson and Steer, 1977).

Butz and Jackson (1977) have claimed that nitrate reductase itself is the carrier protein for nitrate uptake, but current evidence strongly argues against this idea. Apparently nitrate uptake can take place in cells which do not contain active nitrate reductase (Heimer and Filner, 1971; Schloemer and Garrett, 1974), and no close correlation exists between the processes of nitrate uptake and nitrate reduction (Smith, 1973; Chantarotwong et al., 1976; Pearson and Steer, 1977).

The rate of nitrate uptake appears to be an important factor in the regulation of nitrate assimilation. From experiments with barley seedlings, Chantarotwong et al. (1976) have concluded that the rate of in vivo nitrate reduction is more a function of the rate of uptake than of the amount of nitrate reductase present in the tissues, and the same has been suggested for the utilization of nitrate by the green alga *Ankistrodesmus braunii* (Eisele and Ullrich, 1977). On the other hand, it has also been proposed that nitrate reduction has a regulatory role in nitrate uptake (Ben-Zioni et al., 1971; Rao and Rains, 1976b).

The rapid inhibitory effect of ammonia on the rate of nitrate uptake in green algae has led to the suggestion that the nitrate uptake system is the first locus for ammonia action (Pistorius et al., 1976; Syrett and Leftley, 1976; E. Pistorius and E.A. Funkouser, unpublished). Inhibition by ammonia of nitrate uptake has also been reported to occur in higher plants (Minotti et al., 1969; Rao and Rains, 1976a), but exceptions have been observed in many instances (Smith and Thompson, 1971; Orebamjo and Stewart, 1975a; Mohanty and Fletcher, 1976).

12.5.2. Synthesis and degradation of enzymes

In general, the level of nitrate reductase is high in cells or plants grown on nitrate and low if grown on ammonia. Nitrite reductase usually follows a similar pattern. Both reductases thus behave as adaptive enzymes. Nitrate seems to act as an inducer of nitrate and nitrite reductases in higher plants (Beevers and Hageman, 1969, 1972; Hewitt et al., 1976), whereas in green algae de novo synthesis of the enzymes can also take place in a nitrogen-free medium or in the presence of certain amino acids (Vega et al., 1971; Herrera et al., 1972; Diez et al., 1977; Hipkin and Syrett, 1977). Actually, ammonia seems to repress the synthesis of both enzymes of the nitrate-reducing pathway in prokaryotic and eukaryotic algae (Losada et al., 1970; Solomonson and Vennesland, 1972; Morris, 1974; Losada, 1967a, b; Syrett and Leftley, 1976), but this may not be a general phenomenon, and many exceptions have been observed for higher plants (Hewitt et al., 1976; Mohanty and Fletcher, 1976). From experiments with *Lemna*, Orebamjo and Stewart (1975) have suggested that the repressive effect of ammonia is indirect through a product of ammonia assimilation acting at the level of transcription.

Synthesis of nitrate reductase following induction by nitrate has been clearly shown to involve de novo protein synthesis (Zielke and Filner, 1971). Nitrate reductase can also be induced by a wide variety of unrelated substances (Hewitt et al., 1976; Vennesland and Guerrero, in press). In rice seedlings, chloramphenicol and related nitro compounds induce an enzyme which reacts with both NADH and NADPH, whereas the normal nitrate-induced enzyme is NADH specific (Shen et al., 1976). The absence of a clear correlation between the nitrate reductase level and the level of endogenous nitrate in the tissues of higher plants has led to the proposal that nitrate exists within the cell in two pools: a storage or nonmetabolic pool (perhaps in the vacuole) and an active or metabolic pool. Only nitrate in the metabolic pool would be effective for induction of nitrate reductase (see Aslam et al., 1976).

Light seems to be a requirement for the synthesis of nitrate reductase in photosynthetic tissues (Beevers and Hageman, 1969, 1972; Hewitt, 1975; Hewitt et al., 1976), and it has been suggested that light-induced electron flow is required for induction of the enzyme (Sawhney and Naik, 1972; Sawhney et al., 1972). According to Aslam et al. (1976), a light-dependent transfer of nitrate from the vacuole to the metabolic pool would be responsible for the dependence on light of nitrate-reductase induction by nitrate in higher plants. Fast changes of the nitrate reductase activity level on illumination or darkness take place in synchronized *Chlorella* cells (Hodler et al., 1972; Tischner, 1976), and circadian rythms of enzyme activity have been reported in algae and higher plants, the nitrate reductase activity level being maximum in the light period and declining in the dark (Tischner, 1976; Steer, 1976). It is not clear whether such changes are brought about by variations in the level of enzyme protein, in the level of activity, or in both. Plant hormones and phytochrome appear to be involved in these effects (Hewitt et al., 1976; Guerrero and Vennesland, in press). In contrast, both nitrate and nitrite reductases are present at considerable levels in *Chlorella* cells grown in the dark with glucose as a carbon source (Guerrero et al., 1971).

A decay of nitrate reductase activity has been observed after transferring plants to darkness or to noninductive conditions, suggesting the operation of a mechanism for enzyme degradation (Kannangara and Woolhouse, 1967; Beevers and Hageman, 1969, 1972; Zielke and Filner, 1971; Hewitt et al., 1976). Specific proteolysis of nitrate reductase could be implicated in this phenomenon (Wallace, 1975).

It appears that the level of enzyme protein at a given time is a function of the relative rate of synthesis and degradation with nitrate reductase turning over continuously (Zielke and Filner, 1971; Beevers and Hageman, 1972). The fluctuations in the level of this protein can be brought about by changes in the relative rate of synthesis and decay.

It has to be considered that the relative rate of assembly and dissembly of the different components of nitrate reductase might also play a role in determining the level of the holoenzyme (Hewitt et al., 1977).

12.5.3. Activation and inactivation of enzymes

In addition to the long-term changes in the amount of enzyme that might be ascribed to controlled synthesis and degradation, the occurrence of rapid variations of the nitrate reductase level in response to the presence of ammonia or to changes in particular environmental factors has been recently reported (see Losada, 1974, 1976a, b; Hewitt, 1975; Hewitt et al., 1976; Guerrero and Vennesland, in press). Enzyme activation and inactivation processes are thought to be responsible for these effects.

Different types of mechanisms have been proposed for the regulation of nitrate reductase activity. In vivo reversible inactivation of the enzyme has been explained by a reversible protein—protein interaction in *Lemna* (Orebamjo and Stewart, 1975b), whereas in the thermophilic alga *Cyanidium caldarium* it has been suggested that in the enzyme complex there is a labile moiety bearing an inhibition site for ammonia (or a product of its assimilation), which on destruction results in reactivation of the enzyme (Rigano et al., 1974). The existence in higher plants of enzymes which specifically inactivate nitrate reductase and which could play a role in regulating the activity level of the enzyme has also been reported (Kadam et al., 1974; Wallace, 1975; Yamaya and Ohira, 1977). The interconversion of nitrate reductase which occurs in green and blue-green algae seems, however, to be the most interesting and exciting example of regulation of nitrate reductase activity so far observed. In these organisms, the enzyme exists in two metabolically interconvertible forms, either active or inactive, the relative proportion of each being dependent on the intracellular concentration of specific metabolites, which in turn are affected by changes in environmental conditions (Losada, 1974, 1976a, b).

12.5.3.1. Metabolic interconversion of nitrate reductase

The phenomenon of nitrate reductase interconversion was first reported in 1970 by Losada et al., who made the observation that addition of ammonia to Chlorella fusca cells growing on nitrate resulted in a rapid decrease of the level of nitrate reductase activity. Only the terminal moiety of the enzyme complex was inactivated, the NADH-diaphorase activity remaining fully active. Subsequent removal of ammonia from the culture resulted in a rapid and complete reactivation of the enzyme. Since then, this reversible inactivation of nitrate reductase has been extensively studied by the Sevilla group, both in vivo and in vitro, and it has been shown to occur in other green algae, such as Chlamydomonas reinhardii (Herrera et al., 1972; Losada, 1974, 1976a, b) and Ankistrodemus braunii (Diez et al., 1977).

Ammonia can be regarded as an uncoupler of photosynthetic phosphorylation, and inactivation of nitrate reductase by this end-product of nitrate reduction has been interpreted as a result of its uncoupling action, through the increase in the cellular levels of reducing power and ADP (Losada, 1974, 1976a,b). In fact, the effect of other uncouplers, such as methylamine and arsenate, parallel the action of ammonia when added to illuminated cells containing active nitrate reductase. However, if non-cyclic electron flow is inhibited or accumulation of reducing power is prevented, inactivation by these compounds does not take place (Losada et al., 1973; Maldonado et al., 1974; Diez et al., 1977). Moreover, experiments carried out with C. fusca cells have shown that inactivation of nitrate reductase following addition of methylamine to the culture medium is concomitant with an increase in the cellular levels of the reduced forms of pyridine nucleotides and of the uncharged form of adenine nucleotides (Chaparro et al., 1976).

In vitro, the active form of nitrate reductase can be converted into the inactive form by adding NAD(P)H with ADP showing a specific cooperative effect with the reduced pyridine nucleotide for this inactivation. Nitrate and its competitive inhibitors, such as cyanate, protect the enzyme against this in vitro inactivation. Strong oxidants, such as ferricyanide, instantaneously reactivate both the in vivo and the in vitro inactivated enzyme (Herrera et al., 1972; Vega et al., 1972; Maldonado et al., 1973; Losada, 1974, 1976a, b; Diez et al., 1977).

According to this hypothesis, the metabolic interconversion of nitrate reductase in green algae is primarily induced by a change in the redox status of the cell, the active oxidized enzyme becoming inactivated upon reduction and, conversely, the inactive reduced enzyme becoming activated upon oxidation. This phenomenon represents a new type of metabolic enzyme interconversion through chemical modification by reduction and oxidation (Losada, 1974, 1976a,b). The mechanism of light activation—inactivation of Calvin cycle enzymes has been recently shown to involve also reduction-oxidation of the regulatory enzymes (see Losada, 1976a; Montagnoli, 1977).

The same basic mechanism for the regulation of nitrate reductase activity which operates in green algae appears also to function in blue-green algae. Ammonia and other uncouplers inhibit the photosynthetic reduction of nitrate by particles of *N. muscorum*. Furthermore, nitrate reductase is reversibly inactivated by reduction in a ferredoxin-dependent reaction and reactivated by molecular oxygen. Nitrate can protect against this inactivation (Ortega et al., 1977a, b). Interesting enough, addition of ammonia to particles of *Nostoc* not only affects nitrate reductase, but also ferredoxin : NADP reductase, which becomes reversibly inactivated. The reduced inactive enzyme can be reactivated by oxidation (Ortega et al., 1977b). These results mean that ferredoxin—NADP reductase can be added to the list of enzymes subjected to regulation according to the above model of metabolic enzyme interconversion by reduction and oxidation.

A different conclusion about the mechanism of the reversible inactivation of nitrate reductase in algae has been reached by workers in Berlin based on experiments carried out with *Chlorella vulgaris* (see Vennesland and Guerrero, in press). Addition of ammonia to cells of this alga growing on nitrate results also in inactivation of the enzyme, and the inactivated enzyme can be rapidly reactivated by ferricyanide or more slowly by incubation with nitrate and phosphate buffer. Applying these reactivating procedures to the inactive enzyme obtained by addition of ammonia to the cells, stoichiometric amounts of HCN are released as the enzyme becomes activated (Lorimer et al., 1974). The inactive species of nitrate reductase is thus regarded as the cyanide complex of the reduced enzyme, identical to that form of the enzyme which results from the in vitro treatment of active nitrate reductase with NADH and HCN (Solomonson et al., 1973; Solomonson, 1974; Lorimer et al., 1974).

Histidine has been proposed to be a precursor of cyanide in *C. vulgaris* through the action of peroxidase and an amino acid oxidase (Pistorius and Voss, 1977; Pistorius et al., 1977). Glyoxylate oxime is another candidate for the in vivo production of cyanide (Gewitz et al., 1976). Solomonson and Spehar (1977) have recently proposed a model for the inactivation by ammonia of nitrate reductase in *C. vulgaris*, through the formation of glyoxylate oxime and cyanide.

In contrast to the situation with the nitrate reductase of *C. fusca*, *C.*

reinhardii and *A. braunii* (Losada, 1974, 1976b; Diez et al., 1977), incubation with reduced pyridine nucleotide (alone or in conjunction with ADP) does not lead to inactivation of the purified enzyme of *C. vulgaris* (Solomonson, 1974). Differences have also been found in the effect of some factors such as oxygen and CO_2 tension on the in vivo inactivation of nitrate reductase from *C. fusca* and *C. reinhardii* on one side, and from *C. vulgaris* of the other (see Pistorius et al., 1976; Vennesland and Guerrero, in press). The existence of such discrepancies does not allow the results obtained in the two laboratories to be understood in terms of a common interpretation.

It should be pointed out that, at present, the Berlin group regards inhibition of the nitrate uptake system as the primary event in the prompt cessation of nitrate utilization after ammonia addition, the reversible inactivation of nitrate reductase having also a physiological significance in control but coming in a second stage (Pistorius et al., 1976; E. Pistorius and E.A. Funkhouser, unpublished). The experiments by Ortega et al. (1977) with *Nostoc* particles described above speak, however, in favor of a more direct effect of ammonia on the photosynthetic reactions leading to inactivation.

Aparicio et al. (1976) have reported that the inactive form of nitrate reductase from *C. fusca* and spinach leaves can be rapidly activated by blue light but not by red light, and have suggested a physiological role for quality of light in modulating nitrate reductase activity, the flavin prosthetic group being the light-absorbing pigment. In the same context are also the results of Jones and Sheard (1977), who have observed that plants grown under blue light exhibit higher levels of nitrate reductase than those grown under red light, and that transfer of plants from red to blue light results in increased nitrate reductase activity whereas transfer from blue to red brings about a decay in enzyme activity.

12.6. CONCLUSIONS

The long controversy regarding mechanism, pathway, regulation, and localization of nitrate reduction in green cells, as well as the role of light in the process, now seems to be virtually resolved in its most basic aspects.

Photosynthesis consists essentially in the transduction of light energy into electronic energy and subsequently into redox energy leading to the reduction by water of the bioelements carbon, nitrogen and sulfur. Ferredoxin is the universal intermediate electron carrier which connects the photooxidation of water with the assimilation of these bioelements. Carbon is halfway reduced in a 4-electron reaction from its most oxidized state, carbon dioxide (oxidation state +4), to carbohydrate (oxidation state 0), whereas nitrogen and sulfur are totally reduced in an 8-electron reaction from nitrate (+5) to ammonia (−3) and from sulfate (+6) to sulfide (−2). Only in the case of nitrate-nitrogen, the photosynthetically generated reducing power at the level of ferredoxin (E'_0 = —0.42 V) or pyridine nucleotide (E'_0 = —0.32 V) is energetic enough to promote its reduction to ammonia (E'_0 = +0.35 V). By contrast, the reduction of carbon and sulfur by these carriers are endergonic processes which require, in addition, adenosine triphosphate for the corresponding uphill transfer of electrons. Since both the photooxidation of water in the intrathylakoid space and the reduction of the bioelements in the extrathylakoid space involve protons, the light-driven overall redox reaction is accompanied by the generation of an electrochemical potential gradient of protons.

The assimilatory nitrate-reducing system consists of only two metalloproteins, nitrate reductase and nitrite reductase, which catalyze in series the 2-electron reduction of nitrate to nitrite and the 6-electron reduction of nitrite to ammonia. Both enzymes are ferredoxin-dependent and tightly bound to the chlorophyll-containing particles in blue-green algae, whereas in green algae and higher plants only nitrite reductase is directly dependent on ferredoxin as the electron donor, nitrate reductase being a NAD(P)H-enzyme. There seems to be general agreement about the intracellular localization of nitrite reductase in green eukaryotic cells as a soluble chloroplast enzyme, but there is contradictory evidence regarding nitrate reductase; recent well-supported proposals claim, however, for its localization in the outer membrane of the photosynthetic organelle.

In the transfer of electrons from NAD(P)H to nitrate, catalyzed by the enzyme complex NAD(P)H-nitrate reductase from eukaryotic cells, two moieties participate sequentially: the first is a FAD-dependent NAD(P)H diaphorase (or dehydrogenase), and the second is the molybdenum-dependent nitrate reductase proper or terminal nitrate reductase. Cytochrome b-557 seems to act as electron carrier in the interconnection between both activities. Ferredoxin-nitrate reductase from blue-green algae apparently lacks the diaphorase moiety and, accordingly, is a much smaller protein.

Ferredoxin-nitrite reductase from photosynthetic organisms contains an iron—sulfur center (Fe_2—S_2^*) and a special heme (siroheme) as functional prosthetic groups in the transfer of electrons from ferredoxin to nitrite.

Nitrate reduction, being the port of entry for nitrogen assimilation, is a step of outstanding importance in the regulation of photosynthetic metabolism, and both nitrate and ammonia (the end-product of its assimilatory pathway) seem to be very efficient antagonists in controlling the process. Ammonia is not only a nutritional repressor of the enzymes of the nitrate-reducing systems, but acts in vivo through a very sophisticated mechanism to regulate the activity of the first enzyme of the pathway by promoting the reversible conversion of the oxidized active form of the terminal moiety of nitrate reductase into its reduced inactive form. By itself, ammonia does not have any effect on the enzyme in vitro. Although ammonia may play a relevant role in the inhibition of nitrate uptake, it seems that its intracellular accumulation induces an increase in the level of reducing power and a

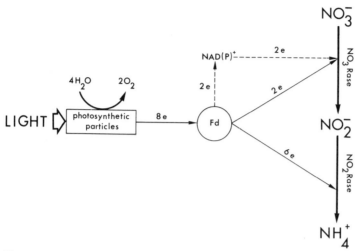

Fig. 12.10. Schematic diagram of the photosynthetic reduction of nitrate to ammonia coupled with the oxidation of water to oxygen. In prokaryotes, the two enzymes of the nitrate-reducing system (nitrate reductase and nitrite reductase) are ferredoxin-dependent and accept electrons directly from this carrier, whereas in eukaryotes the first enzyme of the system is a NAD(P)H-dependent complex that accepts electrons indirectly from ferredoxin through pyridine nucleotide (dashed line).

decrease in the energy charge as a result of its uncoupling action on photophosphorylation, and that nitrate reductase inactivation is a direct consequence of the synergistic effect of NADH and ADP on the overreduction of the enzyme protein. Such an effect is prevented by the presence of nitrate. Alternatively, ammonia might act through the formation of cyanide, a very strong inhibitor of the reduced enzyme.

Although in the last few years great advances have been made in the understanding of the molecular mechanism of nitrate reduction and its regulation, it still remains a fascinating subject for further research. At present, the genetic approach is being successfully applied to the study of nitrate utilization in algae and higher plants.

Besides its biochemical and chemical interest, ammonia is an excellent and powerful fuel and a source of hydrogen, and as such has found prominent application in internal-combustion engines, jet propulsion, and the oxyhydrogen blow-pipe. Its photoproduction by the reduction of nitrate with water may, therefore, be of great interest as an attractive model to be imitated for the conversion and storage of solar energy.

In summary, the photosynthetic reduction of nitrate to ammonia represents one of the simplest and most relevant examples of photosynthesis (Fig. 12.10). Further insight in its knowledge will undoubtedly provide a more solid basis for practical applications in agriculture and in the field of energy utilization.

12.7. ACKNOWLEDGEMENTS

The preparation of this article was aided in part by grants from the Philips Research Laboratories (The Netherlands), the National Science Foundation (U.S.A.) and the Comisaría Asesora de Investigación (Spain). We thank W. Ullrich for a critical reading of the manuscript.

REFERENCES

Åberg, B. (1947) K. Lantbrukshoegsk. Ann., 15, 37—107.

Ahmed, J. and Spiller, H. (1976) Plant Cell Physiol., 17, 1—10.

Alef, K. and Klemme, J.H. (1977) Z. Naturforsch., 32c, 954—956.

Allen, J.F. (1977) in Superoxide and Superoxide Dismutases (A.M. Michelson, J.M. McCord, I. Fridovich, eds.), pp. 417—436. Academic Press, London.

Amy, N.K., Garrett, R.H. and Anderson, B.M. (1977) Biochim. Biophys. Acta, 480, 83—95.

Anderson, J.W. and Done, J. (1977) Plant Physiol., 60. 504—508.

Aparicio, P.J., Knaff, D.B. and Malkin, R. (1975) Arch. Biochem. Biophys., 169, 102—107.

Aparicio, P.J., Roldán, J.M. and Calero, F. (1976) Biochem. Biophys. Res. Commun., 70, 1071—1077.

Aparicio, P.J., Cárdenas, J., Zumft, W.G., Vega, J.M., Herrera, J., Paneque, A. and Losada, M. (1971) Phytochemistry, 10, 1487—1495.

Arnon, D.I. (1977) in Encyclopedia of Plant Physiology, New Series (A. Trebst, M. Avron, eds.), Vol. 5, pp. 7—56. Springer Verlag, Berlin.

Arnon, D.I., Losada, M., Whatley, F.R., Tsujimoto, H.Y., Hall, D.O. and Horton, A.A. (1961) Proc. Natl. Acad. Sci. U.S.A., 47, 1314—1334.

Aslam, M., Oaks, A. and Huffaker, R.C. (1976) Plant Physiol., 58, 588—591.

Barea, J.L., Maldonado, J.M. and Cárdenas, J. (1976) Physiol. Plant., 36, 325—332.

Beevers, L. and Hageman, R.H. (1969) Annu. Rev. Plant Physiol., 20, 495—522.

Beevers, L. and Hageman, R.H. (1972) in Photophysiology (A.C. Giese, ed.), Vol. 7, pp. 85—113. Academic Press, New York.

Ben-Zioni, A., Vaadia, Y. and Lips, S.H. (1971) Physiol. Plant., 24, 288—290.

Bothe, H. (1969) in Progress in Photosynthesis Research (H. Metzner, ed.), Vol. 3, pp. 1483—1491. IUBS, Tübingen.

Bothe, H. (1977) in Encyclopedia of Plant Physiology, New Series (A. Trebst, M. Avron, eds.), Vol. 5, pp. 217—221. Springer Verlag, Berlin.

Butz, R.G. and Jackson, W.A. (1977) Phytochemistry, 16, 409—417.

Buvet, R., Allen, M.J. and Massué, J.-P. (eds.) (1977) Living Systems as Energy Converters. North-Holland, Amsterdam.

Campbell, W.H. (1976) Plant Sci. Lett., 7, 239—247.

Candau, P., Manzano, C. and Losada, M. (1976) Nature, 262, 715—717.

Canvin, D.T. and Atkins, C.A. (1974) Planta, 116, 207—224.

Cárdenas, J., Rivas, J. and Barea, J.L. (1972a) Rev. Real Acad. Cienc. (Spain), 66, 565—577.

Cárdenas, J., Barea, J.L., Rivas, J. and Moreno, C.G. (1972b) FEBS Lett., 23, 131—135.

Castillo, F., De la Rosa, F.F., Calero, F. and Palacián, E. (1976) Biochem. Biophys. Res. Commun., 69, 277—284.

Chantarotwong, W., Huffaker, R.C., Miller, B.L. and Granstedt, R.C. (1976) Plant Physiol., 57, 519—522.

404

Chaparro, A., Maldonado, J.M., Diez, J., Relimpio, A.M. and Losada, M. (1976) Plant Sci. Lett., 6, 335—342.

Coddington, A. (1976) Mol. Gen. Genet., 146, 195—206.

Coombs, J. (1976) in Topics in Photosynthesis (J. Barber, ed.), Vol. 1, pp. 279—313. Elsevier, Amsterdam.

Coupé, M., Champigny, M.L. and Moyse, A. (1967) Physiol. Veg., 3, 271—279.

Cove, D.J. (1976) Mol. Gen. Genet., 146, 147—159.

Cramer, M. and Myers, J. (1949) J. Gen. Physiol., 32, 93—102.

Czygan, F.C. (1963) Planta, 60, 225—242.

Dalling, M.J., Tolbert, N.E. and Hageman, R.H. (1972) Biochim. Biophys. Acta, 283, 505—512.

Davies, D.D., Texeira, A. and Kenworthy, P. (1972) Biochem. J., 127, 335—343.

De la Rosa, F.F. (1975) Ph.D. Thesis. University of Sevilla.

Del Campo, F.F., Paneque, A., Ramirez, J.M. and Losada, M. (1963) Biochim. Biophys. Acta, 66, 450—452.

Diez, J., Chaparro, A., Vega, J.M. and Relimpio, A.M. (1977) Planta, 137, 231—234.

Downey, R.J. (1973) Biochem. Biophys. Res. Commun., 50, 920—926.

Eaglesham, A.R.J. and Hewitt, E.J. (1971) FEBS Lett., 16, 315—317.

Eaglesham, A.R.J. and Hewitt, E.J. (1975) Plant Cell Physiol., 16, 1137—1149.

Eisele, R. and Ullrich, W.R. (1975) Planta, 123, 117—123.

Eisele, R. and Ullrich, W.R. (1977) Plant Physiol., 59, 18—21.

Eisner, T. (1970) in Chemical Ecology (E. Sondheimer, J.B. Simeone, eds.), pp. 157—217. Academic Press, New York.

Elstner, E.F. and Frommayer, D. (1978) Z. Naturforsch., 33c, 276—279.

Garner, C.D., Hyde, M.R., Mabbs, F.F. and Routledge, V.I. (1974) Nature, 252, 579—580.

Garrett, R.H. and Nason, A. (1969) J. Biol. Chem., 244, 2870—2882.

Gewitz, H.S., Pistorius, E.K., Voss, H. and Vennesland, B. (1976) Planta, 131, 145—148.

Grant, B.R. and Canvin, T. (1970) Planta, 95, 227—246.

Grant, B.R. and Turner, I.M. (1969) Comp. Biochem. Physiol., 29, 995—1004.

Grant, B.R., Atkins, C.A. and Canvin, D.T. (1970) Planta, 94, 60—72.

Guerrero, M.G. and Gutierrez, M. (1977) Biochim. Biophys. Acta, 482, 272—285.

Guerrero, M.G. and Vennesland, B. (1975) FEBS Lett., 51, 284—286.

Guerrero, M.G., Jetschmann, K. and Völker, W. (1977) Biochim. Biophys. Acta, 482, 19—26.

Guerrero, M.G., Manzano, C. and Losada, M. (1974) Plant Sci. Lett., 3, 273—278.

Guerrero, M.G., Rivas, J., Paneque, A. and Losada, M. (1971) Biochem. Biophys. Res. Commun., 45, 82—89.

Guerrero, M.G., Vega, J.M., Leadbetter, E. and Losada, M. (1973) Arch. Mikrobiol., 91, 287—304.

Harel, E., Lea, P.J. and Miflin, B.J. (1977) Planta, 134, 195—200.

Hattori, A. (1962) Plant Cell Physiol., 3, 355—369.

Hattori, A. (1970) Plant Cell Physiol., 11, 975—980.

Hattori, A. and Myers, J. (1967) Plant Cell Physiol., 8, 327—337.

Hattori, A. and Uesugi, I. (1968) Plant Cell Physiol., 9, 689—699.

Heber, U. and French, C.S. (1968) Planta, 79, 99—112.

Heimer, Y.M. and Filner, P. (1971) Biochim. Biophys. Acta, 230, 362—372.

Herrera, J., Paneque, A., Maldonado, J.M., Barea, J.L. and Losada, M. (1972) Biochem. Biophys. Res. Commun., 48, 996—1003.

Hewitt, E.J. (1975) Annu. Rev. Plant Physiol., 26, 73—100.

Hewitt, E.J. and Nicholas, D.J.D. (1964) in Modern Methods of Plant Analysis (H.F. Linskens, B.D. Sanwall, M.V. Tracey, eds.), Vol. 7, pp. 67—122. Springer Verlag, Berlin.

Hewitt, E.J., Hucklesby, D.P. and Notton, B.A. (1976) in Plant Biochemistry (J. Bonner, J.E. Varner, eds.), 3rd edn., pp. 633—681. Academic Press, New York.

Hewitt, E.J., Notton, B.A. and Rucklidge, G.J. (1977) J. Less-Common Metals, 54, 537—553.

Hill, R. (1965) in Essays in Biochemistry (P.N. Campbell, G.D. Greville, eds.), Vol. 1, pp. 121—151. Academic Press, New York.

Hipkin, C.P. and Syrett, P.J. (1977) Planta, 133, 209—214.

Ho, C.H., Ikawa, T. and Nisizawa, K. (1976) Plant Cell Physiol., 17, 417—430.

Hodler, M., Morgenthaler, J.J., Eichenberger, W. and Grob, E.C. (1972) FEBS Lett., 28, 19—21.

Hucklesby, D.P., James, D.M., Banwell, M.J. and Hewitt, E.J. (1976) Phytochemistry, 15, 599—603.

Jensen, R.G. and Bassham, J.A. (1966) Proc. Natl. Acad. Sci. U.S.A., 56, 1095—1101.

Jolly, S.O., Campbell, W. and Tolbert, N.E. (1976) Arch. Biochem. Biophys., 174, 431—439.

Jones, K. (1975) The Chemistry of Nitrogen. Pergamon Press, Oxford.

Jones, R.W. and Sheard, R.W. (1977) Plant Sci. Lett., 8, 305—311.

Kadam, S.S. and Naik, M.S. (1975) Indian J. Biochem. Biophys., 12, 317—320.

Kadam, S.S. and Naik, M.S. (1976) Indian J. Biochem. Biophys., 13, 223—227.

Kadam, S.S., Gandhi, A.P., Sawhney, S.K. and Naik, M.S. (1974) Biochim. Biophys. Acta, 350, 162—170.

Kannangara, C.G. and Woolhouse, H.W. (1967) New Phytol., 66, 553—561.

Katoh, T. (1963a) Plant Cell Physiol., 4, 13—28.

Katoh, T. (1963b) Plant Cell Physiol., 4, 199—215.

Kessler, E. (1964) Annu. Rev. Plant Physiol., 15, 57—72.

Kessler, E. (1971) Prog. Bot., 33, 95—103.

Kessler, E. (1974) Prog. Bot., 36, 99—107.

Kessler, E. (1976) Prog. Bot., 38, 108—117.

Kessler, E. and Zumft, W.G. (1973) Planta, 111, 41—46.

Ketchun, P.A., Taylor, R.C. and Young, D.C. (1976) Nature, 259, 202—204.

Klepper, L., Flesher, D. and Hageman, R.H. (1971) Plant Physiol., 48, 580—590.

Kulandaivelu, G., Spiller, H. and Böger, P. (1976) Plant Sci. Lett., 7, 225—231.

Leech, R.M. and Murphy, D.J. (1976) in Topics in Photosynthesis (J. Barber, ed.), Vol. 1, pp. 365—401. Elsevier, Amsterdam.

Lipmann, F. (1976) in Reflections on Biochemistry (A. Kornber, B.L. Horecker, L. Cornudella, J. Oró, eds.), pp. 33—38. Pergamon Press, Oxford.

Lips, S.H. (1975) Plant Physiol., 55, 598—601.

Lips, S.H. and Avissar, Y. (1972) Eur. J. Biochem., 29, 20—24.

Lorimer, G.H., Gewitz, H.-S., Völker, W., Solomonson, L.P. and Vennesland, B. (1974) J. Biol. Chem., 249, 6074—6079.

Losada, M. (1974) in Metabolic Interconversion of Enzymes (E.H. Fisher, E.G. Krebs, H. Neurath, E.R. Stadtman, eds.), pp. 257—270. Springer Verlag, Berlin.

Losada, M. (1976a) in Reflections on Biochemistry (A. Kornberg, B.L. Horecker, L. Cornudella, J. Oró, eds.), pp. 73—84. Pergamon Press, Oxford.

Losada, M. (1976b) J. Mol. Catal., 1, 245—264.

Losada, M. (1977) Invest. Cien. (Spanish Edn. of Sci. Am.), 7, 6—18.

Losada, M. (1978a) Bioelectrochem. Bioenerg. (submitted).

Losada, M. (1978b) International Conference Bio-Energy: Energy from Living Systems. Ruschlikon, Gottlieb Duttweiler Institute, Zurich.

Losada, M. (1978c) International Symposium on Energy Sources and Development. Barcelona Vol. 2, pp. 32—51. Editorial Moneda y Crédito, Madrid.

Losada, M. and Arnon, D.I. (1964) in Modern Methods in Plant Analysis (H.F. Linskens, B.D. Sanwall, M.V. Tracey, eds.), Vol. 7, pp. 569—615. Springer Verlag, Berlin.

Losada, M. and Paneque, A. (1966) Biochim. Biophys. Acta, 126, 578—580.

Losada, M. and Paneque, A. (1971) Methods Enzymol., 23, 487—491.

Losada, M., Herrera, J., Maldonado, J.M. and Paneque, A. (1973) Plant Sci. Lett., 1, 31—37.

Losada, M., Paneque, A., Ramirez, J.M. and Del Campo, F.F. (1963) Biochem. Biophys. Res. Commun., 10, 298—303.

Losada, M., Ramirez, J.M., Paneque, A. and Del Campo, F.F. (1965) Biochim. Biophys. Acta, 109, 86—96.

Losada, M., Paneque, A., Aparicio, P.J., Vega, J.M., Cárdenas, J. and Herrera, J. (1970) Biochem. Biophys. Res. Commun., 38, 1009—1015.

Magalhaes, A.C., Neyra, C.A. and Hageman, R.H. (1974) Plant Physiol., 53, 411—415.

Maldonado, J.M., Pueyo, M.C. and Chaparro, A. (1974) Rev. Real Acad. Cien. (Spain), 68, 633—642.

Maldonado, J.M., Herrera, J., Paneque, A. and Losada, M. (1973) Biochem. Biophys. Res. Commun., 51, 27—33.

Malofeeva, I.V., Kondratieva, E.N. and Rubin, A.B. (1975) FEBS Lett., 53, 188—189.

Manzano, C. (1977) Ph.D. Thesis. University of Sevilla.

Manzano, C., Candau, P. and Guerrero, M.G. (1977) European Seminar on Biological Solar Energy Conversion System, Abstr. 14. Grenoble-Autrans.

Manzano, C., Candau, P., Gómez-Moreno, C., Relimpio, A.M. and Losada, M. (1976) Mol. Cell. Biochem., 10, 161—169.

McDonald, D.W. and Coddington, A. (1974) Eur. J. Biochem., 46, 169—178.

Mehler, A.H. (1951) Arch. Biochem. Biophys., 33, 65—77.

Mellor, G.E. and Tregunna, E.B. (1971) Can. J. Bot., 49, 137—142.

Meyer, V. and Schultze, E. (1894) Ber. Dtsch. Chem. Ges., 17, 1554.

Miflin, B.J. (1974) Plant Physiol., 54, 550—555.

Minotti, P.L., Williams, D.C. and Jackson, W.A. (1969) Planta, 86, 267—271.

Mohanty, B. and Fletcher, J.S. (1976) Plant Physiol., 58, 152—155.

Montagnoli, G. (1977) Phytochem. Photobiol., 26, 679—683.

Morris, I. (1974) in Algal Physiology and Biochemistry (W.P.D. Stewart, ed.), pp. 583—609. Blackwell, Oxford.

Murphy, M.J. and Siegel, L.M. (1973) J. Biol. Chem., 248, 6911—6919.

Murphy, M.J., Siegel, L.M., Kamin, H. and Rosenthal, D. (1973) J. Biol. Chem., 248, 2801—2814.

Murphy, M.J., Siegel, L.M., Tove, S.R. and Kamin, H. (1974) Proc. Natl. Acad. Sci. U.S.A., 71, 612—616.

Nason, A. (1962) Bacteriol. Rev., 26, 16—41.

Neyra, C.A. and Hageman, R.H. (1974) Plant Physiol., 54, 480—483.

Nicholas, D.J.D. and Stevens, H.M. (1955) Nature, 176, 1066—1067.

Notton, B.A., Fido, R.J. and Hewitt, E.J. (1977) Plant Sci. Lett., 8, 165—170.

Orebamjo, T.O. and Stewart, G.R. (1975a) Planta, 122, 27—36.

Orebamjo, T.O. and Stewart, G.R. (1975b) Planta, 122, 37—44.

Ortega, T., Castillo, F. and Cárdenas, J. (1976) Biochem. Biophys. Res. Commun., 71, 885—891.

Ortega, T., Castillo, F., Cárdenas, J. and Losada, M. (1977a) Biochem. Biophys. Res. Commun., 75, 823—831.

Ortega, T., Rivas, J., Cárdenas, J. and Losada, M. (1977b) Biochem. Biophys. Res. Commun., 78, 185—193.

Pan, S.-S. and Nason, A. (1976) Fed. Proc., 35, 1530, Abstr. 887.

Paneque, A., Del Campo, F.F. and Losada, M. (1963) Nature, 198, 90—91.

Paneque, A., Ramirez, J.M., Del Campo, F.F. and Losada, M. (1964) J. Biol. Chem., 239, 1737—1741.

Paneque, A., Aparicio, P.J., Cárdenas, J., Vega, J.M. and Losada, M. (1969) FEBS Lett., 3, 57—59.

Patterson, C.O.P. and Myers, J. (1973) Plant Physiol., 51, 104—109.

Payne, W.J. (1973) Bacteriol. Rev., 37, 409—452.

Pearson, C.J. and Steer, B.T. (1977) Planta, 137, 107—112.

Pistorius, E.K. and Voss, H. (1977) Biochim. Biophys. Acta, 481, 395—406.

Pistorius, E.K., Gewitz, H.-S., Voss, H. and Vennesland, B. (1976) Planta, 128, 73—80.

Pistorius, E.K., Gewitz, H.-S., Voss, H. and Vennesland, B. (1977) Biochim. Biophys. Acta, 481, 384—394.

Plaut, Z. and Littan, A. (1974) in Proc. 3rd Int. Congr. Photosynthesis (M. Avron, ed.), pp. 1507—1515. Elsevier, Amsterdam.

Plaut, Z., Lendzian, K. and Bassham, J.A. (1977) Plant Physiol., 59, 184—188.

Rabinowitch, E.I. (1945) Photosynthesis and related Processes, Vol. 1. Interscience Publishers, New York.

Racker, E. (1976) A New Look at Mechanisms in Bioenergetics. Academic Press, New York.

Radmer, R.J. and Kok, B. (1976) Plant Physiol., 58, 336—340.

Ramirez, J.M., Del Campo, F.F., Paneque, A. and Losada, M. (1966) Biochim. Biophys. Acta, 118, 58—71.

Rao, K.P. and Rains, D.W. (1976a) Plant Physiol., 57, 55—58.

Rao, K.P. and Rains, D.W. (1976b) Plant Physiol., 57, 59—62.

Rao, K.K., Rosa, L. and Hall, D.O. (1976) Biochem. Biophys. Res. Commun., 68, 21—28.

Rathnam, C.K.M. and Das, V.S.R. (1974) Can. J. Bot., 55, 2599—2605.

Rathnam, C.K.M. and Edwards, G.E. (1976) Plant Physiol., 57, 881—885.

Relimpio, A.M., Aparicio, P.J., Paneque, A. and Losada, M. (1971) FEBS Lett., 17, 226—230.

Relimpio, A.M., Vega, J.M., Guerrero, M.G. and Losada, M. (1977) Potenciometría y Bioenergética. Universidad de Sevilla, Sevilla.

Rigano, C., Aliotta, G. and Violante, U. (1974) Arch. Microbiol., 99, 81—90.

Ritenour, G.L., Joy, K.W., Bunning, J. and Hageman, R.J. (1967) Plant Physiol., 42, 233—237.

Robinson, J.M. and Stocking, C.R. (1968) Plant Physiol., 43, 1597—1604.

Roussos, G.G. and Nason, A. (1960) J. Biol. Chem., 235, 2997—3007.

Sawhney, S.K. and Naik, M.S. (1972) Biochem. J., 130, 475—485.

Sawhney, S.K., Prakash, V. and Naik, M.S. (1972) FEBS Lett., 22, 200—202.

Schloemer, R.H. and Garrett, R.H. (1974) J. Bacteriol., 118, 259—269.

Schwenn, J.D. and Trebst, A. (1976) in Topics in Photosynthesis (J. Barber, ed.), Vol. 1, pp. 315—334. Elsevier, Amsterdam.

Shen, T.C., Funkhouser, E.A. and Guerrero, M.G. (1976) Plant Physiol., 58, 292—294.

Smith, F.A. (1973) New Phytol., 72, 769—782.

Smith, F.W. and Thompson, J.F. (1971) Plant Physiol., 48, 219—223.

Solomonson, L.P. (1974) Biochim. Biophys. Acta, 334, 297—308.

Solomonson, L.P. (1975) Plant Physiol., 56, 853—855.

Solomonson, L.P. and Spehar, A.M. (1977) Nature, 265, 373—375.

Solomonson, L.P. and Vennesland, B. (1972a) Plant Physiol., 50, 421—424.

Solomonson, L.P. and Vennesland, B. (1972b) Biochim. Biophys. Acta, 267, 544—557.

Solomonson, L.P., Jetschmann, K. and Vennesland, B. (1973) Biochim. Biophys. Acta, 309, 32—43.

Solomonson, L.P., Lorimer, G.H., Hall, R.L., Borchers, R. and Bailey, J.L. (1975) J. Biol. Chem., 250, 4120—4127.

Steer, B.T. (1976) Plant Physiol., 57, 928—932.

Stevens, S.E. and Van Baalen, C. (1973) Plant Physiol., 51, 350—356.

Stiefel, E.I. (1973) Proc. Natl. Acad. Sci. U.S.A., 70, 988—992.

408

Stoller, M.L., Malkin, R. and Knaff, D.B. (1977) FEBS Lett., 81, 271—274.

Swader, J.A. and Stocking, C.R. (1971) Plant Physiol., 47, 189—191.

Syrett, P.J. and Leftley, J.W. (1976) in Perspectives in Experimental Biology (N. Sunderland, ed.), Vol. 2, pp. 221—234. Pergamon Press, Oxford.

Thacker, A. and Syrett, P.J. (1972) New Phytol., 71, 423—433.

Tischner, R. (1976) Planta, 132, 285—290.

Tromballa, H.W. and Broda, E. (1971) Arch. Mikrobiol., 78, 214—223.

Ullrich-Eberius, C.I. (1973) Planta, 115, 25—36.

Van Niel, C.B. (1941) Adv. Enzymol., 1, 263—328.

Vega, J.M. and Kamin, H. (1977) J. Biol. Chem., 252, 896—909.

Vega, J.M., Cárdenas, J. and Losada, M. (1978) Methods Enzymol. (in press).

Vega, J.M., Herrera, J., Relimpio, A.M. and Aparicio, P.J. (1972) Physiol. Veg., 10, 637—652.

Vega, J.M., Herrera, J., Aparicio, P.J., Paneque, A. and Losada, M. (1971) Plant Physiol., 48, 294—299.

Vennesland, B. and Guerrero, M.G. (1978) in Encyclopedia of Plant Physiology, New Series (M. Gibbs, E. Latzko, eds.), Vol. 6, Springer Verlag, Berlin (in press).

Vennesland, B. and Solomonson, L.P. (1972) Plant Physiol., 49, 1029—1031.

Villalobo, A., Roldán, J.M., Rivas, J. and Cárdenas, J. (1977) Arch. Microbiol., 112, 127—132.

Wald, G. (1966) in Current Aspects of Biochemical Energetics, Fritz Lipmann Dedicatory Volume (D.O. Kaplan, E.P. Kennedy, eds.), pp. 27—32. Academic Press, New York.

Wallace, W. (1975) Biochim. Biophys. Acta, 377, 239—250.

Whatley, F.R. and Losada, M. (1964) in Photophysiology (A.C. Giese, ed.), Vol. 1, pp. 111—154. Academic Press, New York.

Warburg, O. (1948) Schwermetalle als Wirkungsgruppen von Fermenten, 2nd Edn. W. Saenger, Berlin.

Warburg, O. (1964) Annu. Rev. Biochem., 33, 1—14.

Warburg, O. and Negelein, E. (1920) Biochem. Z., 110, 66—115.

Warburg, O., Krippahl, G. and Jetschmann, C. (1965) Z. Naturforsch., 20b, 993—996.

Yamaya, T. and Ohira, K. (1977) Plant Cell Physiol., 18, 915—925.

Zielke, H.R. and Filner, P. (1971) J. Biol. Chem., 246, 1772—1779.

Zumft, W.G. (1972) Biochim. Biophys. Acta, 276, 363—375.

Zumft, W.G. (1976) Naturwissenschaften, 63, 457—464.

Zumft, W.G., Aparicio, P.J., Paneque, A. and Losada, M. (1970) FEBS Lett., 9, 157—160.

Subject Index

416

Author Index